About the Authors

Paul A. Tucci is chief operating officer and partner at iwerk, inc., an innovative software developer and IT services corporation based in Michigan. He is also a partner in the private equity investment banking firm Cranbrook Partners, and is on the board of directors of the Rizlov Foundation, an organization that grants scholarships to students of classical music. Previously, Paul held senior executive management positions at leading global academic information publishing companies, managing global sales and marketing and product innovation for 18 years.

Tucci has traveled to and done business in more than half of the countries of the world, and is also the author of the book *Traveling Everywhere: How to Survive a Global Business Trip.* He has lectured throughout the world and written extensively on global information development issues. He has also guest-lectured in international management, marketing, and culture at such institutions as the University of Michigan, Northwestern University, New York University, and INSEAD in Fountainbleau, France. A graduate of the University of Michigan, Tucci received a B.A. in international politics, with a concentration in international development, foreign relations, and languages.

Matthew Todd Rosenberg studied geography at the University of California at Davis and earned a master's at California State University, Northridge. He is the author of the first edition of *The Handy Geography Answer Book,* and also published *The Geography Bee Complete Preparation Handbook.* He has worked as an adjunct professor, a newspaper columnist, a city planner, and as a disaster manager for the Red Cross. His work in disaster relief took him all across America, as well as Asia, Africa, the Middle East, and Europe. His contributions to the science of geography earned him the Excellence in Media Award from the National Council for Geographic Education.

Also from Visible Ink Press

The Handy Anatomy Answer Book
by James Bobick and Naomi Balaban
ISBN: 978-1-57859-190-9

The Handy Answer Book for Kids (and
*   Parents)*
by Judy Galens and Nancy Pear
ISBN: 978-1-57859-110-7

The Handy Astronomy Answer Book
by Charles Liu
ISBN: 978-1-57859-193-0

The Handy Biology Answer Book
by James Bobick, Naomi Balaban,
    Sandra Bobick and Laurel Roberts
ISBN: 978-1-57859-150-3

The Handy Geology Answer Book
by Patricia Barnes–Svarney and
    Thomas E Svarney
ISBN: 978-1-57859-156-5

The Handy History Answer Book,
    2nd Edition
by Rebecca Nelson Ferguson
ISBN: 978-1-57859-170-1

The Handy Math Answer Book
by Patricia Barnes–Svarney and
    Thomas E Svarney
ISBN: 978-1-57859-171-8

The Handy Ocean Answer Book
by Patricia Barnes–Svarney and
    Thomas E Svarney
ISBN: 978-1-57859-063-6

The Handy Physics Answer Book
by P. Erik Gundersen
ISBN: 978-1-57859-058-2

The Handy Politics Answer Book
by Gina Misiroglu
ISBN: 978-1-57859-139-8

The Handy Religion Answer Book
by John Renard
ISBN: 978-1-57859-125-1

The Handy Science Answer Book™,
    Centennial Edition
by The Science and Technology
    Department Carnegie Library of
    Pittsburgh
ISBN: 978-1-57859-140-4

The Handy Sports Answer Book
by Kevin Hillstrom, Laurie Hillstrom
    and Roger Matuz
ISBN: 978-1-57859-075-9

The Handy Supreme Court Answer Book
by David L Hudson, Jr.
ISBN: 978-1-57859-196-1

The Handy Weather Answer Book,
    Second Edition
by Kevin S. Hile
ISBN: 978-1-57859-215-9

Visit us at www.visibleink.com

THE HANDY GEOGRAPHY ANSWER BOOK

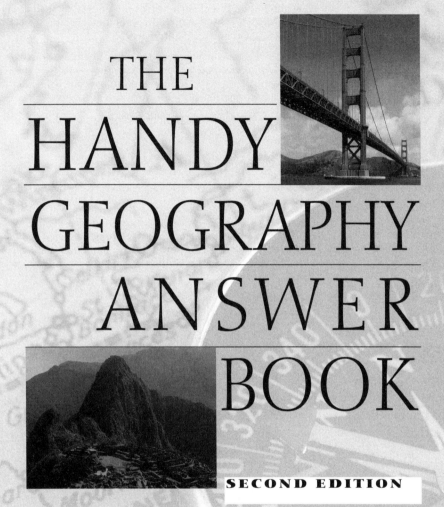

SECOND EDITION

Paul A. Tucci
Matthew T. Rosenberg

VISIBLE
INK
PRESS

Detroit

THE HANDY GEOGRAPHY ANSWER BOOK

Visible Ink Press®
43311 Joy Rd., #414
Canton, MI 48187–2075

Visible Ink Press is a registered trademark of Visible Ink Press LLC.

Most Visible Ink Press books are available at special quantity discounts when purchased in bulk by corporations, organizations, or groups. Customized printings, special imprints, messages, and excerpts can be produced to meet your needs. For more information, contact Special Markets Director, Visible Ink Press, www.visibleink.com, or 734-667-3211.

Managing Editor: Kevin S. Hile
Art Director: Mary Claire Krzewinski
Typesetting: Marco Di Vita
Proofreader: Amy Marcaccio Keyser
ISBN 978-1-57859-215-9

Cover image of Peru by Paul Tucci. All other images from iStock.com. All images in this book that are not otherwise credited are also from iStock.com.
Library of Congress Cataloguing-in-Publication Data

Tucci, Paul A., 1962–
  The handy geography answer book / Paul A. Tucci and Matthew T. Rosenberg. — 2nd ed.
     p. cm.
  Includes index.
  ISBN 978-1-57859-215-9
  1. Geography—Miscellanea. I. Rosenberg, Matthew T. (Matthew Todd), 1973- II. Title.

  G131.R68 2009
  910—dc22

                                           2008052156

Printed in the United States of America

10 9 8 7 6 5 4 3 2 1

Contents

For a list of Internet websites and further resources, visit The Handy Geography Answer Book page at visibleink.com.

Acknowledgments

I am grateful to Roger Jänecke, publisher of Visible Ink Press, for first approaching me and then inspiring me to write the second edition of this book. I would also like to thank our managing editor, Kevin Hile, for his expertise and attention to detail that made the finished book possible; Mathew Rosenberg, the author of the original first edition, for the enormous amount of work and research that went into it; Mary Claire Krzewinski, our designer, for capturing the spirit of the book in her designs; Marco Di Vita for typesetting; and Amy Marcaccio Keyzer, who did the final proofreading.

My interest in international affairs could not have happened without the inspiration of a few of professors at the University of Michigan: Dr. A.F.K. Organski (political science), Dr. Ernest Young (history/Asian studies), and Dr. George Kish (geography). I owe a great debt to the man who hired me for my first position in international business and took a chance on a young kid who wanted to work with the people of the world: Larry Block. Hundreds of people helped introduce and educate me on the front lines of the international publishing world, including Edgar Castillo, Felix Chu, Janet D'Cotta, Jani Dipokusumo, Yoichiro Fudeyasu, Kazuo Hagita, Mark Holland, Dr. Yung Shi Lin, Mani, Mitsuo Nitta, Sue Orchard, Ravichandran, Sunil Sachdev, Tim Smartt, Jae One Son, Simon Tay, Lee Pit Teong, Kelvin Theseira, Sung Tinnie, Takashi Yamakawa, Shinobu Yamashita, Cai Yuniang, and Eve Zhang. Thank you and your organizations for giving me a chance, and for your kindness and patience while I learned.

I would like to thank the staffs of various international organizations, who helped me to learn about the developing world and provided necessary research information to libraries throughout the developing world, including the U.S. Information Agency, U.S Agency for International Development, the World Bank, the Asian Development Bank, the Rockefeller Foundation, and the Soros Foundation.

I wish to also thank two notable graduate business school professors who gave me the opportunity to learn from them and to lecture in their classes on international marketing and management for so many years. They are both my sounding board and sanity check on theories and ideas in international management: Dr. Ann Coughlan (Northwestern University) and Dr. Linda Lim (University of Michigan). I also would like to thank Dr. Evelyn Katz, a great coach and friend who teaches me and inspires me to transform and break through to the next level. Special thanks to Dr. Yung Shi Lin, and all of the people at Jetwin, for the publishing of my first book.

Thanks also to the thousands of people who work at the websites cited in this book, and who give a large part of their lives and passion in getting this information out to the world. Thanks to my friends and colleagues who read my writings over the years and encouraged me to continue. And, of course, I wish to thank my parents, siblings, wife, and daughter for their love and encouragement in writing this book.

—Paul A. Tucci

Introduction

My interest in geography began when I was a little boy, reading whatever I could of my family's collection of *National Geographic* magazines. I still remember actual pictures and stories of far-away places, of distant lands and settlements and modern civilization, of colorful foods that had unimaginable flavors, of people wearing robes and silks, and so many eyes and smiles. I knew at this age that I would want to be a part of the world, and know the whole world.

My geography professor emeritus at the University of Michigan, Dr. George Kish, a noted geographer and cartographer, inspired us with his stories in lectures on Mondays and Wednesdays. I remember he told us what it was like to stand somewhere in Siberia and feel the temperature changes from the thermals on the ground rising up to his waist, creating a gradient of perhaps 30 degrees. I learned that geography was much more than just looking at a globe and naming names on a map. It is about the land, the people on that land, the delicate balance of nature, and our very interdependence upon it, despite the miracles of technology and grocery stores. It's about the effects of nature on places that we may never visit, the stories of human survival and rebuilding, and of renewal.

From the earliest times, mankind has been fascinated with understanding the questions of geography. The Caves of Lascaux, in France, demonstrate the fascination with which our early ancestors—16,000 years ago—had for their surroundings. Their interaction with nature and reverence for where they were, and how they fit into this world we now inherit, is clearly drawn on stone walls.

It is our nature to wonder about places, to try to understand how do we fit in to this great puzzle that we call Earth. When we begin with asking a question about the planet that we live on, we open up a little part of ourselves to that place. Somehow, it becomes less foreign to us. In my travels around the world, I am always amazed at the number of people who know so much about our country. They speak of New Orleans as if they have walked down Bourbon Street. According to a Roper Poll on Geographic Understanding, American kids ranked dead last in their knowledge of the rest of the world. If you know the people, places, and history of the world, you are more likely to promote peace with other lands. You see the differences as well as the plethora of similarities. Quite possibly, you find things about each place that are admirable. Or you see how your country or region compares to some other place and begin to work to solve common problems and inequities.

World change begins with our geographic interest. I hope that this book stimulates your interest and knowledge, perhaps even makes you delve deeper into a particular place, or set foot upon another land and grasp the hands of its people.

—Paul A. Tucci

DEFINING THE WORLD

DEFINITIONS AND HISTORY

What does the word **"geography"** mean?

The word geography is of Greek origin and can be divided into two parts, *geo*, meaning the Earth, and *graphy*, which refers to writing. So geography can be loosely translated to "writing about the Earth." Ancient geography was often descriptions of far away places, but modern geography has become much more than writing about the Earth. Contemporary geographers have a difficult time defining the discipline. Some of my favorite definitions include "the bridge between the human and the natural sciences," "the mother of all sciences," and "anything that can be mapped."

Who **invented** geography?

The Greek philosopher Thales was one of the first to argue about the shape of the world in the sixth century B.C.E. And Chinese texts of the fifth century B.C.E. describe the provinces of China in great detail. However, the Greek scholar Eratosthenes is credited with the first use of the word geography in the third century B.C.E. He is also known as the "father of geography" for his geographical writing and accomplishments, including the measurement of the circumference of the Earth.

What is **geologic time**?

Geologic time is a time scale that divides the history of the planet Earth into eras, periods, and epochs from the birth of the planet to the present. The oldest era is the Precambrian, which began 4.6 billion years ago and ended about 570 million years ago. Next came the Paleozoic Era, which lasted from 570 to 245 million years ago, followed by the Mesozoic Era, from 245 to 66 million years ago. We're now living in the

Cenozoic Era, which began 66 million years ago. The Paleozoic, Mesozoic, and Cenozoic eras are each divided into periods. Additionally, the Cenozoic Era is divided into even smaller units of time called epochs. The last ten thousand years (the time since the last significant Ice Age) is called the Holocene Epoch.

What is the **AAG**?

The Association of American Geographers (AAG) is a professional organization of academic geographers and geography students. The AAG was founded in 1904 and publishes two key academic journals in geography, the *Annals of the Association of American Geographers* and the *Professional Geographer*. The AAG also holds annual conferences and supports regional and specialty groups of geographers.

What is the **NCGE**?

The National Council for Geographic Education (NCGE) is an organization of educators that seeks to promote geographic education. The NCGE publishes the *Journal of Geography* and holds conferences every year.

What is the **National Geographic Society**?

Founded in 1888, the National Geographic Society has supported exploration, cartography, and discovery and publishes the popular magazine *National Geographic,* the fifth most-popular magazine in the United States.

What do **modern geographers** do?

While there are a few jobs with the title of "geographer," many geography students use their analytical ability and knowledge of the world to work in a variety of fields. Geography students often take jobs in fields such as city planning, cartography, marketing, real estate, environment, and teaching.

THE EARTH

How **old** is the **Earth**?

The Earth is approximately 4.6 billion years old.

How was the **Earth formed**?

Scientists believe that the Earth was formed, along with the rest of the solar system, from a massive gas cloud. As the cloud solidified, it formed the solid masses such as the Earth and the other planets.

What is the **circumference** of the Earth?

The circumference of the Earth at the equator is 24,901.55 miles (40,066.59 kilometers). Due to the irregular, ellipsoid shape of the Earth, a line of longitude wrapped around the Earth going through the north and south poles is 24,859.82 miles (40,000 kilometers). Therefore, the Earth is a little bit (about 41 miles [66 kilometers]) wider than it is high. The diameter of the Earth is 7,926.41 miles (12,753.59 kilometers).

Is the Earth a **perfect sphere**?

No, the Earth is a bit wider than it is "high." The shape is often called a geoid (Earth-like) or an ellipsoid. The rotation of the Earth causes a slight bulge towards the equator. The circumference of the Earth at the equator is 24,901.55 miles (40,066.59 kilometers), which is about 41 miles (66 kilometers) greater than the circumference through the poles (24,859.82 miles [40,000 kilometers). If you were standing on the moon, looking back home, it would be virtually impossible to see the bulge and the Earth would appear to be a perfect sphere (which it practically is).

What is a **hemisphere**?

A hemisphere is half of the Earth. The Earth can actually be divided into hemispheres in two ways: by the equator, and by the Prime Meridian (through Greenwich, England) at 0 degrees longitude and another meridian at 180 degrees longitude (near the location of the International Dateline in the western Pacific Ocean). The equator divides the Earth into northern and southern hemispheres.

The Earth is not a perfect sphere but rather an ellipsoid. As the Earth spins, centrifugal forces cause the planet to bulge slightly around the middle.

3

There are seasonal differences between the northern and southern hemispheres but there is no such difference between the eastern and western hemispheres. Zero and 180 degrees longitude divide the Earth into the eastern (most of Europe, Africa, Australia, and Asia) and western (the Americas) hemispheres.

What are the **Arctic and Antarctic Circles**?

The circles are imaginary lines that surround the north and south poles at 66.5 degrees latitude. The Arctic Circle is a line of latitude at 66.5 degrees north of the equator and the Antarctic Circle is a line of latitude at 66.5 degrees south. Areas north of the Arctic Circle are dark for 24 hours near December 21 and areas south of the Antarctic Circle are dark for 24 hours near June 21. Almost all of the continent of Antarctica is located to the south of the Antarctic Circle.

If the Earth is so large, why did **Columbus think that India was close** enough to reach by sailing west from Europe?

The Greek geographer Posidonus did not believe Eratosthenes' earlier calculation, so he performed his own measurement of the Earth's circumference and arrived at the figure of 18,000 miles (28,962 kilometers). Columbus used the circumference estimated by Posidonus when he argued his plan before the Spanish court. The 7,000 mile (11,263 kilometer) difference between the actual circumference and the one Columbus used led him to believe he could reach India rather quickly by sailing west from Europe.

How **fast does the Earth spin**?

It depends on where you are on the planet. If you were standing on the north pole or close to it, you would be moving at a very slow rate of speed—nearly zero miles per hour. On the other hand, those who live at the equator (and therefore have to move about 24,900 miles [40,000 kilometers] in a 24-hour period) zoom at about 1,038 miles (1,670 kilometers) per hour. Those in the mid-latitudes, as in the United States, breeze along from about 700 to 900 miles (1,126 to 1,448 kilometers) per hour.

Why **don't we feel** the Earth moving?

Even though we constantly move at a high rate of speed, we don't feel it, just as we don't feel the speed at which we're flying in an airplane or driving in a car. It's only when there is a sudden change in speed that we notice, and if the Earth made such a change we would certainly feel it.

Does the Earth **spin at a constant** rate?

The rotation of the Earth actually has slight variations. Motion and activity within the Earth, such as friction due to tides, wind, and other forces, change the speed of the

How was the circumference of the Earth determined?

The Greek geographer and librarian at the Great Library of Alexandria, Eratosthenes (c.273—c.192 B.C.E.), was aware that the sun reached the bottom of a well in Egypt only once a year, on the first day of summer. The well was near Aswan and the Tropic of Cancer (where the sun is directly overhead at noon on the summer solstice). Eratosthenes estimated the distance between the well and Alexandria based on the length of time it took camel caravans to travel between the two places. He measured the angle of the sun's shadow in Alexandria at the same time as the well was lit by the sun, and then used a mathematic formula to determine that the circumference of the Earth was 25,000 miles (about 40,000 kilometers)—amazingly close to the actual figure!

planet's rotation a little. These changes only amount to milliseconds over hundreds of years but do cause people who keep exact time to make corrections every few years.

What is the **axis** of the Earth?

The axis is the imaginary line that passes through the north and south poles about which the Earth revolves.

What is **inside the Earth**?

At the very center of the Earth is a dense and solid inner core of iron and other minerals that is about 1,800 miles (2,896 kilometers) wide. Surrounding the inner core is a liquid (molten) outer core. Surrounding the outer core is the mantle, which makes up the bulk of the interior of the Earth. The mantle is composed of three layers—two outer layers are solid and the inner layer (the asthenosphere) is a layer of rock that is easily moved and shaped.

If I **dug through the Earth**, would I end up in China?

If you are in North America and you were able to dig through the Earth (which is impossible due to such things as pressure, the molten outer core, and solid inner core), you would end up in the Indian Ocean, far from land masses. If you were really lucky, you might end up on a tiny island, but you're surely not going to end up in China. The points at opposite sides of the Earth are called antipodes. Most antipodes of Europe fall into the Pacific Ocean.

What is the **mid-Atlantic ridge**?

We don't get to appreciate the beauty of this huge mountain range because it's located at the bottom of the Atlantic Ocean (with one exception: Iceland is a part of the ridge).

The ridge is a crack between tectonic plates where new ocean floor is being created as magma flows up from under the Earth. As more crust is created, it pushes the older crust further away. The new crust at the ridge piles up to form mountains and then begins to move across the bottom of the ocean. Because the Earth can't get larger as more crust is created, the crust eventually has nowhere to go except back into the Earth. This is where subduction occurs.

What is **subduction**?

When two tectonic plates meet and collide, crust must either be lifted up, as in the case of the Himalayas, or it must be sent back into the Earth. When crust from one plate slides under the crust of another, it is called subduction, and the area around the subduction is called a subduction zone.

What is the **North Magnetic Pole**?

The North Magnetic Pole is where compass needles around the world point. It is located in Canada's Northwest Territories at about 71 degrees north, 96 degrees west (latitude and longitude), about 900 miles (1,450 kilometers) away from absolute North Pole. It moves continuously, so to determine true north, look at a recent topographic map for your local area. It should note the "magnetic declination," which means the degrees east or west that you'll need to rotate your compass to determine which way is actually north.

CONTINENTS AND ISLANDS

What are **continents**?

Continents are the six or seven large land masses on the planet. If you count seven continents these include Europe, Asia, Africa, Australia, Antarctica, North America, and South America. Some geographers refer to six continents by combining Europe and Asia as Eurasia, due to the fact that it is one large tectonic plate and land mass. So whether you count Europe and Asia one continent or two (divided at the Ural Mountains in western Russia) is up to the individual. Australia is the only continent that is its own country.

What is the **largest continent**?

The largest continent is Eurasia (Europe and Asia combined) at 21,100,000 square miles (54,649,000 square kilometers). But even if you consider Europe and Asia to be two separate continents, Asia is still the largest, at 17,300,000 square miles (44,807,000 square kilometers).

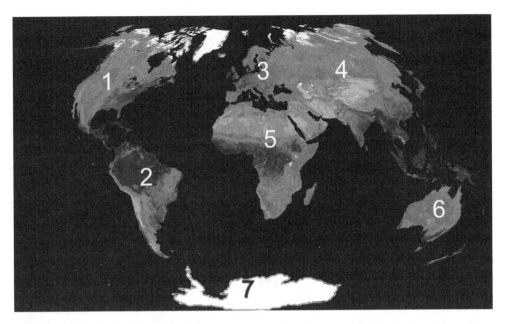

A satellite image showing the continents of North (1) and South America (2), Europe (3), Asia (4), Africa (5), Australia (6), and Antarctica (7). (Image courtesy MODIS Land Group/Vegetation Indices, Alfredo Huete, Principal Investigator, and Kamel Didan, University of Arizona).

What is a **subcontinent**?

A subcontinent is a landmass that has its own continental shelf and its own continental plate. Currently, India and its neighbors form the only subcontinent, but in millions of years, Eastern Africa will break off from Africa and become its own subcontinent.

What was **Pangea**?

About 250 million years ago, all of the land on Earth was lumped together into one large continent known as Pangea. Faults and rifts broke the land masses apart and pushed them away from each other. The continents slowly moved across the Earth to their present positions, and they continue to move today. The Indian subcontinent (composed of India and its neighbors) continues to push into Asia and create the Himalayas.

What is the **world's largest island**?

The world's largest island is Greenland, technically now known as Kalaallit Nunaat. Greenland is located in the North Atlantic Ocean near Canada. It is a territory of Denmark but has locally governed itself since 1979. It is approximately 840,000 square miles (2,175,600 square kilometers). Australia, while it also meets the usual definition

> ## What island did Robinson Crusoe shipwreck on?
>
> **D**aniel Defoe based his novel *Robinson Crusoe* on the story of Alexander Selkirk. Selkirk was an English sailor who had an argument with the captain of his ship and asked to go ashore on the island of Mas a Tierra (also known as Robinson Crusoe Island), about 400 miles (644 kilometers) west of Chile. Selkirk was stranded on the island from 1704 to 1709, when he was rescued by another English ship.

of an island (surrounded by water) and is larger than Greenland, is not considered an island but a continent.

Why is **Greenland** considered an **island** while **Australia** is a **continent**?

Australia is three and a half times larger than Greenland and comprises most of the land on the Indo-Australian plate, while Greenland is distinctly part of the North American plate.

What is an **archipelago**?

An archipelago is a chain (or group) of islands that are close to one another. The Aleutian Islands of Alaska and the Hawaiian Islands are both archipelagos. They are usually formed by plates pushing into one another or by volcanic activity.

What is a **strait**?

A strait is a narrow body of water between islands or continents that connects two larger bodies of water. Two of the most famous straits are the Strait of Gibraltar, which connects the Mediterranean Sea and the Atlantic Ocean, and the Strait of Hormuz, which connects the Persian Gulf to the Gulf of Oman.

HIGH, LOW, BIG, SMALL, AND WONDROUS

What is the **lowest point** in the world on land?

The world's lowest point is at the Dead Sea on the border of Israel and Jordan. It is 1,312 feet (400 meters) below sea level.

What is the **lowest point** on **dry land**?

The world's lowest point is still the Dead Sea shoreline, at approximately 1,378 feet (420 meters) below sea level.

What are the **lowest points** on each continent?

In Africa, the lowest point is Lake Assal in Djibouti, 512 feet (156 meters) below sea level. In North America, California's Death Valley lies at 282 feet (86 meters) below sea level. Argentina's Bahia Blanca is the lowest point in South America at 138 feet (42 meters) below sea level. The Caspian Sea in Europe lies at 92 feet (28 meters) below sea level, and Australia's lowest point is a mere 52 feet (16 meters) below sea level at Lake Eyre.

What is the **highest point** in the world?

At 29,035 feet (8,850 meters), the highest point above sea level in the world is Mt. Everest, which lies on the border of China and Nepal.

Is Mt. **Everest growing taller**?

Because of the shifting plates underneath the surface of the Earth, Mt. Everest is actually growing at a rate of 2.4 inches (6.1 cm.) per year.

What are the **highest points** on each continent?

The highest peak in South America, Aconcagua, lies in Argentina at 22,834 feet (6,960 meters). In North America, Alaska's Mt. McKinley (also called Denali, as it is known indigenously) is 20,320 feet (6,194 meters). The famous Mt. Kilimanjaro (19,340 feet [5,895 meters]) is in Africa's Tanzania. Ice-covered Antarctica's high point is known as Vinson Massif, 16,864 feet (5,140 meters). Europe's Mont Blanc is in the Alps between France and Italy at 15,771 feet (4,807 meters). Australia's high point, Kosciusko, is the lowest of all the continents at 7,310 feet (2,228 meters).

What is the **highest mountain** on **Earth**?

The highest mountain lies on the big island of Hawaii, with a height from the bottom of the sea floor, rising 33,474 feet

Mt. Everest is the tallest peak on Earth, standing 29,035 feet above sea level.

(10,203 meters) to the top of Mauna Kea, which is a volcano that rises 13,680 feet (4,170 meters) above sea level.

What are the **deepest points** in the oceans?

Lying deep below the Pacific Ocean, about 200 miles (322 kilometers) south of the island of Guam, is the Marianas Trench (also known as the Mariana Trench), which is 1,554 miles (2,550 kilometers) long and 44 miles (71 kilometers) wide. The deepest point of the Marianas Trench is 36,198 feet (11,033 meters). In the Atlantic Ocean, the Puerto Rico Trench is 28,374 feet (8,648 meters) below the surface. In the Arctic Ocean, the Eurasia Basin is 17,881 feet (5,450 meters) deep. The Java Trench in the Indian Ocean is 23,376 feet (7,125 meters) deep. Another deep point of note in the Pacific Ocean is Monterey Canyon off the coast of northern California. It is about 95 miles (153 kilometers) long and 11,800 feet (3,600 meters) deep. The cold waters generated in the trench create a perfect environment rich in foods that support a diverse range of wildlife.

In comparison to all of these ocean canyons, the most famous land canyon—the Grand Canyon in Arizona—is 277 miles (446 kilometers) long and 6,000 feet (1,829 meters) deep. The world's ocean canyons are much more impressive, but most people will never see them.

Where is the **farthest point from land**?

In the middle of the Southern Pacific ocean lies a spot that is 1,600 miles (2,574 kilometers) from any land. Located at 47°30' South, 120° West, this spot is equidistant from Antarctica, Australia, and Pitcairn Island.

Where is the **farthest point from an ocean**?

In northern China lies a spot that is over 1,600 miles (2,574 kilometers) from any ocean. Located at 46°17' North, 86°40' East, the land is equidistant from the Arctic Ocean, Indian Ocean, and Pacific Ocean.

What were the **seven wonders of the ancient world**?

While there was often disagreement by ancient and classical scholars as to which major works of art and architecture could be considered wonders, these seven were nearly always on the list: the Pyramids of Egypt (the only remaining wonder), the Colossus of Rhodes (on the island of Rhodes in Greece), the Temple of Artemis at Ephesus (a marble temple in Turkey), the Mausoleum of Halicarnassus (Bodrum, Turkey), the Statue of Zeus at Olympia (an ivory and gold statue in South Western Greece), the Hanging Gardens of Babylon (an enormous garden building, with plants of every kind, near Al Hillah, Iraq) and the Lighthouse of Alexandria (on the island of Pharos, near Alexandria, Egypt).

What are the **seven wonders of the modern world**?

According to the American Society of Civil Engineers, the seven wonders of the modern world include the Channel Tunnel between England and France; the CN Tower in Toronto, Canada; the Empire State Building, New York; the Golden Gate Bridge, San Francisco; the Itaipu Dam between Brazil and Paraguay; the Netherlands North Sea Protection Works; and the Panama Canal.

One of the seven natural wonders of the world is the stunning Grand Canyon in Arizona.

What are the **seven natural wonders of the world**?

These include the Aurora Borealis (northern lights), Mt. Everest (on the border of China and Nepal), Victoria Falls (in eastern Africa), the Grand Canyon (USA), Great Barrier Reef (Australia), Paricutin (volcano in Mexico), and the harbor of Rio de Janeiro (Brazil) with its stunning topography.

HUMAN CIVILIZATION

When did **agriculture** begin?

Agriculture began about 10 to 12 thousand years ago in a time period known as the first agricultural revolution. It was at this time that humans began to domesticate plants and animals for food. Before the agricultural revolution, people relied on hunting wild animals and gathering wild plants for nutrition. This revolution took place almost simultaneously in different areas of human settlement.

Where did **agriculture** begin?

Agriculture simultaneously began in what is known now as the Middle East (Fertile Crescent), the Yangtze River Region of southern China, the Yellow River Region of northern China, Sub Saharan Africa, South-Central Andes near modern day Peru, Bolivia and Chile, Central Mexico, and the eastern United States.

What is the **difference** between **cultivation** and **domestication**?

Cultivation is the deliberate attempt to sow and manage essentially wild plants and seed. Domestication is when people experiment and consciously select the right seeds to grow for various conditions.

When was the second **agricultural revolution**?

The second agricultural revolution occurred in the seventeenth century. During this time, production and distribution of agricultural products were improved through machinery, vehicles, and tools, which allowed more people to move away from the farm and into the cities. This mass migration from rural areas to urban areas coincided with the beginning of the industrial revolution.

What was the **industrial revolution**?

The industrial revolution began in the eighteenth century in England with the transformation from an agricultural-based economy to an industrial-based economy. It was a period of increased development in industry and mechanization that improved manufacturing and agricultural processes, thereby allowing more people to move to the cities. It included the development of the steam engine and the railroad.

What is the **green revolution**?

The green revolution began in the 1960s as an effort by international organizations (especially the United Nations) to help increase the agricultural production of less developed nations. Since that time, technology has helped improve crop output, which is reaching all-time highs throughout the world.

How much of the world's population is **devoted to agriculture**?

In less-developed countries, such as Asia and Africa, a majority of the population is engaged in agricultural activity. In the more-developed countries of Western Europe and North America, less than one tenth of the population relies on agriculture for their livelihood.

How were **animals first domesticated**?

Dogs were probably some of the first animals to become domesticated. Wild dogs probably came close to human villages scavenging for food and were quickly trained as companions and protectors. Over time, early agriculturalists realized the value of domesticating other animals and proceeded to do so. Many different kinds of animals were domesticated in different areas of the world.

PEOPLE AND COUNTRIES

What is the **largest country in the world**?

Russia is by far the largest at about 6.6 million square miles (17.1 million square kilometers). Russia is followed in size by Canada, China, the United States, Brazil, Australia, India, Argentina, Kazakhstan, and Sudan.

How many **people live** on the planet?

As of 2008, approximately 6.7 billion people inhabit Earth. This number is increasing at a rate of around one percent per year.

Which **10 countries** have the **most people**?

Country	2008 Population
China	1.3 billion
India	1.1 billion
United States	303 million
Indonesia	234 million
Brazil	190 million
Pakistan	164 million
Bangladesh	150 million
Russia	141 million
Nigeria	135 million
Japan	127 million

How many **people** are projected to live on the planet in **2040**?

It is estimated that there will be approximately 9.25 billion people on the planet by the year 2040, even with a declining rate of growth.

Why is the **growth rate declining**?

A big reason why the rate of growth is slowly declining is because people are delaying marriage longer. Also, the wide availability of contraception is having a positive impact on decreasing the number of unwanted births.

What top **five countries** will have the **largest populations in 2050**?

Country	2050 Projected Population
India	1.8 billion
China	1.4 billion
United States	420 million
Nigeria	325 million
Indonesia	313 million

Which country has the most neighbors?

China is bordered by 13 neighbors: Mongolia, Russia, North Korea, Vietnam, Laos, Myanmar, India, Bhutan, Nepal, Afghanistan, Tajikistan, Kyrgyzstan, and Kazakhstan. Russia is next in line, as it shares its border with 12 other countries. Brazil is third with nine neighbors.

Which country has the **longest coast line**?

The coastline of Canada and its associated islands is the longest in the world, about 151,400 miles (243,603 kilometers) long. Russia, which is the largest country in the world, has the second longest coastline at about 23,400 miles (37,651 kilometers).

Which countries **have the fewest neighbors**?

All island nations (such as Australia, New Zealand, Madagascar, etc.) have no neighbors. Haiti, Dominican Republic, Papua New Guinea, Ireland, UK, and many others are island nations that share an island.

Which **non-island nations have the fewest neighbors**?

There are 10 non-island countries that share a land border with just one neighbor: Canada (neighboring USA), Monaco (France), San Marino (Italy), Vatican (Italy), Qatar (Saudi Arabia), Portugal (Spain), Gambia (Senegal), Denmark (Germany), Lesotho (South Africa), and South Korea (North Korea).

How does a city get chosen to **host the Olympics**?

The International Olympic Committee chooses a city as an Olympic site through a complex process. Cities (and their countries) are judged on many characteristics, including environmental protection, climate, security, medical services, immigration, housing, and many others. Cities eagerly spend millions of dollars in construction and preparation for possible selection as a host city, as an investment in the city's future.

How is a **capital** different from a **capitol**?

The capital is a city and the capitol is a building. The capitol is located in the capital. To remember the difference, think about the "o" in capitol as being the dome of a capitol building. Capital cities are often the largest cities in a country or region.

Barcelona, Spain, hosted the Olympics in 1992. Hosting the Olympics is a matter of national pride and brings economic benefits, such as the jobs made from building this stadium.

What did the average European know about the world in the **Middle Ages**?

In Europe in the Middle Ages, most individuals' knowledge of the world was quite limited. Geographic knowledge developed by the Greeks and Romans (who knew the Earth was a sphere) was all but lost in Europe. Europeans of the time thought of the world as flat and composed of only Europe, Asia, and Africa.

Where is the **third world**?

Originally, the third world referred to those countries that did not align themselves with the United States (first world) or the Soviet Union (second world) during the Cold War. Over time, the term took on different meanings and has come to refer to less-developed or developing nations, which are the more preferred terms.

What is the **largest landlocked country** in the world?

Kazakhstan, which is the ninth largest country in the world, has no outlet to the ocean. It is over one million square miles (2.59 million square kilometers) in area. While Kazakhstan is located adjacent to the Caspian Sea, the Caspian Sea is a landlocked sea.

Where are cyberspace and the Internet?

Cyberspace is not space in the old-fashioned sense of the word at all. The Internet is composed of millions of computers around the world, which are connected to each other in order to provide information across cyberspace as though there were no global boundaries, mountain passes, or oceans to cross. When you send e-mail to a friend on the other side of the planet, it passes from your computer to that of your Internet service provider and then from computer to computer, making its way to your friend in a matter of seconds. Similarly, when you access a page on the World Wide Web, your computer tells another computer which tells another computer that you want such and such document delivered to your computer, and it arrives in seconds. Some geographers measure and map cyberspace by looking at where most of the Internet's traffic flows though, to, and from.

What **countries** have the most **Internet usage**?

Country	Internet Users
United States	211 million
China	162 million
Japan	86 million
Germany	50 million
India	42 million
Brazil	39 million

Which **countries restrict Internet access** to what is deemed unfavorable content by their governments?

The following countries filter or restrict content that their citizens can access on the Internet: Azerbaijan, Bahrain, Myanmar, China, Ethiopia, India, Iran, Jordan, Libya, Morocco, Oman, Pakistan, Saudi Arabia, Singapore, South Korea, Sudan, Syria, Tajikistan, Thailand, Tunisia, Turkmenistan, United Arab Emirates, Uzbekistan, Vietnam, and Yemen.

What is **geographic illiteracy**?

In 1989, the National Geographic Society commissioned a survey to find out how much Americans and residents of several other countries knew about the world around them. Unfortunately, American youths scored the worst. Swedes knew the most about world geography. The media subsequently reported the "geographic illiteracy" of the American population. Due to the attention given to this problem, geographic education has since become a greater priority for educators.

What is the **National Geographic–Roper Public Affairs Geographic Literacy Study**?

This study assesses the geographic knowledge of young American adults between the ages of 18 and 24. The survey also asks respondents how much they think they know about geography and other subjects, as well as their views on the importance of geographic, technological and cultural knowledge in today's world.

What were the **results** of the **2006 National Geographic–Roper Public Affairs Geographic Literacy Study**?

The 2006 study is the latest in a series of surveys commissioned by the National Geographic Society, with the most recent previous wave being conducted in 2002. The countries that scored highest in the Poll were Sweden, Italy, and France. The United States and Mexico scored the lowest.

MAPS

HISTORY AND INSTRUMENTS

What were the **earliest maps**?

There are many examples demonstrating that humanity has been interested in its surroundings since the beginning of recorded history. In fact, written on the walls of caves in Lascaux, France, are three dots indicating the brightest stars in the sky, which are estimated to have been drawn around 14000 B.C.E. Neolithic wall paintings found in Çatalhöyük, Turkey, show an early city plan from around 7500 B.C.E.

What makes a piece of paper a **map**?

No matter what the medium, all maps must be a representation of an area of the Earth, celestial bodies, or space. Though maps are commonly printed on paper, they can come in a variety of forms, from being drawn in the sand to being viewed on computers. A map should have a legend (a guide explaining the map's symbols), a notation of which way is north, and an indicator of scale. No map is perfect and every map is unique.

What is a good way to **learn where places are**?

The best way to learn about places that you have heard of is to look them up in an updated atlas. An atlas is a collection of maps bound in a book; it may include additional information, such as illustrations, statistical tables, topography, and other important information about a place or places. If you like to use the Internet, you may try to enter the place in the search box at your favorite search engine, and then click on the results. You will be surprised by the amount of information you can find!

Antique maps are much more ornamental than modern-day ones, but they were also considerably less accurate!

Why is a **book of maps** called an **atlas**?

The term "atlas" comes from the name of a mythological Greek figure, Atlas. As punishment for fighting with the Titans against the gods, Atlas was forced to hold up the planet Earth and the heavens on his shoulders. Because Atlas was often pictured on ancient books of maps, these became known as atlases.

How do **cartographers** shape our world?

Cartographers are map makers and cartography is the art of map making. Cartographers map neighborhoods, cities, states, countries, the world, and even other planets. There are as many types of maps to make as there are cartographers to make them.

Who would purposely create **false maps**?

Within the former Soviet Union, incorrect maps were produced as a matter of course. Soviet maps purposely showed the locations of towns, rivers, and roads in incorrect places. Often, in different editions of the same map, towns would disappear from one version to the next. Street maps of Moscow were particularly incorrect and non-proportional. The cartographic deceit of the U.S.S.R. was an effort to keep the geography of the country a secret, not only from foreigners but also from its own citizens. Even official government agencies were not allowed to have accurate maps.

How do I get to the refrigerator in the dark?

Not all maps are written on paper. When trying to reach the refrigerator at night, we do not smell our way to food, we use a map based on our memory of the room. If we stumble on our way, it is usually over a misplaced toy or shoe that we did not remember leaving there. Everyone has these kinds of maps in his or her mind. These mental maps help you find your way not only to the refrigerator in the dark, but also to the grocery store and to work. People not only have mental maps of common trips they make, but also of their city, country, and even the world. Every person's mental map is unique, based on how wide an area that person travels and their knowledge of the world.

Who decides which **names** go **on maps**?

In the United States, the U.S. Board on Geographic Names (BGN) approves the official names and spellings of cities, rivers, lakes, and even foreign countries. If a town would like to change its name, it must petition the BGN for approval. Upon approval, the name is officially changed and updated in federal government gazetteers and records, which official and commercial mapmakers use for their maps.

How can **maps** be used to **start wars**?

Prior to its 1990 invasion of Kuwait, Iraq produced official maps that showed the independent country of Kuwait as Iraq's nineteenth province. Iraq used these maps as justification for its 1990 invasion and attempted annexation of Kuwait (Iraq was actually after Kuwait's oil reserves). Maps have been, and still are, used by a multitude of countries, provinces, and cities to prove ownership of a certain piece of land.

What **other countries** have **disputes** over lines drawn on **maps**?

Japan, China, and Taiwan all have a dispute over the Diaoyu/Senkaku islands in the Pacific Ocean. India and Pakistan have an ongoing dispute over ownership of the northernmost region of Kashmir in India. China has disputed claims of sovereignty by Taiwan. North and South Korea are still in conflict over lines drawn by Korean and American forces that split the country in two.

How does a **sextant** help navigators?

In 1730, the sextant was invented independently by two men, John Hadley and Thomas Godfrey. Using a telescope, two mirrors, the horizon, and the sun (or another

A "compass rose" is the ornamentation surrounding a compass dial that is divided into the four cardinal directions, and then into further divisions.

celestial body), the sextant measures the angle between the horizon and the celestial body. With this measurement, navigators could determine their latitude while at sea.

When was the **compass** invented?

As early as the eleventh century, the Chinese were using a magnetic needle to determine direction. At approximately the same time, the Vikings may have also used a similar device. A compass is simply a magnetic needle that points toward the magnetic north pole.

Have compasses always pointed **north**?

No, they have not. Though compasses always point to the magnetic pole, the magnetic pole has not always been in the north. Every 300,000 to 1 million years, the magnetic pole flips from north to south or from south to north. If compasses had been around before the last time the magnetic pole reversed, their arrows would have pointed south rather than north.

What is a **compass rose**?

On old maps, the directions of the compass were represented by an elaborate symbol, known as a compass rose. Many of the older compass roses displayed 32 points, representing not only the four cardinal directions (north, south, east, and west) but also 28 subdivisions of the circle (south-west, south south-west, etc.). This directional symbol resembled a rose, hence its name. Though compasses are now often drawn with only the four cardinal directions and the resemblance to the flower is minimal, the directional symbol is still called a compass rose.

What is an **azimuth**?

Azimuth is another method for stating compass direction. It is based on the compass as 360 degrees, with north at 0 degrees, east at 90, south at 180, and west at 270 degrees. You can refer to a direction as "head 90 degrees" instead of "head east."

How, why, and how much does **Magnetic North move**?

Scientists aren't sure why the Earth's magnetic pole moves, only that it does. The amount of movement varies, but it's never more than a few miles each year.

How much has the **magnetic north pole moved**?

Since it was documented in 1831 by James Ross to be slightly north of 70 degrees north latitude, it has moved to north of 80 degrees north latitude at an average rate of more than 24.9 miles (40 kilometers) per year.

How old is the **oldest known map**?

Circa 2700 B.C.E., the Sumerians drew sketch maps in clay tablets that represented their cities. These maps are the oldest known maps.

What is the **oldest map** using the word **America**?

The map, acquired by the Library of Congress and dated 1507, incorporates many of the findings of the Italian explorer Amerigo Vespucci. The map was thought to be lost, but was rediscovered in 1901, and was kept in a castle in Wolfegg, Germany, for more than 350 years.

What is the oldest known **map drawn to scale**?

In the sixth century B.C.E., the Greek geographer Anaximander created the first known map drawn to scale. His map was circular, included known parts of Europe and Asia, and placed Greece at its center.

Where is the **equator**?

The equator is the line located equidistant between the North and South Poles. The equator evenly divides the Earth into the northern and southern hemispheres and is zero degrees latitude.

LATITUDE AND LONGITUDE

What are **latitude and longitude** lines?

Lines of latitude and longitude make up a grid system that was developed to help determine the location of points on the Earth. These lines run both north and south and east and west across the planet. Lines of latitude (those that run east and west) begin at the equator, which is zero degrees. They extend to the North Pole and the South Pole, which are 90 degrees north and 90 degrees south, respectively. Lines of longitude (those that run north and south) begin at the Prime Meridian, which is the imaginary line that runs through the Royal Observatory in Greenwich, England. The lines of longitude extend both east and west from the Prime Meridian, which is zero degrees, and converge on the opposite side of the Earth at 180 degrees.

Are lines of longitude and latitude all the **same length**?

No, they are not. Only the lines of longitude are of equal length. Each line of longitude equals half of the circumference of the Earth because each extends from the North Pole to the South Pole. The lines of latitude are not all equal in length. Since they are each complete circles that remain equidistant from each other, the lines of latitude vary in size from the longest at the equator to the smallest, which are just single points, at the North and South Poles.

How wide is a degree of **longitude**?

Though there are only a couple dozen lines of longitude shown on most globes and world maps, the Earth is actually divided into 360 lines of longitude. The distance between each line of longitude is called a degree. Because the lines of longitude are widest at the equator and converge at the Poles, the width of a degree varies from 69 miles (111 kilometers) wide to zero, respectively.

How **wide** is a degree of **latitude**?

Though there are only about a dozen lines of latitude shown on most globes and world maps, the Earth is actually divided into 180 lines of latitude. The distance between each line of latitude is called a degree. Each degree is an equal distance apart, at 69 miles (111 kilometers).

What do **minutes and seconds** have to do with longitude and latitude?

Each degree of longitude and latitude is divided into 60 minutes. Each minute is divided into 60 seconds. An absolute location is written using degrees (°), minutes ('), and seconds (") of both longitude and latitude. Thus, the Statue of Liberty is located at 40°41'22" North, 74°2'40" West.

Latitude lines run horizontally around the globe, while longitude lines run from north to south.

Which comes **first**, **latitude** or **longitude**?

Latitude is written before longitude. Latitude is written with a number, followed by either "north" or "south" depending on whether it is located north or south of the equator. Longitude is written with a number, followed by either "east" or "west" depending on whether it is located east or west of the Prime Meridian.

Why was **computing longitude** so difficult?

It wasn't until the sixteenth century that clocks were fabricated in such a way that they could accurately tell time both on land and at sea. The only way of determining how far east or west one could go is by plotting the stars in two locations and recording the exact time in both locations simultaneously, and then recording the time and position at the destination. As clocks became more accurate, the ability to measure speed and distance became possible.

How can I **remember** which way latitude and longitude run?

You can remember that the lines of latitude run east and west by thinking of lines of latitude as rungs on a ladder ("ladder-tude"). Lines of longitude are quite "long" because they run from the North Pole to the South Pole.

How can a **gazetteer** help me find latitude and longitude?

A gazetteer is an index that lists the latitude and longitude of places within a specific region or across the entire world. Many atlases include a gazetteer, and some are published separately.

How can I find the latitude and longitude of a **particular place**?

To find latitude and longitude of a particular location, you will need to consult either a gazetteer or a computer database that includes longitude and latitude data. Though gazetteers are readily accessible, they don't include as many places as online databases. There are a number of sites on the Internet that have extensive databases of latitude and longitude and even include such specific places as public buildings.

Why was the Prime Meridian established at **Greenwich**?

In 1675, the Royal Observatory in Greenwich, England, was established to study determination of longitude. In 1884, an international conference established the Prime Meridian as the longitudinal line that passes through the Royal Observatory. The United Kingdom and United States had been using Greenwich as the Prime Meridian for several decades before the conference.

READING AND USING MAPS

What is the difference between a **physical and a political** map?

A physical map shows natural features of the land such as mountains, rivers, lakes, streams, and deserts. A political map shows human-made features and boundaries such as cities, highways, and countries. The maps we use in atlases and see on the walls of classrooms are typically a combination of the two.

What is a **topographic** map?

A topographic map shows human and physical features of the Earth and can be distinguished from other maps by its great detail and by its contour lines indicating elevation. Topographic maps are excellent sources of detailed information about a very small area of the Earth. The United States Geological Survey (USGS) produces a set of topographic maps for the United States that are at a scale of 1:24,000 (one inch equaling 2,000 feet [or 1 centimeter equaling about 240 kilometers]). You can purchase these maps online, at sporting goods stores, or through the USGS itself.

Why are road maps so **difficult to fold**?

The problem lies with the multitude of folds required to return the map to its original, folded shape. The easiest way to fold a road map is to study the creases and to fold the map in the order that the creases will allow. But once you've made a mistake, the folds have lost their tell-tale instructions. To fold a road map, begin by folding it accordion style, making sure that the "front" and "back" of the folded design appear on top. Then, once the entire map is folded accordion style, fold the remaining slim, long, folded paper into three sections. And, voilà, your road map is folded!

Why is **color** important on a relief map?

A relief map portrays various elevations in different colors. But, a common color scheme found on relief maps causes a problem. On these maps, mountains are displayed as red or brown while lowlands are shown in shades of green. This is confusing because the green areas on the map are often misconstrued as fertile land while brown areas are mistaken for deserts. For example, an area such as California's Death Valley, which is shown in green on relief maps because it lies below sea level, seems fertile, when actually it is an inhospitable desert.

What does the **scale of a map** tell me?

A scale indicates the level of detail and defines the distances between objects on a map. On a map, scales can be written as a fraction, a verbal description, or as a bar scale.

A fraction, or ratio, using the example of 1/100,000 or 1:100,000, indicates that one unit of any form of measurement on the map is equivalent to 100,000 units of the same measurement in the area being represented. For instance, if you use inches as the unit of measurement, then one inch on the map would equal 100,000 inches in the area represented by the map.

A verbal description describes the relationship as if it were a verbal instruction, such as "one inch equals one mile." This allows the versatility of having different units of measurement.

A bar scale uses a graphic to show the relationship between distance on the map to distance in the area represented. The bar scale is the only type of scale that allows a reduction or enlargement of the map without distorting the scale. This is because when you increase the size of the map, the bar scale is increased proportionally. For a fraction or verbal scale, the proportion (1:1,000) is only true for the map at that size. For example, when enlarging a map, the map might become twice as large but the numbers in a ratio do not change, as they would need to in order to stay accurate.

How can I determine the **distance between two places** by using a scale?

By using a ruler, compare the distance between two points on a map with the information on the scale to calculate the actual distances between the two points. For example, if you measure the distance between two towns as being five inches and the ratio says 1:100,000, then the actual distance between the towns is 500,000 inches (7.9 miles [12.7 kilometers]).

What is the difference between **small and large** scale maps?

A small-scale map shows a small amount of detail over a wide area, such as the world. A large-scale map shows a large amount of detail while representing a limited area, such as neighborhoods or towns.

27

How is the Earth's surface like an orange peel?

All attempts to represent a sphere, like the Earth, in a flat representation result in distortions. The Earth's surface is like an orange peel. If one were able to peel an orange in one piece and then try to flatten the peel, cracks and tears would appear. Attempting to "peel" the Earth and then lay that information on a flat surface in a map creates these same open areas. Map makers attempt to create maps that represent the spherical Earth with as little distortion as possible. The various strategies for this are called projections.

Why is every map **distorted**?

No map is completely accurate because it is impossible to accurately represent the curved surface of the Earth on a flat piece of paper. A map of a small area usually has less distortion because there is only a slight curve of the Earth to contend with. A map of a large area, such as maps of continents or the world, are significantly distorted because the curvature of the Earth over such a large area is extreme.

Why does **Greenland appear larger** on most maps than it actually is?

Because of the distortions that must appear on all maps, many maps place the distortion in the northern and southern extremes of the Earth. In one of the common projections, known as Mercator, Greenland appears to be similar in size to South America, despite the fact that South America is actually eight times larger than Greenland. The advantage of the Mercator projection is that the lines of latitude and longitude remain perpendicular; thus the map is useful for navigation.

How can a **legend** help me read a map?

The legend, usually found in a box on the map, is information that explains the symbols used on a map. Though some symbols seem standard, like a railroad line, even those can be represented differently on different maps. Since there really are no standard symbols, each map's legend should be consulted when reading a map.

Why is there often a **cross** next to the east direction on maps?

On old maps, a cross often sits next to the east direction on a compass rose. This cross represents the direction to Paradise and the Holy Land.

Where can I **buy maps**?

There are many places you can buy maps. Most large bookstores offer an extensive collection of local and foreign travel maps, wall maps, and atlases. Also, many cities have specialty travel and map stores that offer a larger and more varied collection of maps, as well as maps of more exotic locales. Maps are also available at many Web sites on the Internet. Just put in the name of the place, region, city, or country that you wish to find, and you will discover both maps that are either free or available for purchase. Look for the most up-to-date maps when selecting a map from more than one available vendor.

But what if I **can't find** the map I'm looking for?

Not all maps can be found at bookstores or even in specialty stores. If you are looking for an extremely specific and relatively uncommon map, visit a local university's map collection. Their collections are often far greater in size and breadth than any store. If you need help locating a map, you should be able to discuss your map needs with a friendly map librarian. You can also try constructing a search on the Internet that is as specific as possible. Use advanced search options to filter out results so that you may find a precise and detailed map.

What is the difference between **relative and absolute** location?

There are two different ways to describe where a place is located; relative location and absolute location. Relative location is a description of location using the relation of one place to another. For instance, using relative location to describe where the local video store is, you might say that it's on Main Street, just past the high school. Absolute location describes the location of a place by using grid coordinates, most commonly latitude and longitude. For instance, the local video store would be described as being located at 23°23'57" North and 118°55'2" West.

MODERN MAPPING

What are **satellites** photographing?

Satellites capture images of the Earth's weather patterns, the growth of cities, the health of plants, and even individual buildings and roads. Satellites circle the Earth, or remain geostationary (in the same place with respect to the Earth), and send data back to the Earth via radio signals.

How have satellites **changed map making**?

Satellite images, which are accurate photographs of the Earth's surface, allow cartographers to precisely determine the location of roads, cities, rivers, and other features on the Earth. These images help cartographers create maps that are more accurate than ever before. Since the Earth is a dynamic and ever-changing place, satellite images are great tools that allow cartographers to stay up-to-date.

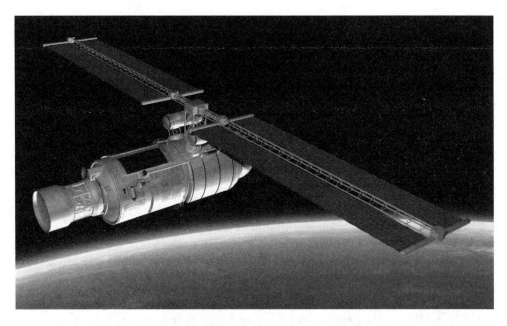
Satellites orbiting the Earth take photographs that make it possible to create extremely accurate maps.

How much **junk** is there in space?

In addition to operational satellites, there are approximately 8,800 pieces of space junk surrounding the Earth, from tiny screws to booster rockets. There are plans for the future to build a radar system that would track every piece of space junk so that space vehicles and satellites can avoid irreparable damage.

How has **GIS** revolutionized cartography?

Geographic Information Systems (GIS) began in the 1960s with the popularity of computers. Though very simplistic in its beginning, new technology and inventions have expanded and enhanced the functions of GIS. GIS has revolutionized cartography by using computers to store, analyze, and retrieve geographic data, thus allowing infinite numbers of comparisons to be made quickly. The program formulates information into various "layers," such as the location of utility lines, sewers, property boundaries, and streets. These layers can be placed together in a multitude of combinations to create a plethora of maps, unique and suitable to each individual query. The versatility of GIS makes it indispensable to local governments and public agencies.

How can GIS **help my town**?

Your community can use GIS on a day-to-day basis and in emergency situations. GIS allows public works departments, planning offices, and parks departments to monitor

How did a map stop cholera?

In the 1850s an outbreak of cholera threatened London. Dr. John Snow, a British physician, mapped the deaths associated with the disease and determined that many deaths were occurring near one water pump. The pump handle was removed and the spread of the disease stopped. Prior to this time, the method by which cholera spread was unknown. Today, medical geographers and epidemiologists frequently use cartography to determine the cause and spread of disease or epidemics.

the status of the community's utilities, roads, and properties. In an emergency, GIS can give emergency teams the information they need to evacuate endangered areas and respond to the crisis.

How does a **GPS** unit know where I am?

Individual Global Positioning System (GPS) units on the Earth receive information from a U.S. military-run system of 24 satellites that circle the Earth and provide precise time and location data. The individual GPS unit receives data from three or more satellites that triangulate its absolute location on the Earth's surface. If you are carrying such a device, your absolute location is the same as that of the device.

How can GPS **keep me from getting lost**?

A GPS unit provides precise latitude and longitude for the location of the device. By using a hand-held GPS unit along with a map that provides latitude and longitude (such as a topographic map), you can determine your precise location on the Earth's surface. This is a valuable tool for those who hike or travel in remote regions and for ships at sea. GPS is now widely available in cars; as stand-alone, portable, pocket-sized devices; on cell phones; and even on the boxes that ship products that you buy. In short, GPS is used in all aspects of our lives.

THE PHYSICAL ENVIRONMENT

THE EARTH'S MATERIALS AND INTERNAL PROCESSES

How thick is the **Earth's crust**?

The thickness of the Earth's crust varies at different points around the planet. Under continents, the crust is approximately 15 miles (24 kilometers) thick, but under the oceans it is a mere five miles (8 kilometers) thick.

What is **continental drift**?

The Earth is divided into massive pieces of crust that are called tectonic plates. These plates lie wedged together like a puzzle. The plates slowly move, crashing into each other to form mountain ranges, volcanoes, and earthquakes. The plates are like rafts floating on water; this is called continental drift.

How many **tectonic plates** are there?

There are a dozen significant plates on the planet. Some of the largest include the Eurasian Plate, North American Plate, South American Plate, African Plate, Indo-Australian Plate, Pacific Plate, and Antarctic Plate. Some smaller plates are located between the major plates. The smaller plates include the Arabian Plate (containing the Arabian Peninsula), the Nazca Plate (located to the west of South America), the Philippine Plate (located southeast of Japan, containing the northern Philippine islands), the Cocos Plate (located southwest of Central America), and the Juan de Fuca Plate (just off the coast of Oregon, Washington, and Northern California).

Earth's tectonic plates are in constant motion, causing geologic fault lines like this one, formed in a strata of volcanic ash that was cut open during road construction.

What was **Pangea**?

Pangea, which existed about 250 million years ago, was one huge continent, including the land of all seven continents. It was located near present-day Antarctica and has slowly drifted and split to form the continents as we know them today. The continents and their tectonic plates continue to move and will one day be in a much different arrangement than they are today.

How are **mountains** formed?

The process of orogeny, or mountain building, is related to continental drift. When two tectonic plates collide, they often form mountains. The Himalayas are the result of the Indo-Australian plate colliding with the Eurasian plate. At these collision zones, volcanoes and earthquakes are common.

How did the **Himalayas** form?

About 30 to 50 million years ago, the landmass of India pressed into the landmass of Asia, pushing up land at the place of impact and creating the Himalayas. Even today, as the Indian subcontinent presses against Asia, the Himalayas continue to grow and change.

What type of rocks are **formed by lava**?

Igneous rocks are formed when liquid magma under the surface of the Earth, or lava on the surface of the Earth, cools and hardens into rock.

What type of rocks are **formed from particles**?

Sedimentary rocks are formed by the accumulation and squeezing together of layers of sediment (particles of rock or remains of plant and animal life) at the bottom of rivers, lakes, and oceans or even on land. The continual accumulation of more and more layers of sediment places a great amount of pressure on the lowest layers of sediment and, over time, compresses them into rock.

What are **recycled rocks**?

Metamorphic rocks are recycled rocks. Metamorphic rocks are rocks that had a prior existence as sedimentary, igneous, or even another metamorphic rock. Underground heat and pressure metamorphose one type of rock into another, creating a metamorphic, or recycled rock.

What type of **sand** do we often see on **beaches near volcanoes**?

Because of the content of the sediment near volcanoes comprised of dark black igneous rock, the resulting beach sand in such places as Hawaii and Indonesia is actually a dark brown, even black in color.

Which is **larger, clay or sand**?

A single grain of sand is 1500 times larger than a grain of clay.

What is a **dike**?

A dike is magma that has risen up through a crack between layers of rock. When this magma solidifies, it becomes very solid rock. If the rock around it is eroded, a dike can form great rock monoliths above the ground.

What are **hot springs**?

Hot springs are created by underground water that is heated and percolates to the Earth's surface. Aside from being natural baths, the steam from hot springs can be used to drive turbines, which create electricity. This type of energy production is called geothermal energy.

Where are **hot springs used by people** around the world?

Hot spring baths have been used in cultures throughout the world since ancient times. They are found in such places as Japan, Taiwan, Australia, the United States, Iceland, and Sicily.

Why does the **ground sink**?

In many places over the world, seemingly solid land lies over vast oil deposits or water aquifers. Without the liquid supporting it, the ground sinks into the space left behind. In some parts of California's Imperial Valley, the land has dropped more than 25 feet (7.6 meters) due to underground water being removed from the area. Unless the pumping of underground water and oil is stopped, the land will continue to sink.

Why do houses fall into **sinkholes**?

Houses that sit upon limestone rock have the proclivity to fall into sinkholes. As underground water wears away the limestone rock, it creates underground caverns. If the water wears away too much limestone, the cavern may collapse, taking anything on the surface with it. A sinkhole is just one of the many reasons to have your home inspected by a geologist.

NATURAL RESOURCES

What is a **renewable resource**?

A renewable resource is one that can be replenished within a generation. Forests, as long as they are replanted, are a renewable resource. Materials such as oil, coal, and natural gas are known as nonrenewable resources because they require millions of years to be created. So, once the world's supply of oil is gone, it's gone for a long, long time.

What is a **perpetual resource**?

Perpetual resources are natural resources—such as solar energy, wind, and tides—that have no chance of being used in excess of their availability. They can be used for power generation and conversion to electric energy.

What are **fossil fuels**?

Underground fuels such as natural gas, oil, and coal are all known as fossil fuels because they are encased in rocks, just like fossils. It takes millions of years of compression and the building up of dead plants and animals to create these fuels.

What is a **fossil**?

The outline of the remains of a plant or animal embedded in rock is called a fossil. Fossils are formed when a plant or animal dies and becomes covered up by sediments. Over time, the layers compress the remains, which are then embedded into the rock.

The increasing scarcity of fossil fuels such as crude oil has forced exploration offshore. Ocean rigs such as this one can be vulnerable to hurricanes and oil spills.

LANDSCAPES AND ECOSYSTEMS

What are **basins and ranges**?

Basins and ranges are sets of valleys and mountains that are spaced close together. Most of Nevada and western Utah is composed of sets of basins and ranges.

What is **permafrost**?

Soil that is permanently (or for a good part of the year) frozen is known as permafrost. Permafrost occurs in higher latitudinal regions, which have cold climates.

Is the **permafrost thawing**?

Yes, it is. In places with great areas of permafrost, such as Alaska, the permafrost has warmed to the highest level in 10,000 years. During the last 50 years, the Arctic regions have been warming to record high temperatures. During this time, Alaska's average temperature has warmed an average of 3.3 degrees above normal.

Why is **permafrost** so **important**?

Aside from the mediating effects on the Earth's temperature, permafrost provides a cover for the habitat that supports wild flora and fauna, which provide sustenance for the entire ecosystem of the Arctic regions, including the many indigenous people who live there.

How much of the Earth's surface is **frozen**?

About one-fifth of the planet is permafrost, or frozen for all or most of the year.

What do **deserts** and **polar regions** have in **common**?

Deserts and polar regions both have average rainfall of less than four inches (10 centimeters) per year; they are both the driest places on Earth.

What is a **jungle**?

A jungle is a forest that is composed of very dense vegetation. The term tropical rain forest can often be used interchangeably with jungle. Jungles occur most often in tropical areas such as the Amazon and Congo river basins.

What is a **rain forest**?

A rain forest is any densely vegetated area that receives over 40 inches (100 centimeters) of rain a year.

A lush tropical rain forest in Puerto Rico.

What is a **tropical rain forest**?

A tropical rain forest is a rain forest that lies between the Tropic of Cancer in the north and the Tropic of Capricorn in the south (i.e. within the "tropics"). Tropical rain forests are known for their very diverse species of plant and animal life. Tropical rain forests exist throughout Central America, northern Brazil, the Congo River basin, and Indonesia.

Where is the **northernmost rain forest**?

Juneau, the capital of Alaska, is in the middle of one of the largest and most northerly rain forests in the world, which is located within the Chugach National Forest and Chugach State Park. Others extend to the northern regions of Japan and Siberia, Russia.

What is a **desert**?

A desert is an area of light rainfall. Deserts usually have little plant or animal life due to the dry conditions. Contrary to the popular image, deserts are not just warm, sand-swept areas like the Sahara Desert, but can also be frigid areas like Antarctica, one of the driest places on Earth.

What was the **highest temperature** ever recorded?

The world's highest temperature was recorded in the desert in Libya in 1922. The temperature was 136° Fahrenheit. California's Death Valley holds the record for the highest temperature in the United States, 134° in 1913.

How can an area be lower than sea level?

Land in the midst of a continent can be lower than sea level because the land is not close enough to the sea to be flooded with water. Movement of the tectonic plates pushes areas like the Dead Sea in Israel and Death Valley in California to elevations lower than sea level.

Do **oases** really exist?

Oases do exist and they are quite prevalent throughout the eastern Sahara Desert in Africa. An oasis has a source of water, often a spring, that allows vegetation to grow. Small towns are located at some larger oases in the desert. Oases have been traditional stopping places for nomads traveling across deserts.

How do **sand dunes move**?

Sand dunes are created and transported by wind. Wind blows sand from the windward side of the dune to the opposite side, slowly transporting it across the landscape.

What causes **erosion**?

Wind, ice, and water are the most common agents of erosion. They wear down and carry away pieces of rock and soil. The process is accelerated when trees that help hold the soil in place have been destroyed by fire or have been chopped down. With fewer trees, the soil is easily eroded and washed away, leaving a barren surface where plants can no longer grow.

What does a **glacier leave behind**?

When a glacier moves across the land, it acts like a giant bulldozer, pushing and collecting rock, dirt, and debris. A moraine is a deposit of rock and dirt carried by a glacier and left behind once the glacier melts and recedes.

What is a **tree line**?

A tree line is the point of elevation at which trees can no longer grow. The tree line is caused by low temperatures and frozen ground (permafrost).

How **high** is **tree line**?

Tree lines vary, depending on where they are in the world and how close they are to a geographic pole. Trees stop growing in mountain regions from around 2,600 feet (800 meters) in places such as Sweden, to over 17,000 feet (5,200 meters) in the Andes Mountains in Bolivia.

How do **forest fires** help forests?

An occasional fire is often necessary for a forest. Forest fires clear undergrowth, giving more room for trees to grow, thus rejuvenating the forest. Since forest fires are usually extinguished by fire fighters as rapidly as possible, the amount of undergrowth in forests has increased. This extra undergrowth can become extremely flammable, making fires even more dangerous to people. A policy of allowing the forest to burn naturally, while protecting human structures, produces a more natural environment.

What is **tundra**?

Tundra is dry, barren plains that have significant areas of frozen soil or permafrost. Tundra is common in the northernmost parts of North America, Greenland, Europe, and Asia. Although rather inhospitable, there is plant life on the tundra. This life consists of low, dense plants such as shrubs, herbs, and grasses. There are even some species of insects and birds that can survive the harsh conditions of tundra.

ASTEROIDS AND NEAR EARTH OBJECTS

Did an asteroid kill the **dinosaurs**?

Approximately 65 million years ago, a six-mile-wide (about 10 kilometer-wide) asteroid struck the Earth. This impact might have started a chain of events that led to the extinction of about two-thirds of the Earth's species of animals and plants, including the dinosaurs. An asteroid that large would have created a layer of dust that would have surrounded the Earth, lowering temperatures, and causing deadly, highly acidic rain.

Will another large asteroid **strike the Earth**?

Yes, asteroids have struck the Earth in the past and are likely to strike again in the future. Small asteroids strike the planet about every 1,000 to 200,000 years. Large asteroids are much less likely to threaten the planet; they strike only about three times every million years. Huge asteroids, such as the one that may have killed dinosaur species, are even less frequent.

How many places on Earth have **evidence of asteroid impacts**?

There are about 160 places on Earth that display evidence of asteroid impacts. Depending on the size of an asteroid, it can have significant effects on the planet. If an asteroid strikes the ocean, it can create huge destructive tsunamis or tidal waves. If an

Meteor Crater near Flagstaff, Arizona, was the first definitive site to provide physical evidence that meteors have crashed into the Earth in the past.

asteroid strikes land, it can create a huge crater, cause earthquakes, and propel debris into the atmosphere, creating major climatic changes.

What is a **Potentially Hazardous Asteroid** or PHA?

A PHA is an object that is large enough to be detected by astronomers on Earth. It has an orbital path that could bring it in close proximity to Earth.

How many **Potentially Hazardous Asteroids** is NASA **tracking** that could impact the Earth in the twenty-first century?

More than 210 PHAs are on NASA's list of most likely objects whose orbital paths may bring them close to Earth's orbit sometime in the next 100 to 200 years. Scientists have not fully calculated the risks because they still need to figure out the effects of gravity of various planetary bodies on the exact course of the asteroids.

What is a **Near Earth Object** (NEO)?

A Near Earth Object is any object in space that is relatively close to Earth and is of any size. The difference between an NEO and a Potentially Hazardous Asteroid (PHA) is that PHAs are anything larger than 500 feet (150 meters) in diameter and have orbits that bring them closer than 4,650,000 miles (7,480,000 kilometers) to Earth. All PHAs are also NEOs, but not all NEOs are PHAs.

What is the Space Guard Survey?

The Space Guard Survey is an initiative by NASA to catalogue all near earth objects that are larger than 0.6 miles (1 kilometer) in diameter. The survey assesses their potential intersection with Earth's orbital path.

How **many Near Earth Objects** is NASA **tracking**?

Approximately 1,100 Near Earth Objects are tracked by NASA.

When and where was the **last great impact** on the Earth from a **Near Earth Object**?

The last big impact was in 1908 in Tunguska, Russia, in Siberia. Scientists do not know for sure what hit the Earth—it could have been a comet or meteoroid—but whatever it was, it had an estimated explosive energy of 15 megatons.

WATER AND ICE

How much of the **Earth is covered by water**?

About 70 percent of the surface area of the Earth is covered by water. The other 30 percent of the Earth is land located primarily in the Northern Hemisphere. If you look at a globe, you'll notice that the Southern Hemisphere has a great deal of ocean.

How much is the **sea level rising**?

Recent estimates show the sea level has increased 7 to 12 inches (20 to 30 centimeters) during the twentieth century.

What **countries are threatened** the most from **rising sea levels**, and may cease to exist in the twenty-first century?

Low-lying island nations of the Pacific and Indian Oceans are most at risk, most notably Tuvalu and the Maldives. Other candidates for severe flooding and reclamation of coastal land by the impending sea include Bangladesh, India, Thailand, Vietnam, Indonesia, and China, affecting the lives of hundreds of millions of people.

How much **land disappears** when the **sea level rises**?

Scientists believe that as the sea water rises by 0.04 inches (1 millimeter), the shoreline disappears by 4.9 feet (1.5 meters). This means that if the sea level rises by 3.28 feet (1 meter), the shorelines will extend another 1 mile (1.6 kilometers) inland.

How does the **hydrologic cycle** work?

The movement of water from the atmosphere to the land, rivers, oceans, and plants and then back into the atmosphere is known as the hydrologic cycle. We can pick an arbitrary point in the cycle to begin our examination. Water in the atmosphere forms clouds or fog and falls (precipitates) to the ground. Water then flows into the ground to nourish plants, or into streams that lead to rivers and then to oceans, or it can flow into the groundwater (underground sources of water). Over time, water sitting in puddles, rivers, and oceans is evaporated into the atmosphere. Water in plants is transpired into the atmosphere. The process of water moving into the atmosphere is collectively known as evapotranspiration.

Where is all the **water**?

Over 97 percent of the world's water lies in the oceans and is too salty to drink or to irrigate crops with (except when the water is cleaned through a desalinization plant, which is not done very often). About 2.8 percent of the world's water supply is fresh water. Of that 2.8 percent, about 2 percent is frozen in glaciers and ice sheets. This leaves only about 0.8 percent of the world's water that is accessible through aquifers, streams, lakes, and in the atmosphere. The water that we use primarily comes from this 0.8 percent.

What is **evapotranspiration**?

Evapotranspiration is the combination of water vapor being evaporated from the surface of the Earth (such as from lakes, rivers, or puddles) into the atmosphere, and transpiration, which is the movement of water from plants to the air.

What is an **aquifer**?

An aquifer is an underground collection of water that is surrounded by rock. The creation and filling of an aquifer is a very slow process, as it relies upon water to percolate through the soil and rock layers and into the aquifer. An aquifer lies above a lower layer of rock that holds the water in place and keeps it from moving further underground.

What is the **Oglala Aquifer**?

The Oglala Aquifer is a huge aquifer that spans an area from Texas to Colorado and Nebraska. The oldest water deposited in the aquifer is over one million years old, and only a very small amount of water is added each year. The Oglala Aquifer is being pumped rapidly by the farms in the region, causing a reduction in the amount of water in the aquifer. Consequently, wells have to be continually deepened so that they can continue to pump water.

Why are we losing **ground water**?

Water is pumped from aquifers around the world for irrigation, industrial, and household needs. Aquifers do not refill as rapidly as water is being pumped out, so in many areas there is a danger that some aquifers may disappear altogether.

Will there be another ice age?

Yes, eventually the Earth will again cool and ice will cover land at higher latitudes and elevations. It may be a hundred years from now or it may be thousands of years away, but the Earth's climate is always slowly changing.

What are **ice ages**?

Throughout the life of the planet, the climate has warmed and cooled many times. During the cooling periods, ice ages have occurred. During the ice ages, large sheets of ice cover large portions of land. In the most recent ice age, which ended about 10,000 years ago, large parts of northern Europe and North America were covered by ice sheets.

What is the **Coriolis** effect?

Due to the rotation of the Earth, any object on or near the Earth's surface will veer to the right in the Northern Hemisphere and to the left in the Southern Hemisphere. This applies especially to phenomena such as ocean currents and wind. Imagine a missile being fired at New York by Los Angeles. As the missile flies over the United States, the Earth continues to rotate under the missile and it strikes New Jersey instead. Missile launchers and pilots need to factor the spinning of the Earth into their trajectories in order to end up in the right place. North of the equator, ocean currents and winds rotate clockwise, but south of the equator, the opposite is true.

Does the Coriolis effect make the water in my **toilet, sink, and bathtub swirl clockwise**?

No, the Coriolis effect has very little effect on such small bodies of water. The flow down the drain is mostly a function of the shape of the container.

If I keep walking in a straight line, will the **Coriolis effect** cause me to **veer**?

If your body were completely symmetrical (and no one's is) and neither leg were longer and you were walking on perfectly flat land then yes, you might start veering due to the Coriolis effect.

What is the difference between **a bay and a gulf**?

Both are bodies of water partially surrounded by land, and a bay is a smaller version of a gulf. Famous bays include the San Francisco Bay (California), the Bay of Pigs (Cuba),

45

Chesapeake Bay (Maryland/Virginia area), Hudson Bay (Canada), the Bay of Bengal (a large bay near India and southeast Asia), and the Bay of Biscay (France). Famous gulfs include the Gulf of Mexico (southern United States), the Persian Gulf (between Saudi Arabia and Iran), and the Gulf of Aden (between the Red Sea and the Arabian Sea).

Where does the **Loch Ness Monster** live?

The fabled monster is supposed to live in Loch Ness. The term "loch" is Gaelic and is used in Scotland to refer to a lake or narrow inlet of the sea. Loch Ness is fully surrounded by land and is therefore a lake.

How are **waves** created?

Waves are created by wind blowing across the surface of the water. Though waves appear to move along the surface of the water, they are simply the movement (oscillation) of water up and down due to the friction of the air. When waves occur near the shore, they may become steeper and "break."

Old Faithful in Yellowstone National Park is the most famous geyser in North America. Its eruptions occur so regularly that you can almost set your watch by them.

How does **Old Faithful** shoot water into the air?

A geyser, such as Yellowstone National Park's famous Old Faithful, is the result of an underground aquifer that is warmed by heated rocks and magma underground. There is a small fissure or crack in the aquifer's surface that allows the steam and heated water to jet from the ground (about every hour at Old Faithful).

How does water **wash away** the **land**?

Drops of rain hit soil and rock and displace grains. When water flows over the surface, it loosens and carries away pieces of rock or soil. There is a tremendous amount of energy in a raindrop. Over days, weeks, months, years, centuries, and millennia, the erosive power of water can cut through even the strongest rocks. The material that the flowing water picks up is eventually deposited when the flow of the stream slows down—this is known as deposition.

> ## How tall is the tallest waterfall in the United States?
>
> Yosemite Falls in California has a height of 2,425 feet (739 meters).

How much does a gallon of **water weigh**?

Water is quite a heavy substance. One gallon of water at room temperature weighs about 8.33 pounds (3.78 kilograms).

How is **water used** in the home?

About 41 percent of household water is used for flushing the toilets; 37 percent is used for bathing; and the remaining 22 percent is used for washing dishes (six percent), drinking and cooking (five percent), laundry (four percent), cleaning (three percent), lawn and garden usage (two percent), and other purposes.

What **waterfall** has the largest flow of water?

Niagara Falls, on the border of Canada and the United States between Lake Ontario and Lake Erie, has 212,000 cubic feet (6,000 cubic meters) of water flowing over its 173- and 182-foot-high (52.7- 55.5-meter-high) set of falls (there are two adjacent waterfalls).

What is the **highest waterfall** in the world?

The highest falls in the world are Angel Falls in southeast Venezuela. It is 3,230 feet (984 meters) high.

How does the **boiling point of water** help determine altitude?

The boiling point of water at sea level is 212° Fahrenheit. The boiling point drops about one degree for every 500-foot (152-meter) increase in altitude. Therefore, in Denver, at 5,280 feet (1,609 meters) above sea level (Denver is called the Mile High city because there are 5,280 feet in a mile), water boils at about 202°. The change in the boiling point is why cooking instructions are modified for higher altitudes.

OCEANS AND SEAS

What is the **difference between an ocean and a sea**?

While a sea can be any body of salt water, it usually refers to a body of salt water partially or completely enclosed by land. Oceans, though they can also be referred to as seas, are large areas of salt water, unobstructed by continents.

How **many oceans** are there?

The seas and oceans of the world are actually one big "world ocean" because they are connected. Most of the time the world ocean is divided into four main parts: the Pacific, Indian, Atlantic, and Arctic. By far the largest is the Pacific Ocean, more than twice the size of the Atlantic. The Indian is third in size, followed by the tiny Arctic Ocean.

How salty is **sea water**?

About 3.5 percent of the weight of seawater is salt (not just sodium chloride or table salt, but also potassium chloride, calcium chloride, and other types of salts). This would equal just over three and one-third cups of salt in one gallon of sea water.

What are **ocean currents**?

The oceans don't remain still; their water is constantly moving in giant circles known as currents. In the Northern Hemisphere, currents move clockwise, while in the Southern Hemisphere they move counterclockwise. Currents help to moderate temperatures on land in places like the British Isles—which are farther north than the U.S.-Canadian border—by sending warm water from the Caribbean northeast across the Atlantic Ocean to northern Europe. A current known as the Antarctic Circumpolar Current circles the southern continent. The North Atlantic and North Pacific oceans each have a large clockwise current, while the South Atlantic and South Pacific oceans each have a large counterclockwise current.

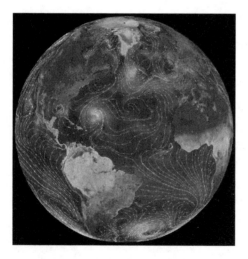

A NASA image measuring temperatures on the Earth reveals the pattern of ocean currents (photo courtesy NASA, Visible Earth Project).

What are the **largest seas**?

Parts of oceans that are surrounded by islands or otherwise partially enclosed are often known as seas. The five largest seas in order are: the South China Sea, the Caribbean Sea, the Mediterranean, the Bering Sea, and the Gulf of Mexico.

Why is the Mediterranean Sea **so salty**?

Due to the high temperatures in the Mediterranean region, evaporation of the Mediterranean Sea occurs more rapidly than in other bodies of water, therefore more salt is left behind. The warm, dense, salty water in the Mediterranean is replaced by less salty and dense Atlantic water in the Strait of Gibraltar. Water that flows into the Mediterranean from the Atlantic Ocean usually remains in the Sea for anywhere from 80 to 100 years before returning to the Atlantic Ocean.

Has the **Mediterranean Sea** always been there?

Salt and sediment found at the bottom of the Mediterranean Sea prove that on several occasions the Mediterranean Sea has dried up, leaving a large layer of salt behind. Scientists speculate that the Strait of Gibraltar has, on occasion, closed up, keeping water from being able to flow back and forth between the Atlantic Ocean and the Mediterranean Sea.

What are the **seven seas**?

The "seven seas" spoken of by mariners from long ago are oceans or parts of oceans. The Atlantic and Pacific Oceans are so large that they were each divided into two "seas." The Antarctic, Indian, and Arctic were also considered seas, thus totaling seven. If a sailor had sailed upon all seven seas, he had sailed around the world. There are not just seven but dozens of seas in the world.

Where are the four **colored seas**: Black, Yellow, Red, and White?

The four colored seas are not geographically associated with one another. The Black Sea is located near the Balkan Peninsula and is bordered by Turkey, Russia, and Ukraine (it is also the home of the port city of Odessa). The Red Sea is located to the south of the Black Sea, between the Arabian Peninsula (Saudi Arabia), and Africa. The Red Sea has been a major trade route for hundreds of years and has been especially useful since the completion of the Suez Canal. The White Sea is in northern Europe. It is part of the Arctic Ocean and is a Russian sea (it lies to the east of Finland). The Yellow Sea is far to the east, between China and the Korean Peninsula.

Is the Black Sea **really black**?

No, it is not. This sea, located to the north of Turkey, is quite deep and has darker looking water than most water bodies, but receives its name from the inhospitableness of the waters for sailing.

49

A desalination plant in the Caribbean pumps salt water out of the ocean and makes it suitable for drinking.

What is a **desalination plant**?

A desalination plant pumps ordinary seawater through a variety of expensive processes, transforming the salty water into fresh water. This process has been used with some success in Texas, the Caribbean, and the Middle East. It is much more efficient and less expensive, however, to clean waste water (water that has been used for bathing, cooking, cleaning, etc.) than it is to clean and desalinate seawater.

How **many desalination plants** are there?

There are approximately 13,000 plants located throughout the world. Saudi Arabia alone accounts for 24 percent of the world output of freshwater from desalination.

Where is the **biggest desalination plant** in the **United States**?

The biggest American desalination plant is in Tampa, Florida. It has an output of 25 million gallons (94.6 million liters) per day. This is a fraction of the largest plant in the world, the Jebel Ali Plant. Located in the United Arab Emirates, it produces over 200 million gallons (757 million liters) per day.

RIVERS AND LAKES

What is the **longest river** in the world?

Egypt's famous Nile River is the longest in the world. It is more than 4,100 miles (6,597 kilometers) long from its sources in the Ethiopian Highlands (the source of the portion of the Nile called the Blue Nile) and Lake Victoria (the source of the White Nile). The Nile Valley is the center of contemporary and ancient Egyptian civilization. Following the Nile in length are the Amazon (in Brazil), the Missouri-Mississippi (United States), the Chang or Yangtze (China), the Huang or Yellow (China), and the Ob (Russia).

What is the **longest river in the United States**?

The Missouri-Mississippi River is the longest in the United States, approximately 3,860 miles (6,211 kilometers) in length.

Where are the **highest rivers** in the world?

The highest rivers are in Tibet, where there are two that stand out from all the rest: the Ating River at 20,013 feet (6,100 meters) and Brahmaputra River at 19,751 feet (6,020 meters).

Why are the **Missouri-Mississippi Rivers** lumped together?

Actually, the Missouri River was incorrectly named. The Missouri River is actually the main feeder river of what is now known as the Mississippi River. Usually, the main feeder bears the same name as the rest of the river. Therefore, the full length of the Mississippi River, including the Missouri River, is known as the Missouri-Mississippi River.

Which river carries the **most water**?

By far, Brazil's Amazon River carries more water to the sea than any other river in the world. The discharge at the mouth of the river is about seven million cubic feet (170,000 cubic meters) per second, which is about four times the flow of the Congo in Africa, the river ranked second in terms of discharge. It would take the Amazon only about 28 days to fill up Lake Erie. The Yangtze, Brahmaputra, Ganges, Yenisy, and Mississippi are other rivers with very high discharges.

What is a **delta**?

A delta is a low-lying area where a river meets the sea. Often, the river divides into many tributary streams, forming a triangular-shaped area. The river deposits a large amount of sediment at its mouth, creating excellent soil for farming once the channel

51

of the stream moves. One of the most famous deltas is where the Nile River meets the Mediterranean Sea. Other major deltas include the Mississippi River delta in Louisiana, the Ganges River delta in India, and the Yangtze delta in China. The word delta comes from the Greek letter delta, referring to its triangular shape when written.

What is a **drainage basin**?

The area that includes all of the tributaries for an individual stream or river is its drainage basin. For example, the drainage basin for the St. Lawrence River includes the area surrounding the Great Lakes. Rivers such as the Platte (which has its own drainage basin) flow into the Missouri, and the Missouri flows into the Mississippi. The combined area drained by the Platte, the Missouri, the Mississippi, and all other Mississippi River tributaries combined create the third largest drainage basin in the world. The Amazon has the largest drainage basin of any river, the Congo has the second largest.

What is a **tributary**?

Any stream that flows into another stream is a tributary. Most major rivers have hundreds of tributaries, which on a map look like branches of a tree. One classification system of rivers is based upon the number of tributaries a river has.

What is a **watershed**?

A watershed is the boundary between drainage basins. It is usually the crest of a mountain where water flows on either side into two different drainage basins.

What is a **wadi**?

Wadi is the Arabic word for a gully or other stream bed that is dry for most of the year. A wadi is a channel for streams that develop during the short rainy season. The channels of wadis were probably initially carved when the desert regions of today had more rainfall.

A perfect example of a meandering river—this one flowing through Alberta, Canada.

What is a **meander**?

Streams and rivers that have carved a flat floodplain commonly flow in curves known as meanders. These S-shaped curves vary by the size and flow of the river. The river flows faster on the outside curve of the meander and therefore continues to cut and create a larger curve.

What is an **oxbow lake**?

An oxbow lake is a crescent-shaped lake that is formed when the meander, or curve of a river, is cut off from the rest of the river during a flood, or when the curve of the meander becomes so large that the river begins flowing along a new path. The curve that remains becomes its own lake. These can commonly be seen along the Mississippi River system.

What is the world's **largest lake**?

The Caspian Sea (which is really a lake) is the largest lake in the world. It is surrounded by Russia, Kazakhstan, Turkmenistan, Iran, and Azerbaijan and is over 143,200 square miles (370,888 square kilometers) in area. The second largest lake, Lake Superior in North America, is a mere 31,700 square miles (82,103 square kilometers).

Where is the **highest lake** in the world?

A yet-to-be-named crater lake found atop Ojos del Salado, a volcanic mountain on the border of Argentina and Chile, has an elevation of 20,965 feet (6,390 meters). Lake Titicaca, a lake sandwiched between Peru and Bolivia, is 12,507 feet (3,812 meters) above sea level and is the highest navigable lake in the world.

PRECIPITATION

How is **rainfall measured**?

Agencies like the National Weather Service use very accurate devices that measure rainfall to the nearest one-hundredth of an inch. The devices, known as rain gauges or tipping-bucket gauges, collect rainwater, usually at a point unaffected by local buildings or trees that may interfere with the rain.

How can I **measure** how much rain falls where I live?

Any container with a flat bottom and flat sides can measure rainfall. The width of the top of the container must be the same as at the bottom of the container, but the diameter does not matter. It could be a device purchased for measuring precipitation or something as simple as a coffee can.

Where does it **rain the most**?

Mt. Waialeale, on the island Kauai in Hawaii, receives a whopping average of 472 inches (1,200 centimeters) of rain a year—that's over 39 *feet* [12 meters] of rain per year!

Where does it **rain the least**?

Northern Sudan's Wadi Halfa (which is in the Sahara Desert) receives an average of less than one-tenth of an inch of rain per year. That's hardly a drop at the bottom of a bucket.

How much **water is in snow**?

When about 10 inches (25 centimeters) of snow melts, it turns into about one inch of water. Snow has pockets of air between snowflakes when they are on the ground, so it takes 10 times the snow to make an equivalent amount of water.

Where was the **most snowfall** ever recorded?

Washington State's Mt. Baker recorded the most snowfall in a single season: 1,140 inches (2,896 centimeters).

Do oceans get more rain than land?

The oceans receive just over their share, percentage-wise, of the world's precipitation, about 77 percent. The remaining 23 percent of precipitation falls on the continents. Some areas of the world receive far more precipitation than others. Some parts of equatorial South America, Africa, Southeast Asia, and nearby islands receive over 200 inches (500 centimeters) of rain a year, while some desert areas receive only a fraction of an inch of rain per year.

What is the difference between **snow and hail**?

Snow is water vapor that freezes in clouds before falling to the Earth. Hail is water droplets (raindrops) that have turned to ice in clouds.

How is **hail formed**?

Hail is ice that is formed in large thunderstorm clouds. Hail begins as droplets of water—normally destined to become raindrops—that are blown upward and subsequently freeze. They then fall lower within the cloud, where they collect more water, are blown upward again, and refreeze. The hailstone grows larger as it collects more and more ice, and eventually falls to the ground.

How big was the **largest hailstone**?

In 2003, a hailstone was recovered near Aurora, Nebraska, with a diameter of 18.75 inches (47.63 centimeters). The previous record was in 1970, when people recovered a hailstone with a 17.5 inches (44.45 centimeters) diameter in Kansas.

GLACIERS AND FJORDS

What is a **glacier**?

A glacier is a mass of ice that stays frozen throughout the year and flows downhill. Glaciers are capable of carving rock with their weight and slow, steady movement. They are responsible for the stunning landscape of Yosemite National Park in California. Large glaciers that cover the land are also known as ice sheets.

Are there **still glaciers** in the United States?

Yes, small glaciers exist throughout Alaska, within the Cascade Range of Washington state, sporadically across the Rocky Mountains, and also in the Sierra Nevadas of California.

How **old** are **glaciers**?

Glaciers present today were created during the last stage of glaciation, the Pleistocene epoch, which lasted from 1.6 million years ago to about 10,000 years ago.

Did glaciers create the **Great Lakes**?

Yes, the Great Lakes are the world's largest lakes formed by glaciers. During the Pleistocene epoch, glaciers inched over the Great Lakes area, moving weak rock out of their way and leaving behind huge, carved basins. As the glaciers began to melt, the basins filled with water and created the Great Lakes.

Athabasca Glacier is located in the Canadian Rockies, in Jasper National Park, outside of Edmonton, Alberta, Canada. Due to global warming, the glacier recedes at a rate of two to three meters per year (photo by Paul A. Tucci).

Are **glaciers** only found in **cold, northern places**?

No, glaciers are found in all six continents.

What is a **tropical glacier**?

Tropical glaciers are those found high in the mountains of tropical regions in the world. The Andes Mountains in South America contain 70 percent of the world's tropical glaciers.

Is **global warming** causing the world's **glaciers to melt**?

Many scientists believe that greenhouse gases from human activities are directly causing glaciers in all parts of the world to melt and recede at an unprecedented rate. It is thought that by 2030 there will be no glaciers in Glacier National Park in Montana. In East Africa, Mt. Kenya's Lewis Glacier in Kenya has lost 40 percent of its size in just the last 25 years.

What are the **consequences** when **glaciers melt**?

Glaciers that have melted in the Himalayas, home to the world's largest mountains, have filled up and burst the banks of nearby glacial lakes, filling rivers and causing widespread flooding and death to nearby populations downstream. Similar consequences will likely befall those now living near other glaciers around the world.

What is a **fjord**?

During the ice ages, glaciers, which were prevalent at higher latitudes and elevations, became so large that gravity drove them to lower elevations, eventually all the way to the sea. On their way, glaciers would carve deep canyons in the surface of the Earth. At the end of the ice age, as the ice melted and the ocean level rose, these glacial troughs filled with seawater. These very dramatic-looking canyons with high cliffs hanging over a thin bay of water are known as fjords. Fjords are very common in Norway and Alaska.

From where does the **word "fjord"** originate?

The word fjord comes from the Norse language and means "where you travel across." It is significant to early Norwegians as a place to travel across to get to the sea when there were no bridges available.

Norway is famous for its scenic fjords, such as Geiranger Fjord, pictured here.

What is the **highest fjord** in **Norway**?

The highest fjord is Sognefjord, which begins at a depth of 4,291 feet (1,308 meters) in the ocean and rises to more than 3,280 feet (1,000 meters).

Are **all fjords** found in **Norway**?

No. In fact, fjords are found throughout the world, wherever glaciers retreated and have cut into the earth, filling in and creating a huge valley of sea water. Notable fjords are found in Alaska and on New Zealand's South Island.

Where is the **longest fjord**?

The longest fjord is in Greenland, Scoresby Dund. It stretches more than 217 miles (350 kilometers).

CONTROLLING WATER

What do **dams** do?

By blocking the flow of a river, a dam allows a reservoir of water to build up. Dams are built in order to minimize floods, to provide water for agriculture, and to provide water for recreational uses. Dams in the United States are somewhat controversial, as the Bureau of Reclamation and the Army's Corps of Engineers battle to build more dams and control more water in the western United States. Many outdoor enthusiasts and environmentalists feel that dams are not always necessary.

What is the **tallest dam** in the world?

Tajikistan is home to the world's two tallest dams—Rogun and Nurek. Rogun is about 1,100 feet (335 meters) tall (about 100 stories!) and Nurek is about 985 feet tall (300 meters). The United States' tallest dam, Oroville (in Northern California), is sixteenth on the world list at 755 feet (230 meters).

How do **farmers water their crops**?

The process of artificially watering crops is called irrigation. In some areas of the world, agriculture can rely on rainfall for all its water needs. In drier areas (usually those receiving less than 20 inches [51 centimeters] of rainfall per year), irrigation is required. Water is pumped from aquifers or delivered via an aqueduct to the fields where it flows through small channels between plants, or is sprayed through sprinklers. In very water-conservative regions such as Israel, water is scientifically dripped onto plants, thereby providing the exact amount of water necessary.

How did the ancient **Romans** get water to their cities?

The ancient Romans and Mesopotamians built aqueducts to transport water between a source and areas where it was needed for agriculture or civilization. The Roman system was very extensive, and was constructed throughout its empire. Some portions of these ancient aqueducts are still in use. Today, modern concrete-lined channels transport water hundreds of miles. The most extensive aqueduct systems in the world today are those that bring water to Southern California from the Colorado River in the east and from the Sacramento River in the north.

Were the **Romans** the **only civilization** to **develop water resources** in an **advanced way**?

No, recent excavations in Henan Province in China uncovered a network of clay pipes built during the Eastern Zhou Dynasty (1122 to 256 B.C.E.); the pipes were connected

Romans constructed this aqueduct in 22 B.C.E. near Caesarea, a town now located in Israel between Tel Aviv and Haifa (photo by Paul A. Tucci).

to many reservoirs around the cities. This use of technology may actually pre-date Roman water works.

Where will the **biggest dam** in the world be **located**?

The Three Gorges Dam, which spans the Yangtze River in China's Hubei province, will be the biggest dam in the world once it is completed. It will be capable of generating 22,500 megawatts of power. Over 1,500,000 people will likely be displaced as result of living in the valleys above the dam. This fact, along with the potential loss of thousands of archaeological sites, has been a source of controversy since the project was conceived in the 1990s.

CLIMATE

DEFINITIONS

What is the difference between **climate and weather**?

Climate is the long-term (usually 30-year) average weather for a particular place. The weather is the current condition of the atmosphere. So, the weather in Barrow, Alaska, might be a hot 70° Fahrenheit, but its tundra climate is generally polar-like and cold.

How are different types of **climates classified**?

The German climatologist Wladimir Köppen developed a climate classification system that is still used today, albeit with some modifications. He classified climates into six categories: tropical humid, dry, mid-latitude, severe mid-latitude, polar, and highland. He also created sub-categories for five of these classifications. His climate map is often found in geography texts and atlases.

What is **global warming**?

Global warming is the gradual increase of the Earth's average temperature—which has been rising since the Industrial Revolution. If temperatures continue to increase, some scientists expect major climatic changes, including the rise of ocean levels due to ice melting at the poles. According to many scientists, global warming is primarily due to the greenhouse effect.

What is the **effect of global warming** and climate change on Earth?

By the year 2100, relative to 1990, world temperatures could rise from 2 to 11.5° Fahrenheit (1.1 to 6.4° Celsius) and sea levels may rise 7.2 to 23.6 inches (18 to 59 centimeters). **61**

How did the **Inca civilization experiment** with **climate**?

In the Urubamba valley in Peru, in a city called Moray, is the remains of a great amphitheater-like terrace system. Archaeologists and scientists now believe that this was a great agricultural laboratory, where each area of the terrace exhibited completely different climates, allowing the Incas to experiment with different climates and growing techniques.

What are the **most polluted cities** in the world in terms of **air quality**?

The most severe air pollution can be found in the following cities: Cairo, Egypt; Delhi, India; Kolkata, India; Tianjin, China; Chongqing, China; Kanpur, India; Lucknow, India; Jakarta, Indonesia; and Shenyang, China.

What are the **sources** of **air pollution**?

Air pollution has two main sources: anthropogenic (man-made) and natural. Man-made sources of pollution include factories, cars, motorcycles, ships, incinerators, wood burning, oil refining, chemicals, consumer product emissions like aerosol sprays and fumes from paint, methane from garbage in landfills, and pollution from nuclear and biological weapons production and testing. Natural sources of pollution may include dust, methane from human and animal waste, radon gas, smoke from wildfires, and volcanic activity.

THE ATMOSPHERE

How much **pressure** does the atmosphere exert upon us?

Average air pressure is 14.7 pounds per square inch (10,335.6 kilograms per square meter) at sea level.

Why is the **sky blue**?

This is one of the world's most frequently pondered questions, and, contrary to what some people believe, the sky's blue color is not due to the reflection of water. Light from the sun is composed of the spectrum of colors. When sunlight strikes the Earth's atmosphere, ultraviolet and blue waves of light are the most easily scattered by particles in the atmosphere. So, other colors of light continue to the Earth while blue and ultraviolet waves remain in the sky. Our eyes can't see ultraviolet light, so the sky appears the only color remaining that we can see, blue.

How many **layers** are in the **atmosphere**?

There are five layers that make up the Earth's atmosphere. They extend from just above the surface of the Earth to outer space. The layer of the atmosphere that we

What is the air made of?

The air near the Earth's surface is primarily nitrogen and oxygen—nitrogen comprises 78 percent and oxygen 21 percent. The remaining 1 percent is mostly argon (0.9 percent), a little carbon dioxide (0.035 percent), and other gasses (0.06 percent).

breathe and exist in is called the troposphere and extends from the ground to about 10 miles (16 kilometers) above the surface. From about 10 miles to 30 miles (16 to 48 kilometers) up lies the stratosphere. The mesosphere lies from 30 to 50 miles (48 to 80 kilometers) above the surface. A very thick layer, the thermosphere, lies from 50 all the way to 125 miles up (80 to 200 kilometers). Above the 125 mile (200 kilometer) mark lies the exosphere and space.

Why can I hear an **AM radio station** from hundreds of miles away at **night** but not during the **day**?

At night, AM radio waves bounce off of a layer of the ionosphere, the "F" layer, and can travel hundreds, if not thousands, of miles from their source. During the day, the same reflection of radio waves cannot occur because the "D" layer of the ionosphere is present and it absorbs radio waves.

Why don't **FM radio waves travel** very far?

FM radio waves are "line of site," which means they can only travel as far as their power and the height of their radio antenna will allow. The taller the antenna, the farther the waves can travel along the horizon (as long as they have enough power).

Does **air pressure change** with elevation?

Yes, it does. The higher you go, the less air (or atmospheric) pressure there is. Air pressure is also involved in weather systems. A low-pressure system is more likely to bring rain and bad weather versus a high-pressure system, which is usually drier. At about 15,000 feet (4,572 meters), pressure is half of what it is at sea level.

What are the different **kinds of clouds**?

There are dozens of types of clouds, but they can all be classified into three main categories: cirriform, stratiform, and cumuliform. Cirriform clouds are feathery and wispy; they are made of ice crystals and occur at high elevations. Stratiform clouds are sheet-like and spread out across the sky. Cumuliform clouds are the ubiquitous cloud—puffy and individual, they can be harmless or they can be the source of torrential storms and tornadoes.

How much of the Earth is usually **covered by clouds**?

At any given time, about one-half of the planet is covered by clouds.

How do **airplanes create clouds**?

When the air conditions are right and it's sufficiently moist, the exhaust from airplanes often creates condensation trails, known as contrails. Contrails are narrow lines of clouds that usually evaporate rather quickly. Contrails can turn into cirrus clouds if the air is close to being saturated with water vapor.

What is the **greenhouse effect**?

The greenhouse effect is a natural process of the atmosphere that traps some of the sun's heat near the Earth. The problem with the greenhouse effect, however, is that it has been unnaturally increased, causing more heat to be trapped and the temperature on the planet to rise. The gasses that have caused the greenhouse effect were added to the atmosphere as a byproduct of human activities, especially combustion from automobiles.

What is the **jet stream**?

The jet stream is a band of swiftly moving air located high in the atmosphere. The jet stream meanders across the troposphere and stratosphere (up to 30 miles [48 kilometers] high) and affects the movement of storms and air masses closer to the ground.

OZONE

What is the **ozone layer**?

The ozone layer is part of the stratosphere, a layer of the Earth's atmosphere that lies about 10 to 30 miles (16 to 48 kilometers) above the surface of the Earth. Ozone is very important to life on the planet because it shields us from most of the damaging ultraviolet radiation from the sun.

Is the ozone layer being **depleted**?

Scientists have recognized that a hole has developed in the ozone layer, a hole that has been growing since 1979. The hole is located over Antarctica and has been responsible for increased ultraviolet radiation levels in Antarctica, Australia, and New Zealand. As the ozone hole grows, it will increase the amount of harmful ultraviolet light reaching the Earth, causing cancer and eye damage, and killing crops and micro organisms in the ocean.

How much of the **ozone layer** is being **depleted**?

Since 1975, scientists believe that more than 33 percent of the ozone layer has disappeared. There is a seasonal factor to the reduction in ozone at any given time during the year, too. At different times, the ozone layer naturally declines or rises. But scientists also know that chlorofluorocarbons (CFCs), which are used for air conditioning, aerosol sprays, halon in fire extinguishers, and the interaction of man-made chemicals with nitrogen in our atmosphere, directly cause ozone depletion. It is a man-made problem that requires a man-made solution.

How do **CFCs** destroy **ozone**?

When CFCs rise up in the atmosphere to the ozone layer, ultraviolet rays break them down into bromine and chlorine, which destroy ozone molecules.

CLIMATIC TRENDS

Why is it **very wet** on one side of a mountain range?

It's much more wet on one side of a mountain than the other because of a process known as orographic precipitation. Orographic precipitation causes air to rise up the side of a mountain range and cool, creating storms. The storms deposit a great deal of precipitation on that side of the mountain and create a rain shadow effect on the opposite side of the range. The Sierra Nevada mountains are an excellent example of orographic precipitation because the mountains of the western Sierras receive considerable rainfall (far more than California's Central Valley), while the eastern Sierras of Nevada are quite dry.

What is a **rain shadow**?

When the moisture in the air is squeezed out by orographic precipitation, there's not much left for the other side of the mountains. The dry side of the mountain experiences a rain shadow effect because they are in the shadow of the rain.

Can people live in a torrid zone?

The ancient Greeks divided the world into climatic zones that are not accurate. The three zones included frigid, temperate, and torrid. They believed that civilized people could only live in the temperate zone (which, of course, was centered around Greece). From Europe northward was part of the inhospitable frigid zone, while most of Africa was torrid. Unfortunately, this three-zone classification system stuck and was later expanded to five zones once the southern hemisphere was explored. People identify everything north of the Arctic Circle (near northern Russia) and south of the Antarctic Circle (near the coast of Antarctica) as frigid, everything between the tropics and the Arctic and Antarctic circles as temperate, and the zone between the Tropic of Cancer and Capricorn as torrid.

What is **El Niño**?

El Niño (also known as ENSO or the El Niño Southern Oscillation), is a large patch of warm water that moves between the eastern and western Pacific Ocean near the equator. When the warm water (about one degree Celsius warmer than normal) of El Niño is near South America, the warm water affects the weather in the southwestern United States by increasing rainfall, and is responsible for changes in the weather throughout the world. El Niño lasts for about four years in the eastern Pacific Ocean and then returns to the western Pacific near Indonesia for another four years. When the warm water is in the western Pacific, it is known as La Niña, the opposite of El Niño. When La Niña is in action, we have "normal" climatic conditions.

Where does the **name El Niño** come from?

The phenomenon of El Niño was discovered by Peruvian fishermen who noticed an abundance of exotic species that arrived with the warmer water. Since this usually occurred around the Christmas season, they called the phenomenon El Niño, which means "the baby boy" in Spanish, in honor of the Christ child. La Niña, the opposite cycle of El Niño, means "the baby girl."

What are **ice core samples** and why are they important?

An ice core sample is a thick column of ice, sometimes hundreds of feet long, that is produced by drilling a circular pipe-like device into thick ice and then pulling out the cylindrical piece. Ice core samples from places like Greenland and Antarctica provide scientists with important clues about past climates. Air trapped in the ice remains there for thousands of years, so when scientists collect ice cores they can analyze the

Climactic changes are making life in places like Ethiopia more difficult. A warming planet means that deserts are expanding, making water resources more scarce.

air to determine the composition of the atmosphere at the time the ice was formed. Sediments and tiny bugs are also found in the ice and provide additional clues to the state of the natural world at the time the ice was deposited.

What is **continentality**?

Areas of a continent that are distant from an ocean (such as the central United States) experience greater extremes in temperature than do places that are closer to an ocean. These inland areas experience continentality. It might be very hot during the summer, but it can also get very cold in winter. Areas close to oceans experience moderating effects from the ocean that reduce the range in temperatures.

What are the **horse latitudes**?

These high-pressure regions, more formally known as subtropic highs, are warm and don't have much wind. Legend has it that the lack of wind sometimes caused sailors of the sixteenth and seventeenth centuries to throw their horses overboard in an effort to conserve water on board. That's how the region, centered around 30° latitude, got its name.

How does land **turn into desert**?

The process known as desertification is complicated and results from such activities as overgrazing, inefficient irrigation systems, and deforestation. It is most widespread in the Sahel region of Africa, a strip of land along the southern margin of the Sahara desert. The Sahara grows larger because of desertification. Desertification can be reversed by changing agricultural practices and by replanting forests.

WEATHER

How do I **convert Fahrenheit** to **Celsius** to **Kelvin**?

Fahrenheit and Celsius are two common temperature scales used throughout the world. Temperature in Fahrenheit can be converted to Celsius by subtracting 32 and multiplying by five; divide that number by nine and you have Celsius. Conversely, you can convert Celsius to Fahrenheit by adding 32, multiplying by nine and finally dividing by five. Kelvin, a system used by scientists, is based on the same scale as Celsius. All you have to do is add 273 to your Celsius temperature to obtain Kelvin. Zero degrees Kelvin is negative 273° Celsius.

What is a **low high** temperature and a **high low** temperature?

When meteorologists look at daily temperature, there is always a low and a high temperature for each day. If the high temperature is the coldest high temperature for that day or for the month, you have a new record—a new low high. Conversely, if the low temperature for a day is quite warm and breaks records, that's a new high low!

What are some **world weather records**?

The following are some amazing weather records. The wettest: Cherranpunji, India, with 500 inches (1,270 centimeters) of rainfall per year; the coldest: Antarctica, with a measurement of −129° Fahrenheit (−89.4° Celsius); the driest: Arica, Chile, which only receives 0.004 inches (0.01 centimeters) of rainfall per year; and the hottest: Azizia, Libya, which has sizzled at 136° Fahrenheit (57.7° Celsius).

What **world weather records** does the **United States** hold?

The United States claims the world's highest surface wind (231 miles [372 kilometers] per hour peak gust in New Hampshire), world's greatest average yearly precipitation (472 inches [1,200 centimeters] in Hawaii), and the world's heaviest 42-minute rainfall (12 inches [31 centimeters] in Missouri).

Why are there so many discrepancies in the world records of weather?

The discrepancies in the data reflect the length of time that we use to measure weather phenomena. Some records were set by observing the weather over decades; others only occurred during the span of a few years or months, or even hours or minutes.

Why is it more likely to **rain in a city** during the week than on the weekend?

Urban areas have an increased likelihood of precipitation during the work week because intense activity from factories and vehicles produce particles that allow moisture in the atmosphere to form raindrops. These same culprits also produce warm air that rises to create precipitation. A study of the city of Paris found that precipitation increased throughout the week and dropped sharply on Saturday and Sunday.

What does a **40 percent chance of rain** really mean?

When the morning weather report speaks of a 40 percent chance of rain, it means that throughout the area (usually the metropolitan area) there is a 4 in 10 chance that at least 0.001 of an inch of rain (0.0025 centimeters) will fall on any given point in the area.

Why is it **hotter in the city** than in the countryside?

Cities have higher temperatures due to an effect known as the urban heat island. The extensive pavement, buildings, machinery, pollution from automobiles, and other things urban cause this warmth in the city. Cities such as Los Angeles can be up to five degrees hotter than surrounding areas due to the urban heat island effect. The term comes from temperature maps of cities where the hotter, urban areas look like islands when isotherms (lines of equal temperature) are drawn.

What is a **thunderstorm**?

Thunderstorms are localized atmospheric phenomena that produce heavy rain, thunder and lightning, and sometimes hail. They are formed in cumulonimbus clouds (big and bulbous) that rise many miles into the sky. Most of the southeastern United States has over 40 days of thunderstorm activity each year, and there are about 100,000 thunderstorms across the country annually. Thunderstorms are different from typical rainstorms because of their lightning, thunder, and occasional hail.

What is **air pollution**?

Air pollution is caused by many sources. There are natural pollutants that have been around as long as the Earth, such as dust, smoke, volcanic ash, and pollens. Humans have added to air pollution with chemicals and particulates due to combustion and industrial activity.

WIND

Where does **wind come from**?

The Earth's atmospheric pressure varies at different places and times. Wind is simply caused by the movement of air from areas of higher pressure to areas of lower pressure. The greater the difference in pressure, the faster the wind blows. Some detailed weather maps show wind speed along with isobars (areas of equal air pressure) indicating the level of air pressure.

In which **direction** does the **west wind blow**?

It blows from the west to the east. Wind is named after the direction from whence it comes.

What are the **westerlies**?

These winds flow at mid-latitudes (30 to 60 degrees north and south of the equator) from west to east around the Earth. The high-altitude winds known as the jet stream are also westerlies.

What are **monsoons**?

Occurring in southern Asia, monsoons are winds that flow from the ocean to the continent during the summer and from the continent to the ocean in the winter. The winds come from the southwest from April to October, and from the northeast (the opposite direction) from October to April. The summer monsoons bring a great deal of moisture to the land. They cause deadly floods in low-lying river valleys, but also provide the water southern Asia relies upon for agriculture.

What is the origin of the **word "monsoon"**?

The word "monsoon" comes from the Arabic word "mausin," meaning season.

What are **dust devils**?

These columns of brown, dust-filled air, which can rise dozens of feet, are not as evil as the name suggests. They are caused by warm air rising on dry, clear days. Winds associated with dust devils can reach up to 60 miles [96.5 kilometers] per hour and cause some damage, but they are not as destructive as tornadoes and usually die out pretty quickly.

What is the **windiest place** on Earth?

Mt. Washington, New Hampshire, is the windiest place. In 1934, regular wind speeds were clocked at 231 miles (372 kilometers) per hour.

Is **Chicago** really the **"windy city?"**

Chicago is not the windiest city in the United States. Chicago's average wind speed of 10.4 miles [16.7 kilometers] per hour is beat by Boston (12.5 [20.1]), Honolulu (11.3 [18.2]), Dallas and Kansas City (both 10.7 [17.2]), and especially in the true windy city, Mt. Washington, New Hampshire (35.3 [56.8]).

Chicago, Illinois, is not really the windiest city in the United States (photo by Paul A. Tucci).

HAZARDS AND DISASTERS

What is a **hazard**?

A hazard is any source of danger that can cause injury or death to humans or that can cause property damage. Hazards range from airline accidents to tsunamis to asteroids smashing into the Earth.

Why is it important to have an **out-of-state contact** in case of a disaster?

It's usually easier to call outside of a disaster area than inside one. By identifying a relative or friend who lives outside of your home state as an emergency contact, your family can ensure communication following a disaster.

What is the difference between a **watch and a warning**?

The U.S. National Weather Service issues watches and warnings for a variety of hazards when they may be imminent. A watch (such as a tornado watch or a flood watch) means that such an event is likely to occur or is predicted to occur. A warning is more serious. It means that a hazard is already occurring or is imminent. Warnings are usually broadcast on television and radio stations via the Emergency Alert System (formerly known as the Emergency Broadcast System).

How should we **prepare for disaster**?

Disasters can and do happen everywhere. You should prepare for disaster by having a disaster supply kit with supplies for you and everyone in your family available at home and work, as well as a mini kit in your automobile. It should include food, water, first aid equipment, sturdy shoes, an AM/FM radio (with batteries kept outside of the radio), a flashlight (with batteries kept outside of the flashlight), vital medication

73

What's the difference between the old Emergency Broadcast System and the Emergency Alert System?

The Emergency Broadcast System (EBS), created in 1964 to warn the country of a national emergency such as nuclear attack, became the Emergency Alert Service (EAS) in 1997. The old EBS system relied on one primary radio station in each region to receive an emergency message and then broadcast it to the public and other media outlets. The new system, which also includes cable television, operates via computer and can be automatically and immediately broadcast to the public. It also allows additional local governmental agencies the opportunity to broadcast emergency messages. Future plans for the EAS include radios and televisions that will automatically turn on when an alert is announced.

(especially prescription medication), blankets, cash (if the power and computers are down, credit and ATM cards won't work), games and toys for children, and any other essentials. Contact your local chapter of the Red Cross for more information about disaster preparedness.

Should we use **candles** after a disaster or power outage?

Many deaths and a great deal of property damage have been caused by fires resulting from people using candles following a disaster. People leave candles burning as a source of light, but these can fall over and start fires. It is strongly advised that people not use candles when the power goes out. There are many flashlights and battery operated lanterns that are available commercially and should be part of your disaster supply kit.

What is the leading cause of **disaster-related death** in the United States?

Lightning is the leading cause of disaster-related death in America. From 1940 to 1981, about 7,700 people died from lightning strikes, 5,300 from tornadoes, 4,500 from floods, and 2,000 from hurricanes. So, it's best to avoid open spaces, high ground, water, tall metal objects, and metal fences during an electrical storm.

How can I learn more about **disasters in my town**?

Each community should have its own disaster plan that includes a history of past disasters (those that have happened in the past are likely to occur in the future) along with plans for dealing with future disasters. You should be able to consult this plan to learn how your community would cope with disaster and to find out the locations of evacuation routes and shelters. Many communities place important disaster planning information in the telephone book for easy reference.

What do medical geographers do?

Medical geographers and epidemiologists (scientists who study disease and health) regularly use maps and spatial information to help control illness and death. Mapping has solved the mystery behind high levels of cancer in small areas and has been used to understand the spread of AIDS. Medical geographers don't just study the distribution of disease, they also investigate the accessibility of people to health services.

What is the best way to **help after a disaster**?

Disaster relief agencies such as the Red Cross are in vital need of money after a disaster to purchase necessary items for victims or provide financial support to them. Call your local chapter of the Red Cross to find out how to help. Donating food or clothing places additional burdens on the agencies in the immediate aftermath of a disaster, as personnel are not available to sort, clean, or distribute donated goods.

How did a **map help stop** the spread of **cholera**?

During an 1854 cholera outbreak in London, a physician named John Show mapped the distribution of cholera deaths. His map showed that there was a high concentration of deaths in an area surrounding one specific water pump (water had to be hand-pumped and carried in buckets at the time). When the handle was taken off of the water pump, the number of cholera deaths plummeted. When it was determined that cholera could be spread through water, future epidemics were curbed. This was the beginning of medical geography.

What are **incidence maps**?

Researchers at such institutions as the Centers for Disease Control and Prevention (CDC) use incidence maps, which plot where and how people have been infected or exposed to such potentially harmful viruses as the influenza, the Ebola Virus, West Nile Virus, and HIV in order to understand the rate of transmission as related to geography. An incidence map may help scientists figure out the origin of a disease and where and how quickly it is spreading. Global incidence maps are of increasing importance in the fight against potentially harmful biological disasters.

Which natural disasters doesn't **Southern California** experience?

While urban southern California is plagued by earthquakes, wild fires, floods, landslides, and tornadoes (yes, even tornadoes), they rarely receive snowstorms or hurricanes.

VOLCANOES

What are **volcanoes**?

Volcanoes are the result of magma rising or being pushed to the surface of the Earth. Hot liquid magma, which is located under the surface of the Earth, rises through cracks and weak sections of rock. The mountain surrounding a volcano is formed by lava (called magma until it arrives at the Earth's surface) that cools and hardens, making the volcano taller or wider or both.

What is the **Ring of Fire**?

If you were to look at a map of the world's major earthquakes and volcanoes, you would notice a pattern circling the Pacific Ocean. This dense accumulation of earthquakes and volcanoes is known as the Ring of Fire. The ring is due to plate tectonics and the merger of the Pacific plate with other surrounding plates, which creates faults and seismic activity (especially Alaska, Japan, Oceania, and the west coasts of North and South America), along with volcanic mountain ranges such as the Cascades of the U.S. Pacific Northwest and the Andes of South America.

Lava erupts from a volcano. Liquid rock, when below the Earth's crust, is called magma.

How many **active volcanoes** are there in the world?

There are about 1,500 active volcanoes around the world. Most are located in the Ring of Fire surrounding the Pacific Ocean. About one-tenth of the world's active volcanoes are located in the United States. A volcano is considered active if it has erupted in the last 10,000 years.

What are some of the **world's most active volcanoes**, in terms of numbers of years of eruptions?

The volcanoes that have been active the most number of years include Mt. Etna in Italy (3,500 years), Mt. Stromboli in Italy (2,000 years), and Mt. Yasur in Vanuatu (800 years).

The ruins of Pompeii, the ancient city destroyed in 79 C.E. by the eruption of Mt. Vesuvius, can still be visited by tourists near Naples, Italy (photo by Paul A. Tucci).

Where are the **active volcanoes** in the **United States**?

Washington, Oregon, and California have many potentially active volcanoes. The most recent eruption in the United States was that of Mt. St. Helens in southern Washington state in 1980. Other volcanoes in the region, such as Mt. Shasta, Lassen, Rainier, and Hood could erupt with little warning.

What is the difference between **magma and lava**?

Magma is hot, liquefied rock that lies underneath the surface of the Earth. When magma erupts or flows from a volcano onto the Earth's surface, it becomes lava. There is no difference in substance; only the name changes.

How was **Pompeii** destroyed?

In the year 79 C.E., the volcano Mt. Vesuvius erupted and buried the ancient Roman town of Pompeii under 20 feet (6 meters) of lava and ash. Pompeii is famous because excavations of the city, which began in 1748 and continue to this day, provide an excellent look at Roman life at the beginning of the millennium. The covering of the city by debris preserved not only the places where people last stood but also paintings, art, and many other artifacts. The nearby city of Herculaneum also was perfectly preserved. Although a much smaller version of Pompeii, it contains some of the best art, architecture, and examples of daily life in Roman times and is only 20 minutes away from Pompeii.

EARTHQUAKES

What causes **earthquakes**?

The tectonic plates of the Earth are always in motion. Plates that lie side by side may not move very easily with respect to one another; they "stick" together and occasionally they slip. These slips (from a few inches to many feet) create earthquakes and can often be very destructive to human lives and structures.

What is an **epicenter**?

An epicenter is the point on the Earth's surface that is directly above the hypocenter, or point where earthquakes actually occur. Earthquakes do not usually occur at the surface of the Earth but at some depth below the surface.

What is a **fault**?

A fault is a fracture or a collection of fractures in the Earth's surface where movement has occurred. Most faults are inactive, but some, like California's San Andreas fault, are quite active. Geologists have not discovered all of the Earth's faults, and sometimes earthquakes occur that take the world by surprise, like the one in Northridge, California, in 1994. When earthquakes occur on faults that were previously unknown, they are called blind faults.

What does an **earthquake feel like**?

Smaller earthquakes or tremors feel disorienting at first. You feel a sense that the room is spinning, as if you are becoming dizzy. Usually preceding an earthquake, when the initial tremors hit, you can hear the sounds of things rattling that you have never heard before, like glasses rubbing against each other and windows vibrating. With larger earthquakes, as the earth nearby tears or opens, you can hear a very loud rumbling sound that is similar to a train driving by.

What is the significance of the **San Andreas fault**?

The infamous San Andreas fault is the border between the North American and the Pacific tectonic plates. This fault lies in California and is responsible for some of the major earthquakes that occur there. Los Angeles is on the Pacific Plate but San Francisco is on the North American Plate. The Pacific Plate is sliding northward with respect to the North American Plate and, as a result, Los Angeles gets about half an inch closer to San Francisco every year. In a few million years, the two cities will be neighbors.

Will California eventually fall into the ocean?

No, it will not. The famous San Andreas Fault, which runs along the western edge of California from the San Francisco Bay Area to Southern California, is known as a transverse fault. This means that the western side of the fault, which includes places like Monterey, Santa Barbara, and Los Angeles, is sliding northward with respect to the rest of the state. In a few million years, the state's two largest urban areas, San Francisco and Los Angeles, will be right next to each other. The fault is moving at about two centimeters (just under an inch) a year.

Was **San Francisco** destroyed by **earthquake** or by fire in 1906?

In 1906 a very powerful earthquake struck San Francisco, California, which sparked a fire that destroyed much of the city. In an effort to preserve San Francisco's image with residents and would-be visitors, official policy regarding the disaster stated that it was not the earthquake but mostly the fire that destroyed the city. Official books and publications produced after the earthquake referred to both the fire and the earthquake as having caused the damage. In fact, the earthquake did considerable damage to the city and killed hundreds.

Which states are **earthquake-free**?

While a 20-year period isn't an excellent indicator, there were four states that had no earthquakes between 1975 and 1995: Florida, Iowa, North Dakota, and Wisconsin.

Is there a high risk of **earthquakes** in the **Midwestern United States**?

Great earthquakes struck the New Madrid, Missouri, area in 1811 and 1812. They caused considerable damage (some areas experienced shaking at the level of XI on the Mercalli scale) and were felt as far away as the East Coast. The potential exists for future earthquakes in the region, since earthquakes have occurred there before. Planning and preparedness continues throughout the region, centered at the junction of Missouri, Arkansas, Illinois, Kentucky, Tennessee, and Mississippi.

What should I do **in the event of an earthquake**?

Duck, cover, and hold! Duck under a table, counter, or any area that can provide protection from falling objects. Cover the back of your head with your hands to help protect against flying debris. Hold on to the leg of the table or anything solid to ride out the shaking.

Is it safe to **stand in a doorway** during an earthquake?

While a doorway is a nice, structurally sound place to be during an earthquake, officials have found that many people are injured when a door swings open and closed during an earthquake, so you may want to avoid standing in a place where your fingers can become crushed.

What is the **Richter scale**?

The Richter scale measures the energy released by an earthquake. It was developed in 1935 by California seismologist Charles F. Richter. With each increase in Richter magnitude, there is an increase of 30 times the energy released by an earthquake. For example, a 7.0 earthquake has 30 times the power of a 6.0, and an 8.0 is 900 times as powerful as a 6.0. Each earthquake only has one Richter magnitude. The strongest earthquakes are in the 8.0 range—8.6 for Alaska's 1964 earthquake, and 8.0 for China's 1976 earthquake in Tangshan.

What is the **Mercalli scale**?

The Mercalli scale measures the power of an earthquake as felt by humans and structures. It was developed in 1902 by Italian geologist Giuseppe Mercalli. The Mercalli scale is written in roman numerals and it ranges from I (barely felt) to XII (catastrophic). The Mercalli scale can be mapped surrounding an epicenter and will vary based on the geology of an area.

The Mercalli Scale of Earthquake Intensity

I	Barely felt
II	Felt by a few people, some suspended objects may swing
III	Slightly felt indoors as though a large truck were passing
IV	Felt indoors by many people, most suspended objects swing, windows and dishes rattle, standing autos rock
V	Felt by almost everyone, sleeping people are awakened, dishes and windows break
VI	Felt by everyone, some are frightened and run outside, some chimneys break, some furniture moves, causes slight damage
VII	Considerable damage in poorly built structures, felt by people driving, most are frightened and run outside
VIII	Slight damage to well-built structures, poorly built structures are heavily damaged, walls, chimneys, monuments fall
IX	Underground pipes break, foundations of buildings are damaged and buildings shift off foundations, considerable damage to well-built structures

X	Few structures survive, most foundations destroyed, water moved out of banks of rivers and lakes, avalanches and rockslides, railroads are bent
XI	Few structures remain standing, total panic, large cracks in the ground
XII	Total destruction, objects thrown into the air, the land appears to be liquid and is visibly rolling like waves

How many **really big earthquakes** occur each year?

On average, there are about 100 earthquakes of magnitude 6.0-6.9, about 20 of magnitude 7.0-7.9, and two huge 8.0-8.9 earthquakes each year. Many of these really big earthquakes occur in the ocean, so we don't hear much about them.

Is a **magnitude 10** the top of the Richter scale?

While the media often refers to the Richter scale as being on a scale of 1 to 10, there is no upper limit, even though the strongest quakes are not as high as 10. It is incorrect to assume that a 7 is on a scale of 1 to 10 because the magnitudes are based on the energy released and it is a logarithmic scale.

TSUNAMIS

What is a **tsunami**?

A tsunami, also known as a seismic sea wave, is usually caused by an earthquake that occurs under the ocean or near the coast. The seismic energy creates a large wave that can cause heavy damage hundreds or even thousands of miles from its source. Hawaii is frequently struck by tsunamis.

How does **Hawaii** protect itself from tsunamis?

There is a sophisticated global monitoring network that provides warnings about possible tsunamis, allowing the islands of Hawaii and other coastal areas to prepare for impending disaster. Hawaii also has a thorough evacuation system to protect lives in the face of tsunami danger.

What caused the **great Indian Ocean Tsunami** of December 2004?

A magnitude 9.0 earthquake in the ocean off the coast of Sumatra, Indonesia, caused a reverberating swell of water to move toward Indonesia, Thailand, Sri

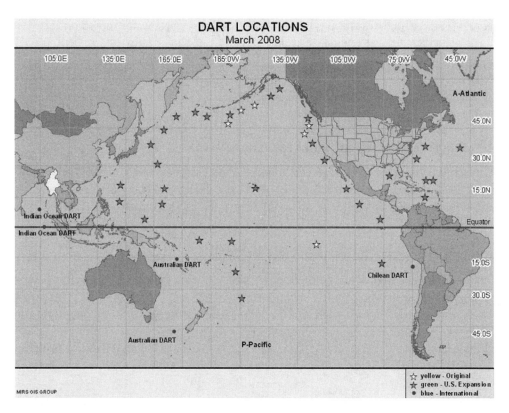

DART LOCATIONS
March 2008

☆ yellow - Original
★ green - U.S. Expansion
● blue - International

MRS GIS GROUP

A map showing the location of DART tsunami monitoring buoys across the world (map courtesy National Oceanic and Atmospheric Administration).

Lanka, India, and the Maldives. Its effects were felt as far away as Africa, before it finally dissipated.

How many people were **killed** during the **2004 tsunami disaster** in Asia?

More than 160,000 people were killed by the initial waves and the resultant flooding and destruction that followed.

What is the **Pacific Tsunami Warning Center**?

Administered by the National Oceanic and Atmospheric Administration, and based in Hawaii and Alaska, the Pacific Tsunami Warning Center is the command center for monitoring and warning all nations which may be affected by a tsunami. With data from a network of 39 detection buoys called the Deep-Ocean Assessment and Reporting of Tsunami array, the center can issue alerts of real time earthquake activity in the Pacific basin, and the tsunamis that may result, giving residents affected time to head to safe ground, away from low lying coastal areas.

What is the **DART array**?

DART stands for the Deep-Ocean Assessment and Reporting of Tsunami array, which consists of 39 buoys that float in critical spots in the Pacific. Each DART system consists of an anchored seafloor bottom pressure recorder (BPR) and a companion moored surface buoy for real-time communications. An acoustic link transmits data from the BPR on the seafloor to the surface buoy. The BPR collects temperature and pressure at 15-second intervals. In normal mode, it transmits the data every 15 minutes. If there is an event, the system reports back data collected in 15-second intervals every minute.

HURRICANES

What part of a **hurricane** is **most damaging**?

Floods caused by hurricanes are the most destructive element. The low-pressure center of a hurricane causes a mound of water to rise above the surrounding water. This hill of water is pushed by the hurricane's fierce winds and low pressure onto the land, where it floods coastal communities, causing significant damage. Hurricanes also spark tornadoes that contribute to the devastation.

What is a **willy-willy**?

Willy-willy is the Australian name for a hurricane.

How fast do **hurricane winds** blow?

The strongest hurricanes have winds that reach speeds well over 150 miles (240 kilometers) per hour.

How are **hurricanes ranked**?

Hurricanes are ranked on a scale of one to five, with category one hurricanes being the weakest and category five being the strongest and most destructive. The damage caused by each category of hurricane ranks from: 1, minimal; 2, moderate; 3, extensive; 4, extreme (such as Hurricane Andrew in 1992); to 5, catastrophic.

Hurricane Type	Winds (MPH/KPH)	Surge Levels
Category One	74-95 mph/119-153 kph	4-5 feet/1.2-1.5 meters
Category Two	96-110 mph/154-177 kph	6-8 feet/1.8-2.4 meters
Category Three	111-130 mph/178-209 kph	9-12 feet/2.75-2.4 meters
Category Four	131-155 mph/210-249 kph	13-18 feet/4-5.5 meters
Category Five	> 155 mph/> 249 kph	> 18 feet/> 5.5 meters

83

What was **Hurricane Katrina**?

Hurricane Katrina was the name given to the hurricane that developed in the Gulf of Mexico and struck New Orleans and many other cities along the southern coast of the United States in late August 2005. Winds from Katrina were initially only a category two hurricane, with a tidal surge ranked as a category three.

How many **people died** as a result of the subsequent **failure of the levees** and flooding after **Hurricane Katrina** struck?

Approximately 1,460 people lost their lives following the landfall of Hurricane Katrina.

Was the **2005 New Orleans disaster caused** by a flood or a hurricane?

The initial cause of the disaster was Hurricane Katrina, which whipped up tides and surged rain and sea water against a very fragile levee system that protected New Orleans. The city is 49 percent below sea level, and so when the man-made levees broke, flood waters moved in and inundated much of the city.

FLOODS

How much rain does it take to make a flood?

The amount varies widely for different areas. In some U.S. western deserts, or in some large urban areas, just a few minutes of strong rain will cause a flash flood in canyons and low-lying areas. In areas prone to greater rainfall amounts, it often takes quite a bit more rain (sometimes a few days' or weeks' worth) to cause rivers to overflow and dams to fill up, raising concerns of those who live downstream. Areas that normally receive more rainfall have better natural drainage systems and are usually home to plants that readily absorb the extra water.

What have been some of the **most destructive floods** in history?

In the United States, the failure of a dam in 1889 upstream from the community of Johnstown, Pennsylvania, killed 2,200 people. Some of the world's most catastrophic flooding takes place in China. A flood on the Huange He River in 1931 killed 3.7 million people.

What is a **floodplain**?

A floodplain is the area surrounding a river that, when unmodified by human structures, would normally be flooded during a river flood. A floodplain can be a few feet or many miles wide, depending on the river flow as well as the local terrain. Even though

Why do people live in floodplains?

People have lived in floodplains for thousands of years. Fertile land for agriculture lines the floodplain, and the nearby water source makes life easier. Unfortunately, when the river does flood, these communities are severely damaged and people suffer. Hazard mitigation, such as levees, dams, dikes and other structures, attempt to limit damage during floods. Sometimes, when the structures fail (such as a levee breaking), large areas are inundated with water. Inhabitants of floodplains must balance the risks with the rewards of living in such an unpredictable environment.

levees and flood walls can be built (with homes and businesses built just behind them), the floodplain does not vanish. If the structures break or are damaged, the water from a flood can fill a floodplain, just as it did before humans occupied it.

What is a **100-year flood**?

A 100-year flood refers not only to the size of a flood, but also to the odds of it occurring. A 100-year flood has a one percent (or 1 in 100) chance of occurring in any given year. It has no relationship to the frequency of occurrence. The magnitude of such a flood is relative to the frequency of occurrence, so a 100-year flood is much larger than any run-of-the-mill annual flood. A 500-year flood only has a one in 500 (0.2 percent) chance of occurring in any given year and would be much larger and more devastating than a 100-year flood.

What is the **National Flood Insurance Program**?

The National Flood Insurance Program (NFIP) was established by the U.S. federal government in 1956 as a subsidized insurance program for home and business owners. The government began the program by creating Flood Insurance Rate Maps (FIRM) showing the boundaries of 100-year and 500-year flood zones. The cost of the insurance is based on the flood risk. The Federal Emergency Management Agency (FEMA) oversees the program and requires the purchase of flood insurance by any owner affected by a disaster before they can be provided with disaster assistance. This way, the next time a flood occurs, they will be insured.

How can I obtain a **flood map** of my community?

The best way to see a Flood Insurance Rate Map (FIRM) for your area would be to contact your local government. Their planning or emergency management agency should have

85

the FIRM maps available. Purchasing them from FEMA is not recommended because the maps change often and are best interpreted by a planning or emergency expert.

What should I do in the event of a flood?

If a flood is expected, turn on your battery-powered radio and listen for information about when and where to evacuate. If a flood or flash flood is coming toward you, move quickly to a higher elevation—but don't ever try to outrun a flood. Also, don't drive through standing water, as it can quickly rise and stall your vehicle, possibly trapping you among swift water.

TORNADOES

What are tornadoes?

Tornadoes are very powerful, yet tiny storms that have destructive winds capable of leveling buildings and other structures. Winds in a tornado form a dark gray column of air, with the center of the tornado acting like a vacuum, picking up objects and moving them along the storm's path. Tornadoes can last from a few minutes to an hour.

What should I do when a tornado approaches?

Try to get to the lowest level of the building (unless you are in a mobile home or outdoors, in which case you should seek a sturdy and safe shelter). Go to the center of the room and hide under a sturdy piece of furniture. Stay away from windows, hold on to the leg of a table or something else stable, and protect your head and neck with your arms.

What is the Fujita scale of tornado intensity?

The Fujita scale measures the strength of a tornado based on observed damage and effects. The scale ranges from F0 (a weak tornado) through F6 (an almost inconceivable tornado, having close to no

A tornado touching down near Bennett and Watkin, Colorado, in 2006.

> ### How many people on average are killed by tornadoes each year in the United States?
>
> **A**pproximately 1,500 people are killed each year due to the destructive power of tornadoes. Many more people are injured and displaced as a result.

chance of actually occurring). About 75 percent of all tornadoes are weak (F0-F1), while only one percent are violent (F4-F5).

Where is **tornado alley**?

Tornadoes occur more frequently in the central United States than anywhere else in the world. Tornado Alley is an area stretching from northwest Texas, across Oklahoma (the tornado capital of the world), and through northeast Kansas. On average, over 200 tornadoes occur across Tornado Alley each year.

What is the **most dangerous state** to live in due to tornadoes?

Massachusetts is considered the most dangerous state to live in due to tornadoes. While Oklahoma receives far more tornadoes than Massachusetts does, the population density and risk of death or severe injury is greater in the New England state.

What were some of the most **destructive tornadoes** in **U.S. history**?

Some of the worst tornadoes in U.S. history include: the Tri State tornado, which struck Missouri, Indiana, and Illinois in 1925, killing 695 people and injuring 2,027; the 1840 Natchez tornado struck Mississippi, killing 317 people and causing injuries to another 109; and the St. Louis/East St. Louis tornado of 1896 killed 296 people and injured 1,000.

Are there **tornadoes in Europe**?

While 90 percent of all tornadoes occur in the United States, there are tornadoes in Europe, especially in western France. Other tornado regions of the world include eastern and western Australia, southern Brazil, Bangladesh, South Africa, and Japan.

LIGHTNING

How many **times** does **lightning strike** the Earth each year?

About 20 million bolts of lightning are generated in the atmosphere every year.

How much **energy** does one **bolt of lightning** contain?

A bolt of lighting contains enough energy to light a 100-watt light bulb for three months.

What different **types of lightning** are there?

There are four types of lightning: cloud-to-cloud, within a cloud, cloud-to-ground, and cloud-to-air. Of course, cloud-to-ground is the most dangerous form of lightning, especially in the spring and summer months, when more people are more likely to be outside.

How many **people are killed** in the **United States** by **lightning**?

Approximately 73 people die each year after being struck by lightning. The highest death rates are in Florida (425 killed between 1959 and 2003). The U.S. total from 1959 to 2003 is 3,696 deaths. This means that about 80 people die each year from lightning, and about 300 are injured annually.

Does lightning ever **strike twice** in the same place?

Lightning can and often does strike in the same place twice. Since lightning bolts head for the highest and most conductive point, that point often receives multiple strikes of lightning in the course of a storm—so stay away from something that has already been struck by lightning! Tall buildings (such as the Empire State Building) often receive numerous lightning strikes during a storm.

OTHER HAZARDS AND DISASTERS

What is **acid rain**?

Motor vehicles and industrial activity release tons of pollutants into the air. When mixed together, the pollutants form sulfuric and nitric acids that later fall to the ground in rain or snow. This precipitation is known as acid rain. Acid rain is responsible for damaging lakes by killing plant and animal life and for killing trees around the world. Canada has been especially hard-hit by acid rain caused by industrial activities in the United States.

Does **radiation** from a **nuclear plant** stop at the 10-mile (16 kilometer) zone?

American nuclear power plants are required to create emergency planning zones within a 10-mile (16 kilometer) radius surrounding their plants. These imaginary 10-mile lines are not walls that hold back the effects of radiation, but simply a distance determined by emergency planners. In the event of an accident, the residents of the

In 1979, the nuclear reactor at Three Mile Island in Pennsylvania caused a panic when radioactive rods broke. However, no radiation was released. A far worse disaster occurred in Chernobyl, Ukraine, in 1986.

10-mile (16 kilometer) zone might not need to be evacuated but could be advised to remain indoors with their windows closed. Nuclear plants also establish smaller zones of two and five miles (three to eight kilometers) surrounding the plants, within which the risk of radiation exposure is much greater.

What happened at **Three Mile Island**?

Three Mile Island, Pennsylvania, was the site of the United States' worst nuclear accident. Luckily, no radiation was released into the environment and no one was killed. In March 1979, the nuclear reactor at the Three Mile Island plant overheated, breaking the radioactive rods. Pennsylvania's governor recommended a voluntary evacuation of pregnant women and preschool children who lived within five miles (8 kilometers) of the plant. It was the unexpected self-evacuation of residents in the area that created major problems. The evacuations yielded surprising information about the lack of preparedness of communities for such an event, and have led to increased planning and preparedness for nuclear accidents and evacuations.

What is **nuclear winter**?

A nuclear winter is what would follow a large-scale nuclear war. Radioactive particles, dust, and smoke released into the atmosphere would create a large cloud over the planet, blocking out sunlight and reducing temperatures worldwide. Plants and animals would die due to the extremely low temperatures. An extended nuclear winter could cause the death of millions of people from starvation, cold, and other problems.

What caused the **Bhopal disaster**?

In December 1984, the U.S.-owned Union Carbide pesticide plant in Bhopal, India, leaked toxic chemicals (methyl isocyanate gas) that killed over 3,800 people. It was the worst industrial accident in history. Union Carbide paid a fine of $470 million to avoid facing criminal charges.

TRANSPORTATION AND URBAN GEOGRAPHY

CITIES AND SUBURBS

What is a **city**?

In the United States, a city is a legal entity, delegated power by a state and county to govern and provide services to its citizens. Cities also have charters, which are somewhat akin to local constitutions, and have specific boundaries.

What was the **first city** to have more than **one million people**?

During ancient times, Rome was the world's first city to have a population larger than one million. Rome's population declined during the fall of the Roman Empire in the fifth century, and a city with a population of one million wasn't again seen until the early nineteenth century in London.

What is an **urban area**?

An urban area consists of a central city and its surrounding suburbs. Urban areas are also known as metropolitan areas. In some cases, urban areas can spread dozens of miles beyond the central city.

How many people in the world are classified as **urban dwellers**?

Forty-eight percent of the world, approximately three billion people, live in urban areas today. Another 52 percent, or 3.3 billion people, live in rural areas.

What are some of the **oldest, continuously inhabited cities** in the world and when were they established?

Some of the oldest cities still inhabited today are Jericho, Palestine (founded 9000 B.C.E.); Byblos, Lebanon (founded 5000 B.C.E.); Damascus, Syria (founded 4300 B.C.E.); Allepo, Syria (founded 4300 B.C.E.); and Susa, Iran (founded 4200 B.C.E.).

Where did some of the **first cities** of the world **begin**?

Scientists generally believe that the Sumer Valley area of ancient Mesopotamia, located in modern-day Iraq in the Middle East, is the origin of the earliest cities.

What is the **most populated urban area** in the world?

The capital city of Tokyo, Japan, still reigns as the most populated urban area in the world, with over 35 million inhabitants. The next largest urban areas include Mexico City, Mexico (18.7 million), New York-Newark, United States (18.3 million), São Paulo, Brazil (17.9 million), and Mumbai, India (17.4 million).

What is a **megalopolis**?

Geographer Jean Gottmann developed the term megalopolis to describe the huge, interconnected metropolitan area from Boston to Washington, D.C. "Boswash," as this original megalopolis has been called, has been joined by such nascent megalopolis as "Chi-Pitts" (from Chicago to Pittsburgh), the Ruhr area in Germany, Italy's Po Valley, and "San-San" (from San Francisco to San Diego).

Are **mega-urban** area populations **growing**?

Despite what we might think, of the top 20 large urban areas in the world, nearly half experience population growth of only 1.5 percent.

What are the **largest metropolitan areas** in the United States?

The New York metropolitan area is America's largest, with 20 million people. Los Angeles is second with 15 million; Chicago is third with 8.5 million; Washington, D.C., is fourth with 7 million; and San Francisco is fifth with 6.5 million.

How does the **population of Tokyo change** each **day**?

It is said that the population of Tokyo City changes each day as 22 million people commute by rail into the city, riding a combination of trains and buses for an average two hours per day.

What **percentage of people** live in the **mega cities** of the world?

Only four percent of the inhabitants of Earth live in gigantic urban areas. The rest live in areas with populations of less than 10 million people. Twenty-five percent of the world's inhabitants live in urban areas with fewer than 500,000 residents.

Large urban areas such as Tokyo, Japan, have made some efforts to add green space to make city living more comfortable for residents (photo by Paul A. Tucci).

What major **American city loses more people** than any other?

Detroit, Michigan, loses more than 25,000 people each year, three times as many people as any other metropolitan area. It leads all other cities in the United States in negative population growth. By comparison, the Dallas/Ft. Worth, Texas, area continues to lead the United States in terms of population growth, adding more than 160,000 people each year.

What is a **central business district**?

A central business district (CBD) of a city is located downtown, often where the city began, and is the primary concentration of commercial buildings.

Who decides where **houses can be built**?

Almost all American city governments have a department that is responsible for planning the layout of the city. The planning department for each city enforces and delineates zoning, which regulates the location of homes, businesses, factories, and even nuclear power plants.

When did **suburbs** become **fashionable**?

Following World War II, the subsequent housing boom and construction of interstate highways led to the development of low-density housing surrounding cities. These areas of low-density housing are known as suburbs. Suburbs have been extremely popular since the 1950s.

What was **Levittown**?

The three Levittowns were large housing developments built from the mid-1940s through the early 1960s by William J. Levitt and his construction company. Levitt invented a process to mass-produce homes by making each home exactly the same. The first Levittown was located in New York and consisted of 17,000 homes. The subsequent Levittowns were built in New Jersey and Pennsylvania. Levittowns were the forerunner of the suburb.

URBAN STRUCTURES

What is the **largest building** in the world, in terms of usable space?

The Boeing Plant in Everett, Washington, has over 4.3 million square feet (398,000 square meters) of space. It is followed by Airbus Industries A380 Assembly Plant near Toulouse, France, with 1,320,000 square feet (122,500 square meters); Aerium in Brandenburg, Germany, with 753,000 square feet (70,000 square meters); and the NASA Vehicle Assembly Building in Brevard County, Florida, at 384,000 square feet (32,374 square meters).

What is the **largest building** in the world in terms of **floor space**?

The Aalsmeer Flower Auction, in Aalsmeer, the Netherlands, is the largest building in the world in terms of floor space. It covers 99 hectares (240 acres) in area.

Why is there such controversy about defining the tallest building in the world?

Most of the controversy about which building is the tallest centers around defining what is a building versus what is a structure. Some buildings, in order to be listed among the tallest buildings, add other non-habitable structures to the top in order to rise in the rankings. These structures may be in the form of communication towers, which can easily add significant height to the building.

What is the **largest office building** in the world?

The United States is home to the largest office building in the world, the Pentagon, which has 3.7 million square feet (343,730 square meters) of space under its roof. The Pentagon is home to the country's Department of Defense.

What is the **largest airport** in the world?

In terms of floor area, China's Beijing Capital International Airport has over 10.6 million square feet (986,000 square meters) of space.

What is the **largest church building** in the world?

The Mormon Church LDS Conference Center in Salt Lake City, Utah, is the largest church building, having over 1.4 million square feet (130,000 square meters).

What is the **longest building** in the world?

Catching a flight from Osaka, Japan's Kansai International Airport could involve a walk of more than a mile; it is 5,580 feet (1.7 kilometers) long.

What is the **largest shopping mall** in terms of gross leasable area?

The South China Mall in Donguan, China, has more than 6.5 million square feet (600,000 square meters) of leasable space. It is the largest shopping mall in the world in terms of total area, consisting of 9.6 million square feet (890,000 square meters).

What is the **tallest self-supporting structure** in the world?

The CN Tower in Toronto, Canada, built as a television transmission tower, is the world's tallest self-supporting structure, at 1,815 feet (553 meters).

What was the **world's first skyscraper**?

Completed in 1885 in Chicago, Illinois, the Home Insurance Company Building was the world's first skyscraper.

What is the **tallest skyscraper** in the world?

The Burj Dubai skyscraper is currently the tallest structure. When completed, it will reach a height of over 2,064 feet (629 meters). It has surpassed the CN Tower in Toronto, Canada, which since 1976 was the world's tallest building at 1,815 feet (553.3 meters).

The Burj Dubai skyscraper is the world's tallest building at 2,064 feet (629 meters).

AIR TRANSPORTATION

What was the **first airplane** flown?

Orville and Wilbur Wright were the first two people to fly in a heavier-than-air vehicle, called the "Flyer." They made their historic flight on December 17, 1903, at Kitty Hawk, North Carolina.

When was the **first flight** achieved?

In 1783 the Montgolfier Brothers flew the first hot air balloon across Paris, France.

What is the **busiest airport** in the world?

The busiest airport in the world, in terms of numbers of passengers annually, is Hartsfield-Jackson Atlanta International Airport in Atlanta, Georgia, with over 78 million passengers annually. The next busiest is Chicago, Illinois's O'Hare International Airport, with 70 million passengers per year.

How many **airports** are there in the **United States**?

There are approximately 19,379 airports in the United States, including 5,233 that are designated for public use.

What is the **busiest airport outside of the United States**?

Heathrow Airport in London, England, has more than 62 million passengers passing through the various terminals annually. This is followed by Charles De Gaulle International Airport near Paris, France, with 54 million travelers; and Schipol Airport in Amsterdam, the Netherlands, with 47 million passengers.

ROADS AND RAILWAYS

When were the **first roads** built?

Archaeologists have found remnants of early roads in Ur, Iraq, and Glastonbury, Scotland, dating back to 4000 B.C.E.

Do all **roads** really **lead to Rome**?

Not any more. During the time of the Roman Empire, the Romans built a massive road network to ensure easy travel in all weather conditions between Rome and the furthest reaches of the Empire. The Romans made their roads as straight as possible and paved large sections of them by precisely piecing together cut rock to make a flat surface. Along the 50,000 miles (80,000 kilometers) of Roman roads, markers were placed every Roman mile (just short of a modern mile) so as to indicate either the distance to Rome or to the city where the road originated. After the fall of Rome, the maintenance of the Roman road system was severely neglected, and during the Middle Ages the roads became overused and dilapidated. Though the Romans built these roads over 2,000 years ago, some segments are still in use today.

What is a **turnpike**?

A turnpike is a toll road. In the late eighteenth century, private companies in the United States and in the United Kingdom built roads and charged users to pass. Beginning in the 1840s, turnpikes had to compete for traffic, and thus profits, with the railroads. The name turnpike is still common on toll highways in the eastern United States, such as the New Jersey Turnpike, the Massachusetts Turnpike, and the Pennsylvania Turnpike.

What road in the United States was known as the **National Road**?

The Cumberland Road, also known as the National Road, was the first federally funded road in the United States. Though construction began in 1811, the Cumberland Road was not completed until 1852. Stretching 800 miles (1,287 kilometers) from Cumberland, Maryland, to Vandalia, Illinois, the road was built to allow settlers to traverse the Appalachian Mountains and settle in the West. With the advent of the

How are interstate highways numbered?

One- and two-digit interstate highways are numbered according to their direction. Highways that run in an east-west direction are even numbered, while highways that run in a north-south direction are odd numbered. The lowest numbers are in the south and west, while higher numbers are in the north and east. For example, Interstate 10 is an east-west highway that runs from Santa Monica, California, to Jacksonville, Florida; thus, it has a even, low number. Interstate 95 is a north-south highway that runs from Houlton, Maine, to Miami, Florida; thus, it has an odd, high number. Three-digit interstate highways are short spur routes connected to a two-digit interstate.

automobile, the road was paved, and in 1926 became part of U.S. Route 40, which stretches across the continent.

What do the Cumberland Road and Cumberland Gap have to do with each other?

Absolutely nothing. The Cumberland Road is more than 100 miles (161 kilometers) from the Cumberland Gap. The Cumberland Gap, which lies near the border of Kentucky, Tennessee, and Virginia, is a pass through the Appalachian Mountains at the Cumberland Plateau. The name "Cumberland" was extremely popular in Colonial America, originating in the name of the British Duke of Cumberland.

What is the difference between a highway and a freeway?

The term highway can be used for any road, but most often describes a paved road connecting distant towns. Freeways are multi-lane highways that use on- and off-ramps, rather than intersections, in order to limit the number of entrance and exit points along the route, hence keeping traffic along the freeway fairly steady.

When was the first freeway built in the United States?

The first freeway (lacking tolls and having limited access) in the United States was the Arroyo Seco Freeway, connecting Pasadena and downtown Los Angeles. It opened in 1940 and is now the Pasadena Freeway, Highway 110.

What are interstate highways?

President Dwight D. Eisenhower signed the Federal-Aid Highway Act of 1956, which
established the system of interstate highways in the United States. Interstate highways

are federally funded freeways that allow the rapid transportation of people, goods, and the military across the country.

Why was **Eisenhower a fan** of **interstate highways**?

In 1919, the young Dwight D. Eisenhower took part in a cross-country military trip from Washington, D.C., to San Francisco. But, due to the state of the highways at that time, the trip took 62 days—far too long to defend the country should the need arise. This experience made Eisenhower realize the need for a faster, more efficient mode of transportation across the country. Because of President Eisenhower's support for the Interstate Highway System, it is now officially known as the Dwight D. Eisenhower System of Interstate and Defense Highways.

Why does **Hawaii** have interstate highways?

Since any freeway funded under the Federal-Aid Highway Act of 1956 is known as an interstate highway, whether it crosses state boundaries or not, Hawaii can have interstate highways. Though they cross no state borders, Hawaii has three interstate highways, H1, H2, and H3.

When was the **last interstate highway** built?

The construction of new interstate highways came to an end in 1993 with the opening of Interstate 105, the Century Freeway, in Los Angeles, 37 years after construction began on the system. The Century Freeway is an inter-city route connecting the coastal community of El Segundo to Interstates 405, 110, 710, and finally 605 in Norwalk.

How many **miles of paved road** are there in the United States?

The United States has more paved road than any other country in the world, with a grand total of 2,335,000 miles (3,757,015 kilometers).

Did Hitler create the **Autobahn**?

Though the first modern freeway system in Germany was begun in 1913, Adolf Hitler did create the Autobahn during the Third Reich, from 1933 to 1945. The Autobahn is a freeway system that includes 6,800 miles (10,941 kilometers) of road across Germany. Though it is widely believed that there are no speed limits on the Autobahn, there are a few segments with marked speed limits.

What is the **longest bridge** in the world?

Completed in 1956, the Lake Pontchartrain Causeway that connects New Orleans with Mandeville, Louisiana, is 24 miles (38.6 kilometers) long—the longest bridge in the world. A parallel bridge adding two more lanes was completed in 1969.

A train enters the Chunnel, an underground tunnel that runs beneath the English Channel and connects France to the United Kingdom.

What is the **Chunnel**?

The Channel Tunnel (or Chunnel) is a railroad tunnel under the Strait of Dover in the English Channel. The Chunnel runs for 31 miles (50 kilometers) between Folkestone (near Dover) in the United Kingdom and Sangatte (near Calais) in France. Opened in 1994, the Chunnel connects England with the rest of continental Europe.

What is the **most common street name** in the United States?

It's not Main Street. Second, Third, First, and Fourth Streets are the most common, followed by Park, Fifth, and Main.

Where is the **longest Main Street** in the United States?

Main Street in Island Park, Idaho, is 33 miles (53 kilometers) long, making it the longest in the United States.

When was the **first automobile** built?

Though it had only three wheels, the world's first gasoline-powered automobile was built by Karl Benz in 1885. Henry Ford built his in 1893.

Which city has the most **taxis**?

Congested Mexico City is home to more than 60,000 taxis among its 3.5 million automobiles. The city with the second greatest number of taxis is Mumbai, India, which has more than 55,000.

What is the world's longest subway system?

The world's first subway system is also the world's longest. In operation since 1863 (when it used steam-powered trains), the London Underground is now 244 miles (393 kilometers) long and uses electricity to power the trains.

How many **taxi cab drivers** are there in the **United States**?

More than 230,000 taxi cab drivers work the streets, with more than 38 percent being immigrants to the United States.

Where was the **first self-service gas station**?

In the automobile city of Los Angeles, George Urich opened the first self-service station in 1947.

Who invented the **traffic signal**?

The red, yellow, and green traffic signal that we are familiar with today was originally invented by Garrett Morgan in 1923. Morgan, who was also the inventor of the gas mask, received numerous awards for his invention.

Who invented the **first train**?

In 1825, British engineer George Stephenson invented the first train, which was powered by steam. Stephenson's train was introduced to North America in the 1830s and was used until the 1940s when diesel-electric locomotives, which didn't run on expensive coal, replaced steam locomotives.

SEA TRANSPORT

What is the world's **busiest seaport**?

Shanghai, China, is the busiest seaport, moving more than 443 million tons of cargo annually. It is followed by Singapore, with 423 million tons, and Rotterdam, the Netherlands, which handles 376 million tons.

What does a **canal lock** do?

Many canals connect two bodies of water that lie at different elevations. Locks are used to gradually move the ships from one elevation to another. Once a ship enters a lock,

The Panama Canal is a remarkable achievement in engineering. Completed in 1914 and costing the lives of over 27,000 workers, the canal cuts shipping routes in half.

doors close in front of and behind it. Water is then added or drained from the area to raise or lower the ship to a different elevation. Then the doors in front of the ship open and the ship sails down the canal to the next lock or to the open sea.

In which **direction** do ships sail **through the Panama Canal**?

Though you would expect them to travel east from the Pacific to the Atlantic Ocean when sailing through the Panama Canal, ships actually travel northwest. Since the Isthmus of Panama lies parallel to the equator, the canal does not lie east-west but rather northwest-southeast.

Why was the **Erie Canal** built?

The 363-mile (584-kilometer) Erie Canal connects the Hudson River to Lake Erie. Opened in 1825, the canal created a new, shorter route from the northern interior of the United States to the Atlantic Ocean. Prior to the opening of the Erie Canal, goods traveled down the Mississippi River and out to the Atlantic. Since New York City lies on the Hudson River, the canal was responsible for the growth of the city as a major port, helping it to become the largest city in the United States. Once the St. Lawrence

Seaway was built in 1959, the Erie Canal became rarely used, since most transportation soon traveled along the Seaway.

What is the **St. Lawrence Seaway**?

Completed in 1959, the 183-mile (294-kilometer)-long St. Lawrence Seaway was built by deepening and widening the St. Lawrence River between Montreal, Canada, and Lake Ontario so that large ships could traverse it. The Seaway consists of a series of locks that allow ships to travel from the Atlantic Ocean to the Great Lakes, and ultimately on to Chicago. A limiting factor of the Seaway is that ships can only use the Seaway between May and November, as it is blocked by ice in the winter.

POLITICAL GEOGRAPHY

How does **geography influence politics**?

Geography is a key component in many political decisions and actions. The borders of countries, location of natural resources, access to ports, and the designation of voting districts are a few of the many geographical factors that affect politics.

What is the difference between a **country and a nation**?

Many people use the terms "country" and "nation" interchangeably. But not all nations are countries, nor are all countries nations. A country is the equivalent of a State, and is a political entity. A nation is a group of people with a common heritage and culture. Some nations have a State and are thus called a nation-state. Nation-states include France, Germany, Japan, China, and the United States. Some nations have no State, such as the Kurds and Palestinians. Some States have multiple nations such as Belgium, which is composed of two nations, the Flemings and Walloons.

What is the difference between a **State and a state**?

A State, with a capital "S," is equivalent to a country. A state, with a lower case "s," is a division of a country, like the states that make up the United States.

Do all countries **have states**?

While most countries are divided into states, provinces, or departments, there are many that have no political divisions. Large countries without political divisions include Mali, Kazakhstan, Saudi Arabia, and Algeria.

The depletion of the world's fisheries begins and ends in fish markets, like this one in Boston, Massachusetts (photo by Paul A. Tucci).

How does a **choke point** "choke" a body of water?

A choke point is a narrow waterway between two larger bodies of water that can be easily closed or blocked to control water transportation routes. Though historically the Strait of Gibraltar (connecting the Mediterranean Sea and Atlantic Ocean between Africa and Spain) has been one of the world's most important choke points, the Strait of Hormuz gained significant attention during the Persian Gulf War of 1991. The Strait of Hormuz, bounded by the United Arab Emirates and Iran, connects the Persian Gulf to the Arabian Sea and, thus, to the Indian Ocean. It was feared that if Iraq controlled the Strait, then most of the oil from the region could not be shipped out.

Who controls the **world's oil supply**?

The Organization of Petroleum Exporting Countries (OPEC) coordinates most of the world's oil production. The members of OPEC meet to coordinate oil policies and prices. Thirteen countries comprise OPEC: Algeria, Angola, Ecuador, Indonesia, Iran, Iraq, Kuwait, Libya, Nigeria, Qatar, Saudi Arabia, United Arab Emirates, and Venezuela. Though Russia, the United States, and Mexico are also leading petroleum producers, the three countries are not members of OPEC.

Who else **controls the price of gas** besides oil producers?

There are many factors that influence the price of gas. Among them are the rising demand for fuel from developed countries such as the United States; new and growing demand from China and India; market speculation; and public policy. The refining industry, which is under government regulation, is also consolidated into the hands of a very few global refiners. This has the potential for removing competitive pricing.

COLONIES AND EXPANSIONISM

Why did the **sun never set** on the **British Empire**?

In the early twentieth century, the United Kingdom included colonies from North and South America (Canada, British Guiana, and Bermuda), Africa (Egypt, South Africa,

Who owns the world's oceans?

The battle over control of the world's oceans has increased over the past few decades due to the discovery of vast mineral and fuel resources located under the sea. In 1958, the United Nations held the first Conference on the Law of the Sea. This conference established territorial seas, measuring 12 nautical miles (22.24 kilometers) from the shore of coastal nations that are under the full control of that country. (The United States, along with such countries as North Korea, Chad, Liberia, and Iran, have refused to sign the treaty.) Additionally, countries have mineral, fuel, and fishing rights in an Exclusive Economic Zone (EEZ) that spans 200 nautical miles (370.6 kilometers) from shore. Problems arise when two countries' zones overlap. Median lines between countries have been drawn in most cases, but there are still many areas of disagreement.

and Nigeria), Asia (India and Burma), and Oceania (Australia and New Zealand). Because the British Empire spanned the globe, there was always at least one portion of the Empire in daylight.

Why would countries want **colonies**?

Colonies are a source of raw materials, new land, wider trading opportunities, and militaristic expansion for the mother country. Colonies were established around the world from the sixteenth century through nineteenth century by powerful western nations. After World War II, the concept of colonization was widely attacked as an exploitive policy. Though most colonies were granted independence, many countries still control colonies around the world.

What were some of the **earliest colonies**?

The Phoenicians, around the year 1000 B.C.E., founded some of the first colonies in Tyre (present-day Lebanon). Colonists from there went on to colonize Carthage (present-day Tunisia) and the coast of Spain. This enabled them to control access to the Atlantic Ocean and trade with the indigenous peoples of what is today Great Britain and France.

How did the **Nazis** use **geopolitics**?

During the Nazi era in Germany, 1933-1945, the "science" of geopolitics was utilized to support Germany's concept of Lebensraum, or living space. The Nazi concept of Lebensraum was based on the idea that there was a racial hierarchy that allowed "superior" races to conquer "inferior" races. Adolf Hitler used this warped sense of geopolitics to invade Czechoslovakia, Poland, and the Soviet Union. For example, Ger-

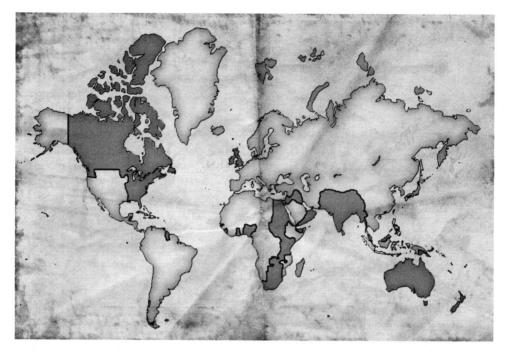

At its height, the British Empire had colonies on every continent except Antarctica (darkened areas of this old illustration indicate Great Britain and its colonies).

many claimed that the ethnic Germans living in the Sudetenland of Czechoslovakia should be included within the German fatherland.

How did **irredentism** help **start World War II**?

Irredentism is a term used to describe a situation in which a minority group in one country shares the culture and heritage of another country. The minority group may attempt to have their region annexed into the mother country or may be happy in the country they are in. Adolf Hitler used irredentism as an excuse to invade and conquer Czechoslovakia in 1938. He claimed that the Germans in the Sudetenland, a part of Czechoslovakia, were being treated unfairly and thus this area should be annexed to Germany. Though Germany's annexation of Czechoslovakia did not start World War II, it was the Nazis' first direct aggressive step toward conquering Europe.

THE UNITED NATIONS

How does the **United Nations preserve peace**?

Established in 1945 at the end of World War II, the United Nations was created for the purpose of maintaining world peace. Its members pledge to work together to solve disputes. The United Nations also oversees many agencies that promote health, welfare, and cooperation around the world.

How many countries are **members of the United Nations**?

There are 192 member states of the United Nations, which is headquartered in New York City. New members include Serbia and Montenegro (from the former Yugoslavia), Tuvalu (an island nation in the Pacific), Timor-Leste (in what was once part of Indonesia), and Switzerland.

Which countries are **not members** of the United Nations?

While almost every country in the world is a member of the United Nations, there is a short list of countries that are not members: Taiwan, Tonga, and Vatican City.

The headquarters of the United Nations is located in New York City.

Why do some countries **choose not to join the United Nations**?

A country must be willing to give up some of its self-rule for the greater good, which often means the good of the larger, western countries.

How did the **League of Nations fail**?

The League of Nations, which was created in 1920 and replaced by the United Nations in 1945, failed in its mission to prevent World War II. Even though the League was essentially the creation of President Woodrow Wilson, the isolationist United States never became a member.

NATO AND THE COLD WAR

Which countries are **members of NATO** (North Atlantic Treaty Organization)?

The 26 members of NATO include Belgium, Canada, Denmark, France, Germany, Greece, Iceland, Italy, Luxembourg, Netherlands, Norway, Portugal, Spain, Turkey, the United

Kingdom, Bulgaria, Czech Republic, Estonia, Hungary, Latvia, Lithuania, Poland, Romania, Slovakia, Slovenia, and the United States.

How did the **Soviet Bloc countries** respond to the creation of NATO?

In 1955, seven Communist countries created the Warsaw Pact to protect against NATO aggression. The Warsaw Pact originally consisted of Albania, Bulgaria, Czechoslovakia, Hungary, Poland, Romania, and the Soviet Union. The Warsaw Pact disbanded in 1991 with the breakup of the U.S.S.R. and the changes in Eastern Europe.

What is the **purpose of NATO** now that the Soviet Union is gone?

The North Atlantic Treaty Organization (NATO) was founded in 1949 as an alliance of European and North American non-Communist countries committed to preventing and protecting against Communist threats. With relative stability in the former Soviet Union (now the Commonwealth of Independent States), NATO sees its role purely as defensive in nature, supporting member states and countries in need of assistance in the surrounding regions.

What did the **domino effect** have to do with United States involvement in the Vietnam War?

American military strategists believed that if one country became Communist, it would begin a never-ending succession of countries converting to Communism (thus the domino metaphor). North Vietnam, at the time of the American invasion, was comprised primarily of communist and communist sympathizers, whereas South Vietnam was more democratic-leaning. Policy makers believed that the United States had to do everything possible to keep every country from falling to communism. This included sending American troops to Vietnam. Though Vietnam fell to the Communists, the theory of the domino effect was proven incorrect because neighboring countries did not fall to communism as predicted. In fact, the U.S. war and policy toward Vietnam caused many adjacent countries to become even more fervently opposed to U.S. policy in the region, and some theorize that this contributed to turmoil and genocide in such countries as Cambodia and Laos.

Who **won** the **Cold War**?

The stalemate of the Cold War effectively ended when the people and governments of the former Soviet Union and Eastern Bloc countries decided that they needed a major change in the way in which they organized their government, societies, and economies. The fall of the Berlin Wall in 1989 was only one indicator of the significance of this trend. Popular demonstrations and changes took place in all parts of the

> ## What criteria must an area meet to be considered an independent State?
>
> **A**State must be an independent country with its own land, government, permanent population, transportation system, and economy, and it must have international recognition.

former U.S.S.R. and Eastern Europe, leading inexorably to the downfall of the Soviet Union and its control of satellite states.

THE WORLD TODAY

What are the world's **newest countries**?

In the 1990s, over *two dozen* new countries appeared on the map. These included 15 new countries that were created when the U.S.S.R. broke up in 1991—Armenia, Azerbaijan, Belarus, Estonia, Georgia, Kazakhstan, Kyrgyzstan, Latvia, Lithuania, Moldova, Russia, Tajikistan, Turkmenistan, Ukraine, and Uzbekistan.

The dissolution of Yugoslavia also created several new countries in 1991 and 1992—Bosnia and Herzegovina, Croatia, Macedonia, Serbia and Montenegro, and Slovenia.

In 1993, Eritrea became independent of Ethiopia, and Czechoslovakia dissolved into the Czech Republic and Slovakia. That same year, the Pacific island countries of the Marshall Islands, Micronesia, and Palau all became independent. In 1990, Namibia split from South Africa to become its own country, while East and West Germany combined to become Germany.

How many countries does the United States recognize?

The State Department is the official U.S. government agency that recognizes independent countries. It maintains an updated list of the official independent States of the world. As of this writing, there are 194 states listed. Taiwan and Vatican City, though commonly considered countries, are not included on this list. Kosovo, which gained independence in February 2008, is the newest country to be recognized.

Why isn't **Taiwan recognized** by the **U.S.** government?

The People's Republic of China has had a dispute with Taiwan since the Nationalists, who were fighting the Communists on the mainland, fled to Taiwan at the end of 111

Taipei is a prosperous, world-class city with a political future that is in doubt because of China's claim on Taiwan.

World War II. Decades later, as China became more powerful in the eyes of various American political administrations, China gave the United States a choice: recognize China or Taiwan, but not both. The United States chose China. Although officially not recognized, Taiwan is one of the United States's strongest allies in the region, and one of its most significant trading partners. Taiwan has very close official and unofficial ties with the United States.

Why are some **borders curvy** while others are **straight**?

There are two primary types of boundaries—geometric and natural. Geometric boundaries are straight and follow lines of latitude, longitude, or a certain compass direction between points. Geometric boundaries were established to divide territories before settlers entered areas. Most of the states in the Western United States have at least a portion of their borders formed by geometric boundaries (especially rectangular-shaped Colorado and Wyoming). Natural boundaries are usually curvy because they follow the crests of mountains or the center of rivers. Natural boundaries are very common in places like Europe, where the region was populated before the countries were created.

Why do **third world countries** no longer exist?

The term "third world" was part of the classification of countries during the Cold War. This classification designated those countries aligned with the United States as "first world," those countries aligned with the Soviet Union as "second world," and those countries that were nonaligned as "third world." Over time, the term "third world" came to mean a poorer or less-developed country. With the dissolution of the Soviet Union, and the acceptance of democracy in Russia and Eastern Europe, the classification no longer exists. The preferred terms are now "developed" countries and "less-developed" (or "developing") countries.

Why was the border between the **Yemen and Saudi Arabia** dashed on maps?

In the sandy desert between Saudi Arabia and Yemen, the border between the two countries was in dispute. A treaty signed in Jedda, Saudi Arabia, in July of 2000 resolved the conflict, and maps no longer show a dashed line between the two countries.

What **other countries** have **border disputes**?

For years, India and Pakistan have had disputes about the borders of Kashmir, a region located in northern India. Other countries in Central America, like Honduras and El Salvador, have had disputes over very small parcels of land because of ill-defined border markings implemented during the countries' colonial period.

Which countries are surrounded entirely by **landlocked countries**?

The two countries surrounded entirely by landlocked countries are Uzbekistan and Liechtenstein. Uzbekistan is surrounded by the landlocked countries of Kazakhstan, Kyrgyzstan, Tajikistan, Afghanistan and Turkmenistan. Liechtenstein is bordered by Switzerland and Austria, neither of which have access to the ocean.

How is **gerrymandering** like a salamander?

In 1812, Massachusetts Governor Elbridge Gerry signed a law that established an oddly shaped congressional district. It was redrawn by political cartoonists into a salamander-type creature and thus the term gerrymander was born. Gerrymandering is the process of establishing oddly shaped congressional districts in order to include voters from dispersed areas. Gerrymandered districts can be helpful or detrimental to minority groups, depending on who draws the borders. The U.S. courts have found gerrymandering to be a legal method of establishing congressional boundaries.

What is the **best shape** for a country?

Though countries come in various shapes for various reasons, the best shape for a country is compact. A compact country, such as Germany or France, is easier to govern than those that are fragmented (such as Indonesia) or elongated (such as Chile). Compact countries are easier to govern because transportation, communication, and internal security are easier to maintain. Also, compact countries have shorter borders to protect. Elongated and fragmented countries are more easily divided and conquered.

THE WORLD ECONOMY

What is the difference between **GNP and GDP**?

GDP, or gross domestic product, is the value of all goods and services produced in a country in a year. GNP, or gross national product, is the total value of GDP plus all income from investments around the world. Per capita GDP is usually compared between countries.

Which country has the **highest GDP**?

GDP is the gross domestic product, which is the total value of all goods and services produced by a country. Technically, the European Community, representing an economic union of the nations of Europe, has the highest GDP at $16 trillion, followed by the United States with $13 trillion and Japan with $4 trillion.

Which country has the **highest per capita GDP**?

Qatar, an oil producing nation on the Arabian coast, has a gross domestic product of $80,600 per person. This is followed by banking giant Luxembourg, with a per capita GDP of $80,457, and Malta, with $53,359.

Which **countries give** the highest proportion of their GNP in the form of **aid to other countries**?

Norway, Sweden, Luxembourg, Denmark, and the Netherlands each give nearly one percent of the value of all of their goods and services to the cause of foreign aid.

Which advanced industrial **democracy ranks last** in giving foreign aid as calculated as a **percentage of its GNP**?

The United States, although giving the highest amount in dollars of any country, contributes the least amount of money in foreign assistance as a percentage of GNP: less than 0.16 percent.

Which country spends the **most money** on the **military**?

The United States leads the world in military spending. American taxpayers commit more than $580 billion per year for military activities. The next closest nation is France, which spends $74 billion.

Which country is the **top exporter** of goods and services?

Germany exports $1.113 trillion in goods and services. It is followed by the United States, with $1.024 trillion, and China, with $974 billion.

CULTURAL GEOGRAPHY

POPULATION

Of all the humans who have ever lived, what proportion of them are **alive today**?

Only a small percentage, anywhere from 5 to 10 percent, of the humans who have ever lived are alive today. Since humans have existed for approximately 100,000 years, the total number that have ever lived is probably between 60 billion and 120 billion.

How many people **live on the Earth**?

There are approximately 6.7 billion people on Earth. The planet's population is expected to grow to 8 billion by the year 2025 and 9.3 billion by 2050.

What has the **world's population** been over time?

Year	Population
0	200 million
1000	275 million
1500	450 million
1750	700 million
1850	1.2 billion
1900	1.6 billion
1950	2.6 billion
1960	3 billion
1975	4 billion
1985	4.85 billion
1990	5.3 billion
1999	6 billion

How long did it take for the **world population to double** in size?

From 1959 to 1999, the population went from 3 billion to over 6 billion.

How **fast** is the **population growing**?

The highest growth rate recorded was around two percent in the late 1950s. This rate has been in decline and is now less than one percent per year.

Which **countries lead** the world in **child poverty**?

Child poverty is defined as the number of children from households who earn 50 percent less than the national median. Given this measurement, the countries with the most child poverty are Mexico (26.4%), the United States (22.4%), Italy (20.5%), and the United Kingdom (19.8%).

Which country has the world's highest **population density**?

Macao, a former Portuguese colony consisting of a peninsula and two islands off the southern coast of China near Hong Kong, has the highest population density in the world. There are 538,000 people living in an 11-square-mile area (28.8 square kilometers).

What is a **census**?

A census is an enumeration, or counting, of a population. The information from a census is used to help governments determine where to provide services, based on the demographics of the population. Information about age, gender, number of children, race, languages spoken, education, commuting distance, salary, and other demographic variables is common in a census. This information is compiled and provided to government agencies and is usually accessible to the general public.

In the United States and most other developed nations, a census takes place once every decade. The Constitution of the United States requires a census to be taken every 10 years, in order to create districts and determine the number of members of Congress each state is able to send to the House of Representatives.

What was the **baby boom**?

Due to post-World War II prosperity, there was a boom in American births between 1946 and 1964, now referred to as the "baby boom." During this time, approximately 77 million babies were born in the United States, a very large number compared to that of other time spans. As the baby-boomers approach retirement age, health and welfare services for the elderly will become high priority as the country prepares for a higher proportion of older people in its population than ever before.

Are there an **equal** number of **boy and girl babies born**?

Though scientists aren't sure why it occurs, there is an average of 105 boys born for every 100 girls.

How many **lesbian and gay people** are there in the world?

Most scientists believe that between 1 and 10 percent of the world's population is homosexual. This means that between 6.7 million to 670 million people are gay or lesbian.

LANGUAGE AND RELIGION

What are the most **commonly spoken languages** of the world?

Mandarin Chinese is spoken by 1.05 billion people. This is followed by English, with 1 billion people; Hindustani (Hindu and Urdu), with 650 million speakers; Spanish, with 500 million speakers; and Arabic, with over 400 million speakers.

What is the difference between a **lingua franca** and a **pidgin**?

A lingua franca is a language used between people who do not have a common language. English is often used as a lingua franca in international business transactions. A pidgin is a language that has a small vocabulary and is a combination and distortion of two or more languages. For example, pidgin English, a combination between English and indigenous languages, is used in Papua New Guinea between English-speaking and indigenous people. Most pidgins are lingua francas but not all lingua francas are pidgins.

Dancers performing in a Buddhist temple on Sukhumvit Road in the capital city of Bangkok, Thailand. Buddhism is the world's fourth most-practiced religious belief (photo by Paul A. Tucci).

119

The Wailing (or Western) Wall (foreground) in Jerusalem, and the Dome of the Rock (background) are holy sites of Judaism and Islam, respectively (photo by Paul A. Tucci).

The Sultan Omar Ali Saifuddin Mosque, built in 1958, stands in the capital city of Bandar Seri Begawan, Brunei. It is a beautiful example of the domed architecture seen in many mosques (photo by Paul A. Tucci).

Which **religions** are practiced by the **most people**?

Christianity has the most adherents, with 33.32 percent of the world's population practicing some form of this faith. This is followed by Islam (21.01 percent), Hinduism (13.26 percent), and Buddhism (5.84 percent).

How many religions have **holy sites** in **Jerusalem**?

Judaism, Islam, and Christianity all regard Jerusalem as a very holy city. The Western Wall, the remaining wall of the Second Temple, is the holiest site in Judaism. Islam's third-holiest site is the Dome of the Rock and the mosque, both located in Jerusalem. The Church of the Holy Sepulcher is a holy Christian site.

Why do **mosques have domes**?

The onion-shaped domes of Islamic mosques and other religious buildings of eastern religions were an architectural style borrowed from the Byzantine Empire. One of the world's most famous onion-domed buildings, St. Basil's Cathedral in Moscow's Red Square, was built in the mid-sixteenth century.

DEALING WITH HAZARDS

Is **childbirth** still a significant cause of death for women?

From the beginning of time until the mid-twentieth century, a leading cause of death for young women was complications during childbirth. Now, in developed countries, there is almost no risk of death during pregnancy and labor.

How did the **Black Plague** affect the world's population?

Spread by fleas, the bubonic plague, also called the Black Plague, raged through Europe, Asia, and North Africa between the years 1346 and 1350. Though cities attempted to curb the spread of this highly infectious disease by quarantining cities, the fleas easily spread from city to city. Estimates of those killed reach into the tens of millions. In Europe and Asia, more than half of the population died of the Black Plague during those four years. Many more died of starvation in the famine that followed because of the staggering depletion of the work force.

Is there **enough food** to feed the world?

Though there is enough food produced in the world to feed everyone, logistical and political problems make its distribution inefficient. At the current rate of world population growth, we may soon have to change our eating habits and eat more grain and less meat. There is a limited amount of grain that the Earth can produce. Currently, much of this grain is consumed by cattle, rather than humans. If humans were to eat the grain instead of eating the cattle, the calories from the grain would be twenty times more efficient than those from beef.

What were **Thomas Malthus' ideas** on **population growth**?

In 1798, English clergyman Thomas Malthus wrote "An Essay on the Principle of Population," in which he described the problems of population growth. Malthus argued that the world's population grows faster than the food supply, but there are such checks as war, famine, disease, and disaster that limit the population.

How widespread was the **influenza pandemic of 1918**?

In 1918, a deadly flu spread quickly around the world. Within just two years, this influenza pandemic had sickened over a billion people and killed more than 21 million. Half a million people died in the United States alone.

How does **medical geography** help control the spread of diseases?

Medical geographers and epidemiologists (scientists who study diseases and epidemics) use mapping to monitor the spread of diseases and locate the source of a dis-

What is a refugee?

A refugee is a person who leaves his or her home country for fear of persecution. There are approximately 15 to 20 million refugees in the world today. Most refugees come from developing countries where society is in flux or even chaotic. Refugees usually flee to the closest stable country, so different countries see great variation in the number of refugees based on the political climate of their neighbors. Thus, developed countries consistently have a large number of refugees arriving. Refugees are problematic, just as any mass immigration is.

ease. For example, by mapping a group of inordinately high numbers of cancer patients in a city, we may find that all live close to a factory that has been releasing toxins into the ground water. By identifying the source and spread of a disease, the disease can often be combated.

What **revolution** attempted to stop **world hunger**?

Begun in the 1960s, the "Green Revolution" was an attempt by developed countries and such international organizations as the United Nations to transfer agricultural technology to less-developed countries. While the Green Revolution increased agricultural yields, it modified the ecology of traditional agricultural systems (such as through the use of chemical fertilizers) and has yet to cure world hunger.

What is a **UNESCO World Heritage Site**?

The United Nations Educational, Scientific and Cultural Organization's stated purpose is to build peace in the minds of mankind. It is one of the most important agencies within the U.N. A World Heritage Site represents our natural or cultural wealth, the most important landmarks in the world that should be shared with everyone. Because of the special significance of these places, they must be protected for us to see, so that we may pass on knowledge of our shared culture to future generations. They can be both physical places built by man, or natural wonders.

How **many refugees** are there?

There are approximately 8.3 million people who are classified as refugees, living somewhere outside of their home countries, because of deplorable conditions at home, warfare, or government and societal discrimination and economic oppression.

How many **people** are abducted and **sold into slavery** each year?

About 600,000 to 800,000 people, mostly women and children, are trafficked across national borders each year and forced to work as virtual slaves. Millions more are trafficked within their own countries every year. Seventy-five percent of these people are women who are used for sexual exploitation. About 280,000 of these people are sent to Asia, and another 210,000 people are trafficked to Europe and Russia.

Many countries participate in human trafficking, both as accomplices in providing the people to abduct by not enforcing laws, or by allowing the people to enter their borders. Some of the biggest offenders include Saudi Arabia, Bahrain, Kuwait, Malaysia, Qatar, United Arab Emirates, Ukraine, Russia, Moldova, Mexico, India, Egypt, and China.

In the past, how many people were **sold into slavery** in the **United States**?

Until the nineteenth century and the end of the Civil War, the United States had a long history of slavery. During the African Diaspora, beginning in 1619, approximately 12 to 13 million Africans were taken from their homes and sold into slavery in the Americas and the Caribbean. The majority of the slaves who survived the voyage were sent to Brazil. By the beginning of the American Civil War, nearly 240 years after the first slaves arrived in the American colonies, four million people were held as slaves, principally in the South, where slave ownership was legal and was a main source of labor for agriculture.

CULTURES AROUND THE WORLD

What are **nomads**?

Nomads are tribes that migrate in a seasonal circuit over a large region. Though nomadic people often build temporary homes, they consider migratory life within their tribe to be home. Nomadic tribes are located in marginal areas around the world, from the Sahara Desert to northern Siberia. The nomadic way of life is threatened because of general cultural prejudice against unsettled peoples.

What is a **Gypsy**?

Gypsies are nomadic tribes that travel throughout Europe. Though they once were thought to have originated in Egypt (hence the European name for them), linguistic studies have placed their origin in India. These traveling tribes have been subject to centuries of persecution, including "Gypsy hunts" and extermination at the Auschwitz Death Camp.

What is **brain drain**?

When highly educated or highly skilled individuals leave their home countries to go to countries where opportunities are better, the home country experiences "brain drain."

Many Arabic people live nomadic lives, such as this man living in the Al Khatim desert in the United Arab Emirates.

This occurs especially in Asian countries, as highly educated Asians move to the United States, Canada, and Australia for higher-paying jobs.

How do people cope with **continual light or darkness** in high latitudes?

Murmansk, Russia, is the largest city north of the Arctic Circle. The city receives no sunlight for several months out of the year, making it one of the most psychologically extreme environments on the planet. Residents of the city (about 470,000) walk along artificially lit streets that give the appearance of sunlight, undergo artificial sun treatments (much like tanning booths), and often suffer from the condition known as Polar Night Stress. Polar Night Stress symptoms include fatigue, depression, vision problems, and susceptibility to colds and flus.

What is a **long lot**?

Long lots are long and narrow pieces of property. This type of division of land is common in Europe and places in North America that were initially settled by the French (such as Québec and Louisiana). Each lot has a narrow access to a stream or road but is several hundreds of feet deep.

Why do some cultures **kill infants**?

Infanticide is the practice of killing an infant. For centuries, various cultures around the globe used infanticide as a form of population control, most commonly because their limited food supply could only feed a certain number of humans. Because of cultural biases, female infants were more often victims of infanticide. The practice of infanticide still occurs today.

Why don't Americans eat **horse meat**?

Most religions and cultural groups have some kinds of food taboos. Foods may be avoided entirely or may be avoided on certain days or during certain festivals. Religious food taboos include the avoidance of pork by Muslims and Jews and the avoid-

Why do some people eat dirt?

The practice of eating dirt, called geophagy, is most commonly practiced by pregnant and lactating women. Since women's bodies require additional nutrients during pregnancy and lactation, the body craves clays and dirts that carry these additional minerals. This practice is most common in Africa, but the practice spread to the United States with the forced migration of Africans during slavery. Geophagy is now also practiced in the southern United States, but the practice has become a cultural rather than physiological action.

ance of beef by Hindus. Cultural food taboos also play an important role. For example, Americans don't eat horse meat because it is a cultural food taboo, despite the fact that horse meat is a nutritional and edible type of food.

How many **McDonald's restaurants** are there in the world?

To some people McDonald's represents the extreme example of cultural imperialism, an import of one's culture supplanted into another. Today, there are 31,000 McDonald's restaurants located in 119 countries.

Why does a **first-born son** get everything?

Primogeniture is the system of inheritance in which all inheritable land and property is passed on to the first-born son. A common worldwide tradition, primogeniture enabled a family's possessions and status to remain intact as they were passed from generation to generation. This practice of the entire inheritances benefiting only the first-born son resulted in subsequent sons needing to find alternative livelihoods.

Can a woman have **multiple husbands**?

While some cultures allow men to have multiple wives, there are other cultures that allow women to have multiple husbands. This practice, known as polyandry, is presently observed by only two cultures, the Tibetans and the Nair people of southwestern India. Polygyny, when men have multiple wives, remains legal in Islamic and many African countries. The collective name for multiple spouses, both polyandry and polygyny, is polygamy.

When did people start eating with **forks and spoons**?

Though introduced in the fifteenth century, forks and spoons did not come into common use in Europe until the seventeenth century. Prior to that time, people ate with their hands and a knife.

What is the most common **last name** in the world?

Zhang (or Chang) is the surname used by more than 100 million people.

TIME, CALENDARS, AND SEASONS

Why did early humans have **no need for hours, days, weeks, or months**?

Because early humans were hunters and gatherers, they had little need to know exact time. What *was* essential was an understanding of the seasonal migration of animals and varieties of plant life.

What time is it at **12:00 A.M.**?

12:00 A.M. is midnight. In the middle of the night, 11:59 P.M. is followed by 12:00 A.M., not 12:00 P.M., which is noon.

What do **A.M. and P.M.** mean?

The distinctions A.M. and P.M. are abbreviations for "ante meridiem" and "post meridiem," which mean before midday and after midday, respectively.

How does **military time** work?

Often known as military time, 24-hour time is used by many countries. It begins at midnight with 0000 and the day ends with 2359. The first two digits represent the hour and the last two digits represent the minutes. Since there are 24 hours in a day, each hour is numbered 00 through 23. For instance, 0100 is 1:00 A.M., 1200 is noon, 1300 is 1:00 P.M. , and 2043 is 8:43 P.M.

How **long is a day**?

A day is the time it takes the Earth to make one rotation, which is 23 hours, 56 minutes, and 4.2 seconds. We round this to an even 24 hours for convenience.

Does the Earth always **rotate and revolve around the sun** at the same speed?

No, the Earth's revolution around the sun and rotation on its axis are not perfect. Its daily rotation may vary by approximately four to five milliseconds, and is slowing down at a rate of one millisecond each century, due to tidal friction. Additionally, the axis wobbles slightly, and the length of revolution around the sun varies by a few milliseconds.

How can I find the **exact time**?

Today, you can go to many Web sites on the Internet and find the exact time. Just search the word "time," and you will see links for the exact time. Your computer also is set to automatically update time, so your desktop time is usually accurate to within a few milliseconds.

TIME ZONES

When were **time zones established** in the United States?

In 1878, Sir Sanford Fleming proposed dividing the world into 24 time zones, each spaced 15 degrees of longitude apart. The contiguous United States was covered by four time zones. By 1895, most states had begun to institute the standard time zones of Eastern, Central, Mountain, and Pacific on their own. But it wasn't until 1918 that Congress passed the Standard Time Act, establishing official time zones in the United States.

How did **trains** help establish time zones?

Before trains, many cities and regions had their own local time, which was set based on the sun at their location. The great variation of local times made train schedules confusing. In November 1883, railroad companies across the United States and Canada began to use standard time zones—years before they came into general use across the United States.

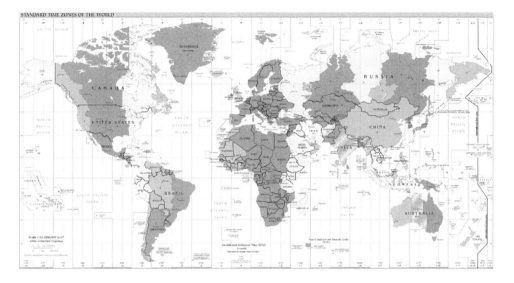

Most—but not all—countries around the world are divided into one-hour time zones (U.S. Central Intelligence Agency, The World Factbook 2008).

How many time zones does the United States have?

The United States spans nine time zones: Eastern, Central, Mountain, Pacific, Alaska, Hawaii-Aleutian, Samoa, Wake Island, and Guam.

Which **states are split** into multiple time zones?

Florida, Indiana, Kentucky, and Tennessee are split into Eastern and Central time. Kansas, Nebraska, North Dakota, South Dakota, and Texas are split between Central and Mountain time. Idaho and Oregon are split between Mountain and Pacific time.

Are **all countries** divided into **one-hour time zones**?

No. China and some other countries do not divide their territory by time zones at all. Several countries use half-hour time zones, including Afghanistan, Australia, Burma, India, Iran, the Marquesas, and Venezuela; also, the Canadian province of Newfoundland uses half-hour zones. Quarter-hour divisions are used in Nepal, as well as in some provinces, including the Chatham Islands.

How many time zones does **China** have?

Since China is such a large country, it should span five time zones, but the entire country uses one time—eight hours ahead of UTC.

What time is it at the **North and South Poles**?

Because time zones get narrower the farther you get from the equator, time zones would be very thin near the North and South Poles. To simplify things, researchers living in Antarctica use Coordinated Universal Time (UTC), which is the time at Greenwich, England.

What happens when I cross the **International Date Line**?

If you fly, sail, or swim across the International Date Line from east to west, such as from the United States to Japan, you add a day (Sunday becomes Monday). When you travel from west to east, such as from Japan to the United States, you subtract a day (Sunday becomes Saturday).

How fast do you have to travel west to **arrive earlier** than when you left?

Normally, when flying between London and New York, the trip takes seven hours. Thus, with the five-hour difference in time between the two cities, you arrive two hours "later" than when you left London. If you could have flown on the now decommissioned Concorde, which traveled at Mach 2 (1,300 miles [2,092 kilometers] per hour—two times the speed of sound), the trip between London and New York would have taken only three hours. Thus, with the five-hour difference in time zones, you would have arrived two hours "earlier" than when you left!

Why is **Russia** always **one hour ahead**?

In an effort to take advantage of the limited amount of light available in winter months, each of Russia's time zones are one hour ahead of the standard time for those zones. Russia also follows Daylight Saving Time and adds an additional hour during spring and summer months.

When was Daylight Saving Time instituted?

Though Benjamin Franklin suggested the concept of Daylight Saving Time in 1784, it was not implemented in the United States until World War I. Between World Wars I and II, states and communities were allowed to choose whether or not to observe the change. During World War II, Franklin Roosevelt again implemented Daylight Saving Time. Finally, in 1966, Congress passed the Uniform Time Act, which standardized the length of the Daylight Saving Time period. But states and territories can choose not to observe Daylight Saving Time. Arizona, Hawaii, parts of Indiana, Puerto Rico, and some island territories have chosen not to observe Daylight Saving Time.

DAYLIGHT SAVING TIME

Why do we have **Daylight Saving Time**?

By moving our clocks forward one hour between spring and fall, we more effectively utilize the light of the sun to keep homes and businesses lit, saving electricity.

When did **Daylight Saving Time move** from the end to the **beginning of April** in the United States?

The shift of the beginning of Daylight Saving Time from the last Sunday in April to the first Sunday in April took place in 1987, when the Uniform Time Act was amended. In 2007, Daylight Saving Time was moved again to the second Sunday in March and the first Sunday in November.

When do countries in the **Southern Hemisphere** observe Daylight Saving Time?

Because Daylight Saving Time is an effort to save daylight during the summer months, Daylight Saving Time in the Southern Hemisphere occurs from October through March.

KEEPING TIME

What is a **sundial**?

A sundial is an instrument that uses the sun to measure time. A sundial consists of an angled marker, called a gnomon, that casts a shadow on a plate, called a dial plane. On the dial plane there are marks indicating the hours of the day. During the day, as the

131

One of the first instruments to measure time was the sundial.

sun moves across the sky, the shadow from the gnomon moves across the dial plane, indicating the hour. Sundials were used to measure time before clocks and watches were invented.

When were the **first clocks made**?

Around 3500 B.C.E. clocks began to appear in a variety of cultures, most notably in Sumeria in what is now the Middle East.

What is a **water clock**?

Water clocks were the first clocks that didn't rely upon sunlight (as with sundials) to tell time. They operated by dripping water from containers at measured intervals. There were two key types of water clocks: those that measured time by the amount of water remaining in the clock and those that measured time by how much water dripped from the clock and filled a measuring device.

When was the **first watch** made?

In the early sixteenth century, German locksmith Peter Kenlein began to produce portable clocks called Nürenberg eggs. Advances in later centuries led to timepieces that could be worn on the wrist.

What is an **atomic clock**?

An atomic clock uses measurements of energy released from atoms to precisely measure time. The current model of the atomic clock, created in 1957 by Norman Ramsey, uses measurements from cesium atoms. Atomic clocks are used by NASA, physicists, astronomers, and other scientists who need extremely precise time.

CALENDARS

What do B.C. and A.D. stand for?

In our modern calendar, the year 0 represents the year of the birth of Jesus Christ. Years before his birth are known as B.C., or "before Christ." Years since his birth are known as A.D. or "Anno Domini," the "Year of Our Lord."

> ## What was the longest year in history?
>
> The year 46 B.C.E. was decreed by the Roman Emperor Julius Caesar to be 445 days long in order to correct the calendar, which was 80 days off, based on the seasons.

What do B.C.E. and C.E. stand for?

To secularize the calendar, the terms B.C.E. and C.E. have come into use to replace B.C. and A.D. respectively. B.C.E. means "before common era" and C.E. stands for "common era."

What is the problem with a **calendar based** on the **cycles of the moon**?

The time between two new moons is 29 and one half days. After 12 lunar months, a calendar based on the cycles of the moon falls short of a solar year—and thus the cycle of seasons—by 11.25 days. To compensate, the Hebrew Calendar, which is based on the moon cycle, has a regulated 19-year cycle in which an extra month is added every two or three years.

How did **Julius Caesar** fix the calendar?

For years, the Romans had been using a calendar based on lunar cycles. Since each lunar month is 29.5 days, 12 months only adds up to 354 days. But seasons do not follow a lunar cycle, they follow a solar one. A solar year lasts 365 days, 5 hours, 49 minutes. Julius Caesar implemented a solar calendar so that the seasons would occur at the same times every year. Additionally, Caesar made each year 365 days long, with every fourth year being 366 days long (a "leap" year). Unfortunately, each calendar year was still 11 minutes longer than a solar year, a problem that Caesar did not feel was a big concern at the time.

When was **January 1** chosen as the beginning of the year?

In Caesar's calendar modifications of 46 B.C.E., he decreed that the year would begin on January 1 instead of March 25, as it had in the past. At that time Caesar also designated number of days in each month, unchanged to the present day.

Why were **10 days lost** from the year in 1582?

In 46 B.C.E., Julius Caesar implemented the Julian calendar, which was 11 minutes longer than a solar year. By 1582, those 11 minutes each year had added up to 10 days. Pope Gregory XIII aligned the calendar with the solar year by declaring October 5,

133

1582, to be October 15, 1582, in the Catholic regions of the world, thus correcting for the 10 lost days.

What is the **Gregorian calendar**?

In addition to moving the calendar forward by 10 days in 1582, Pope Gregory XIII also corrected the error of the Julian calendar. He declared that years ending in "00" would not be leap years, except those divisible by 400 (such as the year 2000). The Julian calendar, with Pope Gregory's correction, is known as the Gregorian calendar, which most of us use today.

When was the **Gregorian calendar adopted in the United States**?

Though Catholic countries switched to the Gregorian calendar in the sixteenth century, Protestant countries, such as England and its colonies, refused to switch from the Julian to the Gregorian calendar at that time. It wasn't until 1752 that Britain and its colonies, including the colonies that soon thereafter became the United States, switched to the Gregorian calendar. By that time, there was an 11-day difference in time, so September 3, 1752, became September 14, 1752.

Is the Gregorian calendar **accurate**?

Almost! It is still 25 seconds longer than the solar year. Therefore, after about 3,320 years we will be a full day ahead of the solar year. The keepers of time will have to deal with this problem when the time comes.

What type of calendar did the **French** use between 1793 and 1806?

In 1793, during the French Revolution, an entirely new calendar was established by the National Convention. The calendar was designed to help rid French society of its Christian influences. Within this new calendar there were 12 months, each consisting of three decades. Each decade was composed of 10 days. Five days (six during leap year) were added at the end of the year to add up to 365 (or 366) days. Napoleon reinstated the Gregorian calendar in 1806.

What type of calendar was used in the **Soviet Union** between 1929 and 1940?

The Soviets created the Revolutionary calendar, which had 5 days in a week (four for work, the fifth as a day off) and six weeks in a month. Five days (six during leap year) were added at the end of the year to add up to 365 (or 366) days.

Which **celestial bodies** are the days of the week named after?

The names of the days of the week come from Roman or Norse names for the planets:

Day	Celestial Body (Roman/Norse)
Sunday	Sun/Sol
Monday	Moon
Tuesday	Mars/Tui
Wednesday	Mercury/Woden
Thursday	Jupiter/Thor
Friday	Venus/Frygga
Saturday	Saturn

When did the **twenty-first century** begin?

The twenty-first century began at 12:00 A.M. on January 1, 2001. Since the first century, which spanned the years 1 to 100, centuries have been counted beginning with the year ending in "01" rather than "00." For instance, the twentieth century consists of the years 1901 through 2000.

THE SEASONS

How does the **tilt of the Earth** affect the seasons?

Since the Earth is tilted 23.5 degrees, the sun's rays hit the northern and southern hemispheres unequally. When the sun's rays hit one hemisphere directly, the other hemisphere receives diffused rays. The hemisphere that receives the direct rays of the sun experiences summer; the hemisphere that receives the diffused rays experiences winter. Thus, when it is summer in North America, it is winter in most of South America, and vice versa.

The ancient Mayans were fascinated with measuring time, and even their pyramids were constructed to mark the time of equinoxes. The pyramid of Chichen Itza cast a shadow that reached the snakehead (foreground) at equinox.

Where on the planet is it **light 24 hours a day** in the summer?

In the extreme north and south parts of the Earth (north of 66.5 degrees north, and south of 66.5 degrees south latitudes) it is light 24 hours a day during the summer and dark 24 hours a day during the winter. Northern cities like Reykjavik, Iceland, and Murmansk, Russia, have nearly 24 hours of daylight for a short period of time during the summer months.

What are the **Tropics of Cancer and Capricorn**?

The two Tropics are the lines of latitude where the sun is directly overhead on the summer solstices. The Tropic of Cancer is at 23.5 degrees north and passes through central Mexico, northern Africa, central India, and southern China. The Tropic of Capricorn is at 23.5 degrees south and passes through central Australia, southern Brazil, and southern Africa.

What are the **solstices**?

There are two solstices—one on June 21 and the other on December 21. On June 21, the sun is directly above the Tropic of Cancer at noon and heralds the beginning of summer in the Northern Hemisphere and the beginning of winter in the Southern Hemisphere. On December 21, the sun is directly above the Tropic of Capricorn at noon and heralds the beginning of winter in the Northern Hemisphere and the beginning of summer in the Southern Hemisphere.

Can you stand an **egg on end** only on the spring equinox?

It is a common legend that an egg can be balanced on its end only on the spring equinox (March 21). Actually, there's nothing magical about gravity on the spring equinox that would allow an egg to stand on end—it can happen at any time of the year with patience and perseverance.

What are the **equinoxes**?

There are two equinoxes—one on March 21 and the other on September 21. On both equinoxes, the sun is directly over the equator. March 21 heralds the beginning of spring in the Northern Hemisphere and the beginning of fall in the Southern Hemisphere. September 21 heralds the beginning of fall in the Northern Hemisphere and the beginning of spring in the Southern Hemisphere.

Where are the **Arctic and Antarctic Circles**?

The Arctic Circle is located at 66.5 degrees north of the equator and the Antarctic Circle is located at 66.5 degrees south of the equator. Areas north of the Arctic Circle and south of the Antarctic Circle have 24 hours of light during the summer and 24 hours of darkness during the winter.

EXPLORATION

EUROPE AND ASIA

Who were the **earliest explorers from Asia**?

Many scientists believe that around 14000 B.C.E. small groups of peoples, originating from what is now Siberia in Russia, crossed a land bridge to what is now Alaska. From there, they moved as far south as South America. Considerable genetic evidence, archaeological finds, and skeletal remains support this theory.

When did the **Chinese Empire** begin **naval exploration**?

During the Song Dynasty (960–1270 C.E.), the Chinese began to build sea-faring trade ships. It was during the Yuan Dynasty (1271–1368 C.E.) that Chinese traders began to appear in the ports of Ceylon (Sri Lanka), India, and as far west as Africa. They were seeking goods for the royal kingdom, such as spices, ivory, medicine, and tropical woods.

Who **disguised himself** as a Muslim **to travel to Mecca**?

Since non-Muslims are not allowed into the sacred city of Mecca, British explorer Sir Richard Francis Burton disguised himself as an Afghan pilgrim in order to enter the city in 1853. Burton, having learned various languages in the military, explored India, the Middle East, Africa, and South America. Also a prolific writer, Burton published many accounts of his journeys and is perhaps most famous for translating *1001 Arabian Nights* into English.

What were **Marco Polo's contributions** to exploration?

An illustration of Marco Polo, the great explorer who introduced Europe to the Asian world.

Though Marco Polo did not actually discover anything, his writings in *Travels of Marco Polo* served as Europe's introduction to the East, and spurred interest in exploration. Marco Polo, born in the mid-thirteenth century in Venice, traveled with his father and uncle to China. During his stay, Polo served the Emperor Kublai Khan as an ambassador, as a governor, and in a host of other diplomatic positions. In his 30s he returned to Venice and fought against the city-state of Genoa and was eventually captured. While imprisoned in Genoa, he dictated the story of his travels to a fellow prisoner, creating the somewhat exaggerated memoir *Travels of Marco Polo*.

What did **Marco Polo** note in his journals about the **Chinese fleet**, when he arrived in the thirteenth century?

Marco Polo noted that Chinese ships had crews of more than 300, cabins for 60 people, and four sailing masts.

Which explorer was named the **Grand Imperial Eunuch** by the Emperor of China?

The Chinese explorer Cheng Ho helped Emperor Yung-lo come to power in 1402, and in 1404 the Emperor named Cheng Ho the Grand Imperial Eunuch. In 1405, Cheng Ho set sail on the first of his seven voyages, which spread Chinese influence and knowledge throughout South Asia and Africa. China moved toward isolationism after Yung-lo died.

Who was **Alexander von Humboldt**?

Alexander von Humboldt (1769–1862) was a German geographer who explored South America, Europe, and Russia. Von Humboldt traveled deep within the Amazon rain forests, developed the first weather map, and wrote a five-volume encyclopedia in which he sought to describe all of human knowledge about the Earth. Von Humboldt was the world's last great polymath (one of encyclopedic learning).

Where was the world's first geographic research institute?

In 1418, Prince Henry the Navigator created the Institute at Sagres, Portugal, to study navigation, cartography, and advances in ship building. Though Prince Henry was not an explorer himself, he commissioned many explorations that sailed south along the coast of Africa.

How early were the **islands of the Pacific explored**?

Polynesians, the people who settled many of the islands in the South Pacific, brought their people, culture, and knowledge to the islands before 1500 B.C.E. Experts at celestial navigation, reading the currents, flight patterns of birds, and how to travel in light canoe-like crafts, the ancient Polynesians left New Guinea and traveled first to the areas we know of as the Solomon Islands, then on to present-day Vanuatu. As the distance between islands became even greater, they refined their boat-making ability to create ships with double hulls that were able to carry animals, people, and trading supplies as far east as Hawaii and even Easter Island, which is 2,237 miles (3,600 kilometers) west of Chile. By the year 1000 C.E., Polynesian culture could be found in a gigantic triangle—the Polynesian Triangle—spreading across thousands of miles of ocean.

Who was the greatest **explorer of the Arab world**?

Known as the "Muslim Marco Polo," Ibn-Batuta (1304–1369) explored much of Africa and Asia. In his lifetime, Ibn-Batuta traveled over 75,000 miles (120,000 kilometers), gaining the reputation of the most-traveled man on Earth.

What did **Genghis Khan** conquer?

Genghis Khan, the ruler of the Mongol Empire, conquered an area stretching from China to western Russia to the Middle East. Khan created the world's largest empire, which began to dissolve following his death in 1227.

Who was **Prester John**?

In the twelfth century, a letter arrived for the Pope claiming to be from "Prester John," the leader of a Christian kingdom in the east that was in danger of being over-run by infidels. Prester John reportedly asked for help from European brethren. Though Prester John and his kingdom were never discovered, his mysterious letter sparked travels and explorations for centuries in an attempt to rescue the kingdom.

Which explorer fought for both the Union and Confederate armies in the U.S. Civil War and went on to discover a famous missing African explorer?

Though born in Britain, Sir Henry Morton Stanley sailed to the United States and worked there for several years before the start of the Civil War. He joined the Confederate Army but was captured in 1861 at the Battle of Shiloh. He then joined the Union Army. Stanley is best known for his search for the missing African explorer, David Livingstone, and his greeting upon finding him: "Dr. Livingstone, I presume?"

When did **Phoenician explorers** begin discovering and colonizing **Europe**?

Around 800 B.C.E., the Egyptian Pharaoh Necho began commissioning naval ships to go beyond the Mediterranean Sea. They traveled along the coast of France, Spain, Great Britain, and Africa.

What were the **Crusades**?

From the eleventh through the fourteenth centuries, groups of armed Christian Europeans invaded the Middle East to take the Holy Land from the Muslims and reclaim it for Christianity. The Crusaders ruthlessly murdered and pillaged throughout their long journey to the Middle East, and continued their brutality once there. Though the Crusades were a horrific era, the knowledge of the world gained by the Crusaders spurred a better geographic understanding.

AFRICA

Who discovered the **source of the Nile River**?

In 1856, John Hanning Speke was sent by the British Royal Geographical Society to discover lakes believed to exist in eastern Africa. In 1858, Speke and fellow explorer Sir Richard Francis Burton discovered Lake Tanganyika. Speke and Burton split, and while traveling on his own, Speke discovered Lake Victoria and claimed it to be the source of the Nile River. Though many did not believe the lake was the source, Speke returned to the lake in 1860 and proved Lake Victoria was indeed the source of the Nile River.

Whose **body was preserved** and then hand-carried for nine months to the coast of Africa?

Dr. David Livingstone, the world-famous explorer of Africa, died while exploring the area now known as Zambia. His body was embalmed with sand and his heart was buried under a nearby tree. His body was then wrapped in cloth and covered with tar to waterproof it. Loyal servants carried his body for nine months, all the way to the eastern coast of Africa, where the body was then transported to Britain on the *HMS Vulture*. On April 18, 1874, his body was buried in Westminster Abbey.

Who was **Captain Kidd**?

Though Captain William Kidd was hired by the British to fight pirates, he himself soon became a pirate. After he set sail for Africa, an area swarming with pirates, reports returned to Britain that Kidd himself had captured several ships. Upon learning of a warrant for his arrest, Kidd

Famous African explorer Dr. David Livingstone was buried at London's Westminster Abbey (photo by Paul A. Tucci).

sailed to Boston to meet a benefactor, who then had Kidd arrested and sent back to Britain. On May 23, 1701, Kidd was hanged for piracy.

THE NEW WORLD

Who was the **first European** to reach **North America**?

In the early eleventh century, Leif Ericsson was the first European to set foot in North America. According to legend, Leif Ericsson, a Norse explorer, visited Helluland, Markland, and Vinland, which are believed to be Baffin Island, Labrador, and Newfoundland, respectively.

Who was **Ponce de Leon**?

The Spanish conquistador Juan Ponce de Leon searched for the fountain of youth. In 1513 during his travels in search of the mythical fountain, reportedly located on the legendary island of Bimini, Ponce de Leon discovered Florida.

143

Who was the first European to see the **Pacific Ocean** from its eastern shore?

In 1513, Vasco Náñez de Balboa crossed the Isthmus of Panama and became the first European to see the Pacific Ocean from its eastern side. Wearing full armor, he walked straight into the ocean and claimed it and all the land it touched for Spain. He named the discovery the "Mar del Sur" (the South Sea). Only six years later, Balboa was beheaded by a jealous rival.

What is the **Strait of Magellan**?

Ferdinand Magellan discovered the strait that bears his name in 1520 during his voyage that circumnavigated the Earth. Magellan used the strait as a shortcut around the southern tip of South America. The waters of the Strait of Magellan are violent and surrounded by dangerous rocks.

What was the intent of **Magellan's Expedition**?

Ferdinand Magellan left Europe in 1519 hoping to circumnavigate the globe. He succeeded in reaching the Philippines in 1521, where he was later killed by natives. Though five ships and 241 men left Europe on September 20, 1519, only one ship, *The Victoria,* returned to Spain with 18 men on September 6, 1522. Despite Ferdinand Magellan's death on April 27, 1521, during a war in the Philippines, the Magellan Expedition successfully circumnavigated the globe.

Why is **Christopher Columbus** credited with discovering the New World?

Despite the fact that Christopher Columbus was not the first person nor the first European to reach the Americas, Columbus's discovery was important in that it prompted mass exploration and colonization of the New World. Though the idea of Columbus "discovering" the New World is extremely Eurocentric, Europeans have credited Columbus with the discovery of, and the enthusiasm surrounding the exploration of, the New World.

A statue of Christopher Columbus is prominently displayed in New York City's Columbus Square.

Did everyone during Christopher Columbus's time think that the **world was flat**?

No, they did not. Though the common perception is that Christopher Columbus had to convince King Ferdinand and Queen Isabella that the world was a sphere, Columbus actually had to convince the King and Queen of the circumference of the world. Though the ancient Greeks had discovered that the Earth was a sphere, centuries passed before this was generally accepted. By the fifteenth century, however—the time of Columbus's sailing—most educated people believed the world to be round. But the question remained: how far was it to travel around the world?

Is it true that Columbus deliberately **fudged the circumference measurement** so as to make a better case for his trip?

Though most scholars believed that the circumference of the Earth was approximately 25,000 miles (40,000 kilometers), Columbus used an estimate of 18,000 miles (29,000 kilometers) to push his case, in order to make his trip seem more achievable and the costs more reasonable. Columbus used Posidonus' smaller estimate, rather than Eratosthenes' larger and more accurate estimate, to make the trip appear shorter.

What did the **Mason-Dixon line** originally divide?

While the Mason-Dixon line commonly refers to the division between the "North" and "South" in the eastern United States, it was originally a boundary between Pennsylvania and Maryland surveyed by Charles Mason and Jeremiah Dixon in 1763. During the Civil War, the boundary between Pennsylvania and Maryland was extended westward to represent the line between the slave and non-slave states.

Who was **Vancouver, Canada,** named after?

In the 1790s, George Vancouver, who had previously accompanied James Cook on his explorations for "Terra Australis Incognito" and the Northwest Passage, explored and mapped the Pacific coast of North America. Vancouver circumnavigated Canada's Vancouver Island, and it and the city of Vancouver (founded 1881) were named for him.

What were **Lewis and Clark** looking for?

President Thomas Jefferson sent Meriwether Lewis and army officer William Clark to search for a Northwest Passage, a waterway that would connect the Atlantic and Pacific Oceans. Beginning in May 1804 and lasting through September 1806, the two men and their expedition party traveled through the uncharted Louisiana Territory and the Oregon Territory. Though they did not locate a Northwest Passage, Lewis and Clark documented the geography of the West.

Who was **John Wesley Powell**?

Though he lost an arm in the Civil War, John Wesley Powell became one of the leading surveyors of the nineteenth century. In 1869, Powell explored the Grand Canyon.

Traveling on a boat along the Colorado River, he faced dangerous rapids, hostile Native Americans, and weather extremes. In 1880, Powell was appointed the second director of the United States Geological Survey.

How did **America** get its **name**?

Though Christopher Columbus discovered the New World, he always believed that he had reached Asia, not realizing that he had encountered new continents. The Italian explorer Amerigo Vespucci, who had explored the New World and published accounts of his travels, was the first person known to have distinguished the New World from Asia. The German cartographer Martin Waldseemüller, who had read of Amerigo Vespucci's travels, published a map of the New World in 1507 with what is now known as South America named "America," in honor of Amerigo. The name stuck.

What did **James Cook** not discover?

During the eighteenth century, James Cook was sent on several expeditions of discovery, one of which was to the southern Pacific Ocean in search of the legendary landmass "Terra Australis Incognita." Though the continent that is now known as Australia had already been discovered, a centuries-old belief foretold of another huge continent in that area. Cook traveled to the southern Pacific Ocean and disproved the legend of "Terra Australis Incognita." On another expedition, Cook was sent to find a water route north of North America from Asia to Europe. As Cook sailed, he discovered the Sandwich Islands (Hawaiian Islands) and determined that a Northwest Passage was not feasible because of ice. On his way back from this "un"-discovery, Cook was killed in the Sandwich Islands during a struggle over the theft of one of his boats.

A statue of Captain James Cook was erected at London's Admiralty Arch in 1914.

How fast did the **Mayflower** sail?

In 1620, the Pilgrims sailed on the Mayflower from Plymouth, England, to the New World in 66 days. Though the Mayflower relied upon intermittent wind for propulsion, it averaged two miles (3.2 kilometers) per hour across the Atlantic Ocean.

Was George Washington a geographer?

George Washington manifested the abilities of a cartographer and a surveyor at an early age. When he was 13, Washington made his first map, which was of his father's property, Mt. Vernon. At the age of 17, Washington was appointed surveyor of Culpepper County, Virginia. At age 21, Washington entered military service, and the rest is history.

What is a **nautical mile**?

Used for measuring ocean-based distances, a nautical mile is equivalent to approximately 6,076 feet (1,852 meters) or 1.15 miles (1.85 kilometers). The speed of ships is measured in knots. One knot is equivalent to one nautical mile per hour.

What **monetary unit** is used in Panama?

The currency in Panama is called the Balboa, after the explorer Vasco Náñez de Balboa, because Balboa established the first European settlement in Panama.

THE POLES

Who was the first person to reach the **North Pole**?

Though American explorer Robert Edwin Peary is credited as the first to reach the North Pole, it is likely that he only came within 30 to 60 miles (48 to 80 kilometers) of 90° North during his expedition in 1909. Who did actually reach the Pole first is still being debated.

Who was the first person to reach the **South Pole**?

In 1911, the Norwegian explorer Roald Amundsen and the British explorer Robert Scott were racing against each other to be the first to reach the South Pole. On December 4, 1911, Amundsen and his crew of four reached the South Pole at 90° South. Approximately one month later, Scott and his team arrived at the pole. Depressed from their defeat and with inadequate supplies of food, Scott and his team died while trying to return to their base camp.

Who was one of the **great African American explorers**?

Mathew Henson, an African American born in 1866, joined Peary's first Arctic expedition and spent seven years there, covering more than 9,000 miles. Henson arrived at the North Pole 45 minutes ahead of expedition leader Peary, and he actually is the first person to find and stand on the North Pole.

UNITED STATES OF AMERICA

PHYSICAL FEATURES AND RESOURCES

Where is the **center** of the contiguous United States?

The geographic center of the lower 48 states is located at 39°50' North, 98°35' West, approximately four miles (6.5 kilometers) northwest of Lebanon, Kansas.

What is the **highest point** in the United States?

Alaska's Mt. McKinley (also known as Denali) is the highest point in the United States at 20,320 feet (6,194 meters). In the contiguous 48 states, the highest point is California's Mt. Whitney at 14,495 feet (4,418 meters), which is less than 100 miles (161 kilometers) from North America's lowest point, Death Valley (282 feet [86 meters] below sea level).

What is the highest point **east of the Mississippi**?

North Carolina's Mt. Mitchell is the tallest point east of the Mississippi River at 6,684 feet (2,037 meters).

What is the **highest lake** in North America?

Yellowstone Lake in Yellowstone National Park is the highest lake in North America, at 7,735 feet (2,358 meters) above sea level.

What is the **deepest lake** in the United States?

Crater Lake in Oregon, lying within the collapsed crater of an ancient volcano, is the nation's deepest lake at 1,932 feet (589 meters). Crater Lake has no feeder streams—it is filled solely by precipitation.

149

The biologically rich Everglades marshes in Florida cover over 2,000 square miles (5,180 square kilometers) of land, but this is only about 20 percent of this endangered ecosystem's original area.

What is the **largest island in the United States**?

The Island of Hawaii is the largest island in the United States at 4,021 square miles (10,414 square kilometers). Puerto Rico is the second largest at 3,435 square miles (8,897 square kilometers).

Where is the world's **largest marsh**?

The Everglades in Florida is the world's largest marsh, consisting of 2,185 square miles (5,659 square kilometers). The water across this southern Florida marsh averages six inches (15 centimeters) in depth. The Everglades is an endangered ecosystem, threatened by excess drainage and the introduction of exotic plants.

Where is the world's **largest mountain**?

Hawaii's Mauna Kea is the world's largest mountain. It begins on the sea floor and rises 33,480 feet (10,205 meters). Mt. Everest only rises 29,000 feet (8,839 meters). Mauna Kea's peak reaches 13,796 feet (4,205 meters) above sea level.

How many **Great Lakes** are there?

There are five Great Lakes: Huron, Ontario, Michigan, Erie, and Superior. The acronym "HOMES" can help you to remember the names of the five Great Lakes. All of the lakes except Michigan lie on the U.S.-Canada border.

Where do the Great Lakes rank in terms of size?

Lake Superior is 31,700 square miles (82,103 square kilometers) and is the world's largest freshwater lake; Lake Huron is 23,000 square miles (59,570 square kilometers) and the third largest freshwater lake; Lake Michigan is 22,300 square miles (57,757 square kilometers) and the fourth largest freshwater lake; Lake Erie is 9,900 square miles (25,641 square kilometers), the world's tenth largest freshwater lake; and Lake Ontario, 7,300 square miles (18,907 square kilometers) is the twelfth largest freshwater lake.

What is the **largest freshwater lake** in the world?

Lake Superior is the largest freshwater lake in the world. It is 31,700 square miles (82,103 square kilometers) in area and is approximately 350 miles (563 kilometers) long.

What is the world's **shortest river**?

The world's shortest river is Oregon's D River, which is a mere 120 feet long (36.6 meters) . It connects Devil's Lake to the Pacific Ocean near Lincoln City, Oregon.

Why is **Coney Island** called an island even though it's not?

Though now a peninsula of Long Island, Coney Island was actually an island at one time. The popular amusement park of the early twentieth century is now attached to Long Island due to the silting up of Coney Island Creek, which once separated the two islands.

THE STATES

What are the five **largest states**?

The five largest states are Alaska (591,000 square miles [1,530,690 square kilometers]), Texas (266,800 square miles [691,012 square kilometers]), California (158,700 square miles [411,033 square kilometers]), Montana (147,000 square miles [380,730 square kilometers]), and New Mexico (121,600 square miles [314,944 square kilometers]).

What are the five **smallest states**?

The five smallest states are Rhode Island (1,200 square miles [3,108 square kilometers]), Delaware (2,000 square miles [5,180 square kilometers]), Connecticut (5,000 square miles [12,950 square kilometers]), Hawaii (6,500 square miles [16,835 square kilometers]), and New Jersey (7,800 square miles [20,202 square kilometers]).

What are the five **most populous U.S. states**?

The five most populous states are California (33.8 million), Texas (20.8 million), New York (18.9 million), Florida (15.9 million), and Illinois (12.4 million).

What are the five **least populous U.S. states**?

The least populous U.S. states are Wyoming (493,000), the District of Columbia (572,000, though not technically a state). Vermont (608,000), Alaska (626,000), and North Dakota (642,000).

Which state has the **most lakes**?

Though Minnesota is known for its "10,000 lakes" (the slogan on its license plates), it is not the state with the most lakes. Neighboring Wisconsin has even more, about 14,000 lakes, but Alaska is the definite winner, with over three million lakes!

How many states have **four-letter names**?

There are three states that tie for having the shortest name: Utah, Ohio, and Iowa.

How many **state names end** in the **letter "a"**?

Twenty-one of the 50 states end in the letter "a."

How was the **Delmarva Peninsula** named?

The Delmarva Peninsula, located on the East Coast, contains all of Delaware and portions of Maryland and Virginia. The 180-mile (290-kilometer) long peninsula was named by combining the abbreviations of each of those three states—Del., Mar., and Va.

How many **islands** does **Hawaii** include?

There are a total of 122 islands in the Hawaiian chain. The southernmost island, Hawaii, is the largest (4,000 square miles [10,360 square kilometers]). The two westernmost islets are known as the Midway Islands and, while they are part of the United States, they are not part of the state of Hawaii.

Which Hawaiian island was once a **leper colony**?

On the island of Molokai in Hawaii, authorities established a secluded leper colony on an inaccessible cove. Its founder, Father Damian, served and cared for the inhabitants. He eventually died of the disease and the Catholic Church made him a saint for his work.

A catamaran shares the beach with bathers in Waikiki, Honolulu, Hawaii. This fiftieth American state consists of 122 islands altogether (photo by Paul A. Tucci).

Which state has the **highest divorce rate**?

Nevada, with 6.4 divorces per 1,000 people, leads the nation. The District of Columbia has the lowest rate at 1.7 per 1,000, and is followed by Massachusetts with 2.2 divorces per 1,000 people.

What do the **Sandwich Islands** and Hawaiian Islands have in common?

The Sandwich Islands and Hawaiian Islands are actually the same set of islands. In 1778, when Captain James Cook discovered the islands, he named them the Sandwich Islands. Gradually, the islands began to be known by their indigenous name of Hawaii. Cook named the islands after his supporter, John Montagu, the fourth Earl of Sandwich.

What is the only state with a **diamond mine**?

Arkansas is the only state with a diamond mine. Located in southwestern Arkansas, the mine is no longer in commercial production but is now a park, Crater of Diamonds State Park, that allows visitors to take any diamonds they find.

What is the **driest state**?

Nevada averages only 7.5 inches (19 centimeters) of rainfall annually, making it the driest state in the union.

153

What is the **official neckwear of Arizona**?

Like an official flower or animal, each state can add additional official symbols for their state. The official neckwear of Arizona is the bolo tie.

Which state **borders only one other state**?

Bordering New Hampshire, Maine is the only state that borders just one other state.

Which state **borders the most** others?

Tennessee borders eight states: Arkansas, Missouri, Kentucky, Virginia, North Carolina, Georgia, Alabama, and Mississippi. Missouri also borders eight states: Iowa, Nebraska, Kansas, Oklahoma, Arkansas, Tennessee, Kentucky, and Illinois.

Which state has the **longest coastline**?

Alaska has more than 5,500 miles (8,851 kilometers) of coastline. The next largest is actually Michigan, which has 3,288 miles (5,291 kilometers) of coastline, far longer than both Florida (1,197 miles [1,926 kilometers]) and California (840 miles [1,352 kilometers]).

What state has only **one legislative body**?

Nebraska has a unicameral legislature; all other 49 states have two houses.

How many states have land **north of Canada's** southernmost point?

Twenty-seven of the 50 states have land north of Canada's southernmost point, Pelee Island in Lake Erie.

Is **Puerto Rico** a state?

No, Puerto Rico is a commonwealth of the United States. In 1898, the United States acquired the island of Puerto Rico from Spain, but Puerto Rico did not become a commonwealth until 1952. Though Puerto Ricans pay no federal income tax and cannot vote for President, they are citizens of the United States and can move freely around

A view of Old San Juan in Puerto Rico, which is still a commonwealth of the United States and not a state.

the country. Puerto Rico is allowed one representative to the House of Representatives, but that member cannot vote. With approximately 3.6 million residents, Puerto Rico is currently the largest colony in the world.

CITIES AND COUNTIES

How many cities are in the United States?

There are approximately 18,440 cities and towns in the United States, the vast majority of which have fewer than 25,000 people.

Measured by area, what is the largest city in the United States?

Juneau, Alaska, the capital of the state, is the largest city in the United States, with over 3,100 square miles (8,000 square kilometers).

What are the fastest growing U.S. cities in terms of population growth from 2000 to 2006?

From 2000 to 2006, the fastest-growing U.S. cities have been McKinney, Texas (97.3%); Gilbert, Arizona (73.9%); and North Las Vegas, Nevada (71.1%).

No. Louisiana is divided into 64 parishes rather than counties. These parishes are no different than counties, other than in name. The word parish comes from the parish system of the Catholic church, and thus shows the French and Catholic influence on Louisiana.

How much did the population of **New Orleans decline** after Hurricane Katrina struck?

Although many residents returned in the aftermath of Hurricane Katrina, New Orleans's population has declined by more than 53 percent since the 2005 disaster.

What **cities** have the **most crime** in the United States?

According to major rankings, Detroit, Michigan; Camden, New Jersey; St. Louis, Missouri; and Oakland, California have the most crime.

What are the **northernmost, southernmost, easternmost, and westernmost** cities in the United States?

Barrow, Alaska, is the northernmost; Hilo, Hawaii, is the southernmost; Eastport, Maine, is the easternmost; and Atka, Alaska, is the westernmost city in the country.

What is the **highest settlement** in the United States?

Climax, Colorado, is located at an elevation of 11,302 feet (3,445 meters) and is the highest settlement in the United States.

What is the oldest continually **occupied city** in the United States?

Saint Augustine, Florida, was founded in 1565 by Spanish explorer Pedro Menendez de Avilés. Saint Augustine is located on Florida's eastern (Atlantic) coast and now has a population of approximately 12,000. It is not only the oldest continually occupied city in the United States, but also in all of North America.

Which is **further west**—Los Angeles, California, or Reno, Nevada?

Though Nevada is California's eastern neighbor and Los Angeles sits on the Pacific Coast, Reno is farther west than Los Angeles. Reno is located at 119°49' West, while Los Angeles is located at 118°14' West.

What city is **named for a game show**?

The popular radio game show "Truth or Consequences" offered to host its 10th anniversary show in a city that would change its name to the show's title. In 1950 Hot Springs, New Mexico, changed its name and the 10th anniversary show was broadcast from Truth or Consequences, New Mexico. The city still holds the name today.

What city is known as the **Earthquake City**?

Charleston, South Carolina, claims the nickname "Earthquake City." On August 31, 1886, Charleston suffered from the largest earthquake in history to strike the east coast of the United States. Sixty were killed in the quake, which had an estimated Richter magnitude of 6.6.

Where does **Los Angeles** get its **water**?

Not much of the water in Los Angeles comes from local sources—most of it is brought from hundreds of miles away. Large aqueducts, man-made channels used to transport water, were built to carry water from Owens Valley (in East-Central California), from the Colorado River, and from the rivers of Northern California, to Los Angeles. Though this method has brought fresh water to a region that desperately needs it, it has also drained and damaged the ecologies that once depended on the water now being sapped from its supply.

How many **counties** are in the United States?

There are 3,043 counties in the United States. The state with the least number of counties is Delaware, with three, and Texas has the most, with 254.

What are the **largest and smallest counties** in the United States?

San Bernardino County in California is the largest, at 20,000 square miles (51,800 square kilometers). It stretches from metropolitan Los Angeles to the Nevada/Arizona border. The smallest county in the United States is Kalawao, Hawaii, at 13 square miles (33.7 square kilometers), which has a very small population of only 130.

In the nineteenth century, most immigrants to the United States—like this Italian couple—came from Europe. Today, the majority come from Mexico and Asia (photo courtesy Paul A. Tucci).

PEOPLE AND CULTURE

What is the **population** of the United States?

The U.S. population is approximately 304 million people, with one person being born every seven seconds, one person dying every 13 seconds, and one international migrant arriving every 30 seconds, yielding a net gain of one new person every 11 seconds.

Where do most **legal immigrants** to the United States come from?

Of the approximately 1.2 million immigrants who come to live in the United States legally, Mexico is the top country of origin (14.4%), followed by the People's Republic of China (7.2%), and the Philippines (6.2%).

How many **illegal immigrants** are there in the United States?

Some estimates take into account responses to the U.S. Census Survey, pegging the number at approximately 11.3 million people.

Where do the **illegal immigrants come from**?

Approximately 57 percent of illegal immigrants come from Mexico, 11 percent from Central America, 9 percent from East Asia, 8 percent from South America, and 4 percent from Europe and the Caribbean.

What is the **oldest college** in the United States?

Harvard University is the oldest college or university in the United States. It was founded in 1636 in Cambridge, Massachusetts, just outside of Boston.

How **valuable** is a **college education** in the United States?

Workers in the United States with at least a bachelor's degree earn an average of $51,206 per year. High school graduates can expect to earn $27,915 per year, while those who did not complete high school earn $18,734 on average. Workers who earned an advanced degree can make on average $74,602 per year.

What percentage of American adults use a cell phone?

More than 89 percent of American adults have cell phones, and 14 percent of American adults use only a cell phone and have no land line. Only nine percent of Americans use only a land line to make phone calls.

How many **educational institutions** are there in the United States?

There are more than 130,400 primary and secondary schools, and more than 6,400 universities, colleges, and post-secondary schools in the United States.

How many **foreign students** attend **universities and colleges** in the United States?

There are more than 564,000 foreign students attending schools in the United States. Asia contributes 58 percent, Europe 15 percent, Latin America 11 percent, Africa 6 percent, Canada 5 percent, the Middle East 3 percent, and Oceania 0.8 percent.

How many **Internet users** are there in the United States?

There are more than 211 million Internet users in America, making up approximately 72.5 percent of the population. Fifteen percent of all Internet users in the world are located in the United States.

What percentage of **American households** regularly give to **charitable organizations**?

Approximately 89 percent of American households give to charitable organizations. On average, they give $1,620, donating $295 billion annually all together.

What percentage of **Americans volunteer** each year?

More than 26 percent of Americans volunteer in some form each year, which means nearly 61 million people donate their time.

What is the most popular **national park** in the United States?

North Carolina and Virginia's Blue Ridge Parkway is the most popular national park, drawing over 17 million visitors each year. Located in the southern Appalachian Mountains, the Parkway is a highway through beautiful scenery and includes nearby hiking trails, ranger talks, and bird-watching. The other two most-visited national parks in the

159

United States are Golden Gate National Recreation Area, which gets about 14 million visitors annually, and Lake Mead National Recreation Area, with 9.4 million.

What is the **oldest public park** in the United States?

In 1634, William Blackstone sold 50 acres of his land to the town of Boston. This pastureland was set aside for common use and called the Boston Common, making it the oldest public park in the United States. The Common is situated in front of Massachusetts' State House and has always been open public land.

What is the most popular **theme park** in the United States?

Walt Disney World in Orlando, Florida, has more than 17 million visitors annually.

Where was the world's first **monument to an insect** established?

On December 11, 1919, Enterprise, Alabama, dedicated a monument to the boll weevil. This tall statue of a woman with raised arms holding a boll weevil declares, "In profound appreciation of the boll weevil and what it has done as the herald of prosperity." The boll weevil, a beetle that attacks bolls of cotton, spread across the South at the beginning of the twentieth century, wiping out cotton crops. Residents of Enterprise switched from cotton crops to peanut crops, thus discovering a new era of prosperity. The monument to the boll weevil is to remind residents and visitors alike of the resourcefulness of the community and the ability of man to diversify.

How many **automobiles** are there in the United States?

There are approximately 151 million automobiles in use in the United States, about one car for every two people.

Where was the **first commercial air flight**?

On January 1, 1914, the first scheduled commercial air flight took passengers from Tampa Bay, Florida, to St. Petersburg, Florida. The service, which lasted only a few weeks, took one person at a time over the 22-mile (35-kilometer) route.

The impressive, mountain-sized sculpture of four American presidents at Mt. Rushmore was created by Gutzon Borglum.

Where is the **sunbelt**?

The sunbelt, known for its warm temperatures, is a geographical area that spreads across the southern and southwestern states of the United States. It has been an area of high population growth over the past few decades, as more families and individuals have moved to states like California, Arizona, Texas, and Florida.

Where is the **rustbelt**?

The rustbelt is a term used to describe the United States' declining manufacturing region of the northeast and Midwest. Factory closures, especially those of steel and textile mills, have resulted in massive unemployment and declining population in rustbelt cities. The rustbelt runs from about Minnesota to Massachusetts.

Where is the **Bible Belt**?

The Bible Belt, a region noted for its high proportion of fundamentalist Christian beliefs, is located in the southern and Midwestern United States, running from about Oklahoma to the Carolinas.

Who carved **Mt. Rushmore**?

Gutzon Borglum, an American sculptor, designed this national memorial located in the Black Hills of South Dakota. Construction began in 1927 and was nearly complete when Borglum died in March 1941. Borglum oversaw construction of the 60-foot-tall (18-meter-tall) heads of Presidents George Washington, Thomas Jefferson, Abraham Lincoln, and Theodore Roosevelt. After Borglum's death, his son completed the work on the unfinished Roosevelt by the end of 1941.

Why was there a **Russian outpost** in California?

Established in 1812, Fort Ross was a Russian outpost located in what is now Sonoma County, California. The outpost was started so that Russian fur traders could explore and exploit the area. A Russian presence was maintained at the fort until 1841.

What is the **oldest continuously published newspaper** in the United States?

In 1764, Thomas Green founded the *Hartford Courant* in Hartford, Connecticut, which is the oldest continuously published newspaper in the United States.

Where was the **first shopping mall** in the United States?

In 1922, Country Club District opened in the suburbs of Kansas City, Kansas, and was the first shopping mall in the country.

How many **businesses are started** each year in the United States?

Each year, more than 600,000 new companies are started in America, joining the more than 24.7 million companies that make the American economy the largest in the world. Most companies (99.9%) have fewer than 500 employees.

What is the **leading cause of death** in the United States?

Approximately one-third of all Americans die from cardiovascular disease, making it the country's leading cause of death. Cancers are the second leading cause of death in America, killing over one-quarter of the population.

Where is **Acadiana**?

In 1755, the British took control of Acadie, New Brunswick, Canada, and exiled the city's inhabitants. Forced to leave their homes, these people, known as Acadians, moved to Louisiana. The Acadians are still present in Louisiana and their culture has produced famed food and music, called Cajun. The area in southern Louisiana where the Acadians still live is termed Acadiana.

Why are **graves above the ground** in Louisiana?

In Louisiana, the water level is so close to the surface of the ground that coffins, rather than being buried, are placed in tombs above ground to avoid the possibility of coffins floating out of place. A Louisiana cemetery looks like a miniature city, with tombs and alleyways.

HISTORY

How did the **United States** reach its present form?

The United States began as 13 British colonies on the Atlantic Coast. In 1783, the United States gained the Northwest Territory, the area encompassing what is now Ohio, Indiana, Illinois, Michigan, and Wisconsin. Spain and the United States agreed on the northern boundary of Florida in 1798 and the United States then took control of the Mississippi Territory. In 1803, the Louisiana Purchase (which included most of the area west of the Mississippi) doubled the size of the country. In 1845 the independent Republic of Texas was annexed and Spain ceded Florida to the United States. In 1846, the Oregon Territory (which included Oregon, Washington, and Idaho) was officially designated with a treaty between the United States and the United Kingdom. The Mexican War of 1846–1848 led to the secession of California, Utah Territory, and New Mexico territory to the United States. The Gadsden Purchase of 1853 added southern Arizona. Alaska was purchased from Russia in 1867 and, finally, Hawaii was annexed by the United States in 1898.

What was **Manifest Destiny**?

First used by John Louis O'Sullivan in an 1845 editorial, Manifest Destiny was the phrase used to describe the assumption that American expansion to the Pacific Ocean was inevitable and ordained by God. The phrase was used to defend the annexations of Texas, California, Alaska, and even Pacific and Caribbean islands.

What were **Lewis and Clark** looking for?

President Thomas Jefferson sent Meriwether Lewis and army officer William Clark to search for a Northwest Passage, a waterway that would connect the Atlantic and Pacific Oceans. Beginning in May 1804 and lasting through September 1806, the two men and their expedition party traveled through the uncharted Louisiana Territory and the Oregon Territory. Though they did not locate a Northwest Passage, Lewis and Clark documented the geography of the West.

What was **Seward's Folly**?

Seward's Folly, also known as Seward's Icebox, was the derogatory nickname given to the area known as Alaska, purchased by the United States from Russia in 1867. The

Sitka is a prosperous fishing town in Alaska, a state with immense natural resources that was bought by the United States for a little over seven million dollars back in 1867.

$7.2 million purchase, heavily encouraged by Secretary of State William Seward, was criticized by many, thus dubbed "Seward's Folly." The Alaskan Gold Rush of 1900 proved Seward to be a very wise man. In 1959, Alaska became the forty-ninth state.

When was the most **territory added** to the United States at one time?

In 1803, the United States purchased over 800,000 square miles (2,000,000 square kilometers) of land from France for $15 million. This territory, known as the Louisiana Purchase, extended from the Mississippi River to the Rocky Mountains and doubled the size of the United States.

Aside from the **Louisiana Purchase**, how was the **American West** obtained?

There were several other purchases and wars fought to gain the land west of the Louisiana Purchase. These included the Gadsden Purchase from Mexico, sections of the west ceded to the United States by Mexico, Texas, and the Oregon Country.

What was the **Oregon Trail**?

The Oregon Trail was a 2,000-mile-long (3,200-kilometer-long) pioneer trail that extended from Independence, Missouri, to Portland, Oregon. Migrants traveled along this trail in an effort to reach and settle the sparsely populated American West. It took

> ## What happened to the Arawaks
> ## after Christopher Columbus arrived in the New World?
>
> It is said that when Columbus arrived on the shores of the islands of the Caribbean there were between 250,000 and 1 million Island Arawaks. By the middle of the sixteenth century, disease, slavery, and outright killing brought the number down to fewer than 500. Today, about 30,000 Arawaks live in Guyana, and there are small populations in French Guyana and Suriname.

migrants approximately six months to traverse the Oregon Trail and reach Oregon. The trail was heavily used from the 1840s and subsequent decades. Portions of the Trail are still visible today in places such as the Whitman Mission National Historic Site in Washington.

How many **Native American tribes** are there in the United States?

There are 550 recognized tribes in the United States.

How many **Native Americans** were **killed** during the period of **European colonization** in the Americas?

The great majority of Native Americans who had made what is now America their home for tens of thousands of years were killed outright by European colonists. Diseases brought to the Americas from Europe, including small pox and the plague, also decimated the indigenous tribes. Furthermore inter-tribal warfare, spurred on by the political interests of Europeans, enslavement, and the mass killing and forced relocation in the nineteenth century of the hundreds of thousands of Native Americans who still managed to survive all conspired to nearly destroy native tribes and their cultures.

How many **Native American reservations** are there in the United States?

There are 310 reservations in the United States. Although the land is owned and administered by the federal government under the Department of the Interior's Bureau of Indian Affairs, each reservation has limited sovereignty, including its own legal system.

What was the **Trail of Tears**?

In 1838, the United States rounded up approximately 15,000 members of the Cherokee Nation and forced them from their homes in Tennessee to a reservation in Okla-   **165** homa. The removal of the Cherokee Nation from their land was done so that citizens of the United States could use the fertile lands in Tennessee. Along the "trail," approximately one-fourth of the Cherokees died from malnutrition, disease, and government inefficiency.

When was the **first permanent British settlement** established in the United States?

In 1607, the colony of Jamestown was established in Virginia by British colonists transported there by the London Company. Though subjected to many attacks by Native Americans, the colony was ultimately destroyed in 1676 by its own rebelling colonists in Bacon's Rebellion.

How did the **Monroe Doctrine** protect the Americas?

In 1823, President James Monroe gave a speech that declared the Americas off limits to European powers. These policies became known in the 1840s as the Monroe Doctrine. Since 1823, the United States has used the Monroe Doctrine not only to prevent intervention by Europeans but also to further its own expansionist goals.

How many **countries** has the **U.S. military attacked or occupied** since the beginning of the year 2000?

The U.S. has invaded or occupied 14 foreign lands since 2000, including Macedonia, Afghanistan, Yemen, Philippines, Colombia, Iraq, Liberia, Haiti, Pakistan, Somalia, Georgia, Djibouti, Cote d'Ivoire, and Sierre Leone.

How did the United States obtain the **Virgin Islands**?

The United States purchased the Virgin Islands, which includes three main islands (St. Croix, St. John, and St. Thomas) along with 50 smaller islands, from Denmark in 1917 to help defend the Caribbean Sea. Just under 100,000 people live on the islands, which are now a territory of the United States.

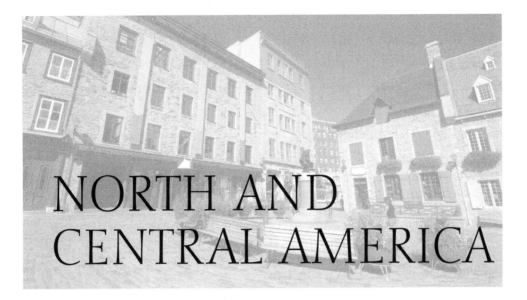

NORTH AND CENTRAL AMERICA

What is the **center of North America**?

The geographic center of North America (including Canada, the United States, and Mexico) is located six miles (10 kilometers) west of Balta, North Dakota, at 48°10' North, 100°10' West.

What's **NAFTA**?

In 1994, Canada, the United States, and Mexico entered into the North American Free Trade Agreement (NAFTA), which reduces tariffs and economic controls between the three countries.

Has **NAFTA helped trade** between Canada, Mexico, and the United States?

Yes, from 1993 to 2007 trade between the NAFTA countries more than tripled, rising from $290 billion to $930 billion.

What is the **continental divide**?

The continental divide is simply the line that divides the flow of water in North America. Precipitation east of the divide flows towards the Atlantic Ocean while precipitation west of the divide flows towards the Pacific Ocean. The divide follows the line of the highest ranges of the Rocky Mountains and is not a distinct mountain range.

GREENLAND AND THE
NORTH POLE REGION

Is **Greenland** really green?

In 982 C.E., Greenland was named by its first colonizer, Eric the Red. Having been banished from Iceland, Eric the Red established a colony on Greenland and gave it a pleasant-sounding name, in order to attract other colonists. In reality, Greenland is not very green. Though small coastal areas are habitable, most of the island is mountainous and covered by ice sheets. Greenland is an autonomous territory of Denmark.

How **thick** is **Greenland's ice sheet**?

Greenland's ice sheet is more than two miles thick.

What is the **northernmost landmass** in the world?

Cape Morris Jesup, located in Greenland's Peary Land, is the northernmost point of land in the world. The cape is at 83°38' North.

Is there land at the **North Pole**?

Since the North Pole lies in the Arctic Ocean, which is mostly covered by a large ice-cap year-round, there is no land near the North Pole. Though there is no land, animals such as the polar bear live upon the icecap.

How big is the **area** of the **North Pole ice** region?

The arctic ice mass covers an area of 1.59 million square miles (4.1 million square kilometers).

Is global warming **melting** the **North Pole**?

Yes, from 1979 to 2005 the North Pole region lost 2.05 million square miles of total area (5.31 square kilometers), which is equivalent to an area the size of California and Texas combined. These changes have not been seen since between 6,000 and 125,000 years ago.

How many **indigenous people live** in the **Arctic**?

More than four million people live in the Arctic. They are spread across national boundaries in such places as Alaska, Canada, Greenland, Scandinavia, and Siberia.

Old Québec in Montréal, Canada, reflects the Old World charm of its French influences.

CANADA

How many **provinces** are in **Canada**?

Canada is divided into 10 provinces and three territories. The 10 provinces are Alberta, British Columbia, Manitoba, New Brunswick, Newfoundland, Nova Scotia, Ontario, Price Edward Island, Québec, and Saskatchewan. Canada's territories are Yukon Territory, the Northwest Territories, and Nunavut.

Where are the **Prairie Provinces**?

The grassy, central Canadian states of Manitoba, Saskatchewan, and Alberta are called the Prairie Provinces.

What is **Nunavut**?

Nunavut, Canada's third territory, whose capital is Iqaluit, is home to Canada's indigenous people, the Inuit. This new territory, which entered the dominion in 1999, covers approximately one-fifth of Canada's land area, but contains less than one percent of Canada's population, with just over 29,000 people.

Over half of Canada's population lies in southern Ontario and southern Québec in eastern Canada. This area, dubbed "Main Street," stretches from Windsor, Ontario, to Québec City, Québec, and includes the major cities of Toronto, Ottawa, and Montréal.

What are the largest **Canadian urban areas**?

Toronto (5.1 million people), Montréal (3.6 million), and Vancouver (2.1 million) are Canada's largest urban areas.

When did Canada have a **transcontinental railway**?

After nearly a decade of setbacks, Canada completed its first transcontinental railroad in 1885. The Canadian Pacific Railway opened western Canada to settlement and greatly helped the city of Vancouver to grow.

How is northern Canada **gaining elevation**?

During the ice ages, thick ice sheets covered northern Canada, including Hudson Bay, pushing the continent down by the sheer force of the extreme weight of the ice. Ever since the melting of the ice sheets at the end of the ice age, the ground has been rising a few inches each year.

Why do they speak **French in Québec**?

Most of Québec speaks French because the French founded the city of Québec, one of the oldest cities in North America, in the seventeenth century. Québec served as the capital of the surrounding New France until the British took control of the territory in 1763. Even though it was ruled by the British from the eighteenth century onward, the region that was once held by France has remained a center of French culture and language in North America.

The differences in culture between the province of Québec and the rest of Canada have been so extreme that there are strong secessionist forces within Québec that want Québec to become its own country. Though two referendums for Québec's independence have been voted upon in Canada-wide elections, both the 1980 and 1995 referendums failed to pass. But these failures have not quieted Québec's secessionist forces, and it is likely that future referendums will be held. If passed, many believe that the secession of Québec would dissolve the loose Canadian federation.

What did the phrase **"fifty-four forty or fight"** mean?

In the mid-nineteenth century, many Americans wanted to see the United States' territory expand northward into the area now known as Canada. The phrase "fifty-four forty or fight" referred to the desire to move the boundary northward to 54°40' North, which would have encompassed much of southern Canada. Ultimately, the boundary was fixed at 49° North, where it sits today.

The Canadian Rockies , off of Icefields Parkway, inside Jasper National Park, Alberta, Canada. The Rockies stretch from the Yukon in Canada all the way down to the American Southwest (photo by Paul A. Tucci).

What is the **deepest lake** in North America?

Northwestern Canada's Great Slave Lake is the continent's deepest, at 2,015 feet (614 meters). The lake is named after the indigenous people who live near the lake, the Slave.

How **long** are the **Rockies**?

The Rocky Mountains (Rockies) extend over 2,000 miles (3,200 kilometers), from the Yukon Territory in Canada to Arizona and New Mexico in the southern United States.

How was **Niagara Falls stopped**?

The water from the Niagara River falls over two waterfalls, divided by Goat Island. Only six percent of the water from the Niagara River falls over American Falls, while Horseshoe Falls (or Canadian Falls) carries the majority of the water. In 1969, a temporary dam was built to divert the water from American Falls to Horseshoe Falls for several months in order to study the erosion endemic to both waterfalls.

What was the **Klondike Gold Rush**?

In 1896, gold was discovered in an area of western Canada known as the Klondike, located in the Yukon Territory where the Klondike and Yukon Rivers meet. Once the news of the discovery spread, tens of thousands of people headed west, creating the Klondike Gold Rush.

MEXICO

What was the **first major city** in the Western Hemisphere?

From the first through the seventh centuries, the city of Teotihuacan flourished. Located northeast of modern Mexico City, Teotihuacan had a maximum population of approximately 200,000 and was the first major city in the Western Hemisphere. The city was graced with the Pyramid of the Sun and the Pyramid of the Moon. Teotihuacan should not be confused with the Aztec city of Tenochtitlan, which was built nearly six centuries later in the area that is now known as Mexico City.

Where are **Sierra Madre Occidental and Sierra Madre Oriental**?

Sierra Madre Occidental and Sierra Madre Oriental are two mountain ranges in Mexico. The ranges' names stem from the meanings of "occidental" (western) and "oriental" (eastern); thus the Sierra Madre Occidental lies along Mexico's west coast and the Sierra Madre Oriental lies along the east coast.

How many states are in Mexico?

Mexico is divided into 31 states plus the federal district of Mexico City. The largest state is Chihuahua, located in northern Mexico, with 95,400 square miles (247,086 square kilometers). The most populous state is the state of Mexico, located west of Mexico City, with 14 million people.

Where is the **Yucatan**?

The Yucatan is a large peninsula in southern Mexico. The Yucatan Peninsula "points" toward Florida and separates the Gulf of Mexico from the Caribbean Sea.

How much of **Mexico's population** lives in Mexico City?

Approximately one-fourth of Mexico's population lives in Mexico City's metropolitan area. Mexico City, Mexico's capital, is the largest metropolitan area in the Western Hemisphere, with a population of 19 million.

How do **maquiladoras** help clothe the United States?

Maquiladoras are Mexican factories owned by foreign (usually U.S.) corporations. Most often located along the Mexican-U.S. border, maquiladoras receive raw materials from the United States and produce finished goods for export. Maquiladoras commonly produce clothing and automobiles.

How much does an **employee** of a **maquiladora earn**?

Border factories prefer to employ women, who tend not to challenge poor working conditions and can be paid as little as one-sixth of what workers can make north of the border in the United States. This is approximately $1.00/hour, with most workers forced to work on average 48 hours per week.

CENTRAL AMERICA

What is the difference between **Central and Latin America**?

Central America includes the countries that connect North and South America and are located between Mexico and Colombia. The seven countries of Central America are Guatemala, Belize, El Salvador, Honduras, Nicaragua, Costa Rica, and Panama. Latin America is a much broader term, and includes Central America as well as Mexico and all of the countries of South America.

What was the **last Central American country** to obtain its **independence**?

In 1981, Belize, formerly known as British Honduras, became the last North American country to obtain independence.

Are there lots of mosquitoes on the **Mosquito Coast**?

The Mosquito Coast is an area 64 miles (103 kilometers) wide and approximately 250 miles (400 kilometers) long along Nicaragua's eastern shore. Though the Mosquito Coast receives an average of 250 inches (635 centimeters) of rain annually, making it a perfect breeding place for mosquitoes, the Mosquito Coast was named after the indigenous people of the area, the Mosquito Indians.

Off the coast of Belize are beautiful shallow reefs.

173

Who owns the **Panama Canal**?

In 1903, the U.S.-backed revolutionaries in western Colombia revolted and created the independent country of Panama, which was immediately recognized by the United States. The newly independent Panama gave the United States use of a 10-mile (16-kilometer)-wide strip of land across the Isthmus of Panama, where the United States built the Panama Canal. The United States maintained control of the Panama Canal and its surrounding land, called the Canal Zone, until 1999, when a 1977 agreement between the United States and Panama officially took effect and turned the canal over to the Central American nation.

Who is a **peon**?

A peon is a farm laborer in Central America who works on large farms known as haciendas.

What is the world's **second-longest barrier reef**?

The second-longest barrier reef in the world lies just off the Atlantic coast of Belize, on the northeastern corner of Central America, and consists of Lighthouse Reef and Glovers Reef. Belize's reefs are only a few dozen miles long while the Great Barrier, the longest reef in the world, is hundreds.

THE WEST INDIES

Where are the **East and West Indies**?

The East and West Indies are separated by half the planet. The West Indies are islands in the Caribbean, including the Greater Antilles, the Lesser Antilles, and the Bahamas; while the East Indies include islands that encompass Indonesia, Malaysia, and Brunei. When Christopher Columbus reached the New World in 1492, he believed that he had actually found a shorter route to the East Indies. Thus, Columbus thought the islands he had reached made up a portion of the Indies and considered the islands' inhabitants to be "Indians."

Where is the Western Hemisphere's oldest university?

Founded in 1538, the Autonomous University of Santo Domingo in the Dominican Republic is the oldest university in the Western Hemisphere.

Where are the **Windward Islands**?

The Windward Islands are located in the Caribbean and are exposed to the northeast trade winds (northeasterlies) of the Atlantic Ocean. Because of their vulnerability to these winds, the islands were named the Windward Islands. The Windward Islands include Martinique, St. Lucia, St. Vincent, the Grenadines, and Grenada.

Where are the **Leeward Islands**?

The Leeward Islands are also located in the Caribbean and are less exposed to the northeasterlies. Because these islands are "lee," or away from the wind, they were named the Leeward Islands. The Leeward Islands include Dominica, Guadeloupe, Montserrat, Antigua, Barbuda, St. Kitts, Nevis, Anguilla, and the Virgin Islands.

Was **Cuba** ever a **part** of the **United States**?

The United States went to war against Spain in 1898 to assist Cubans who were rebelling against Spanish rule. The United States took control of Cuba during the Spanish-American War in 1898 and held it until 1902, when Cuba was granted independence. The three-year military occupation by the United States ended with an agreement that the United States would be allowed to lease Guantanamo Bay, which the United States still uses as a naval base.

How did Cuba become a **Communist** country?

Having been an independent country for 57 years, the Cuban government, run by the dictator Fulgencio Batista y Zalvidar, fell to the Communist leader Fidel Castro in 1959. Because of Cuba's Communist government, the United States severed its relationship with Cuba, forcing the island to ally itself with the Soviet Union. In October 1962, the presence of this nearby Communist country caused extreme terror in the United States when the U.S.S.R. attempted to place nuclear missiles within Cuba. The "Cuban Missile Crisis" is thought to be the closest the Cold War ever got to a real nuclear war.

Where is the **Bay of Pigs**?

The Bay of Pigs is a bay in southwestern Cuba. In 1961, the bay became the location of an attempted coup against the Cuban government by revolutionaries trained and financed by the U.S. Central Intelligence Agency. After the attempted coup failed, the United States abandoned the revolutionaries, most of whom were killed or captured in the days following the coup attempt.

Where is the **oldest church** in **the Americas**?

The oldest church, the Cathedral Basilica Menor de Santa, was built by Columbus' son Diego. The first stone was set in 1514 in Santo Domingo, the Dominican Republic.

Do things really disappear in the **Bermuda Triangle**?

The "Bermuda Triangle," or "Devil's Triangle," is a popular legend that suggests a supernatural or paranormal reason for a supposedly large number of missing aircraft and sea-going vessels within its area. The legend generally places the area of the Bermuda Triangle in the Atlantic Ocean, with its three corners located at Bermuda, Puerto Rico, and Miami, Florida. But you won't be able to find the Bermuda Triangle on a map since it is not a geographically or politically defined area, and its location is solely designated by the legend.

Though the legend has circulated for at least a century, there seems to be little evidence that this area is subjected to anything but natural hazards and human error. Most of the evidence for the phenomena in the Bermuda Triangle stems from the disappearance of the five aircraft of Flight 19 in December 1945, as well as a search plane that was sent to find them. Though the popular version of the disappearance of Flight 19 assumes a mysterious end, a mixture of missing navigational apparatus, human error, low fuel, and choppy seas most likely led to the squadron's disappearance and demise.

What was the **first independent country** in the **Caribbean**?

Haiti was the first independent country in the Caribbean. In 1791, the slaves in Haiti revolted, which led to Haiti's independence from France in 1804. Though Haiti once occupied the entire island of Hispaniola, Haiti now shares the island with the Dominican Republic.

How many **people visit the Caribbean** each year?

Approximately 35 million people visit the Caribbean's beautiful countries and islands.

Which Caribbean country leads the region in **tourism**?

The Dominican Republic is the Caribbean country most visited by tourists, attracting one and a half million people annually. The beautiful beaches, clear seas, and tropical climate lure tourists from around the world, but especially from Europe. Jamaica is the second-most visited Caribbean country, with 850,000 tourists annually.

SOUTH AMERICA

PHYSICAL FEATURES AND RESOURCES

Which is **farther east, Santiago, Chile,** or **Miami, Florida**?

Even though it lies on the west coast of South America, Santiago, Chile, is actually farther east than Miami, Florida. Though it is common to envision South America as directly south of North America, South America actually lies southeast of North America.

Which **river** carries **more water** than any other in the world?

Though the Amazon River is the second longest in the world (4,000 miles [over 6,400 kilometers]), it carries more water to the ocean than does any other river in the world.

What are the **Andes**?

The Andes are a mountain chain that runs along the entire west coast of South America, from Panama (at the southern tip of Central America) to the Strait of Magellan (at the southern tip of South America). This chain is about 4,500 miles (7,240 kilometers) long and contains high plateaus and one of the driest deserts on the planet, the Atacampa. The tallest mountain in South America, Aconcagua (22,834 feet [6,960 meters]), is located in the southern Andes, on the border between Chile and Argentina. The ancient Inca city of Machu Picchu is located in the Andes of Peru.

What are the four **climatic regions of the Andes**?

The Andes are known for their four defined climatic zones, which are based on elevation. The lowest zone, tierra caliente (hot lands), is ascribed to the area from the

Incan terraces on Isla Del Sol by Lake Titicaca in Bolivia.

plains to 2,500 feet (762 meters) and is where most of the population resides. The second zone is tierra templada (temperate land), which is from 2,500 to 6,000 feet (762 to 1,829 meters). The third zone is tierra fria (cold land), which is from 6,000 to 12,000 feet (1,829 to 3,658 meters). Above 12,000 feet (3,658 meters) is the fourth zone, tierra helada (frozen land).

What is the **highest navigable lake** in the world?

Lake Titicaca, located on the border between Peru and Bolivia, is the highest navigable lake in the world, with an elevation of 12,500 feet (3,810 meters). Though there are higher lakes in the world, Lake Titicaca is the highest one in which boats can sail. Lake Titicaca was the center of Incan civilization.

Where is the world's **tallest waterfall**?

Angel Falls, in Venezuela, is the world's tallest waterfall at 3,212 feet (979 meters). American pilot Jimmy Angel discovered the waterfall and named it after himself in 1935. At the time of the discovery, the falls were known to indigenous peoples such as the Pemon for thousands of years. They called the falls *Kerepakupai merú,* which means "waterfall of the deepest place."

What are the Cordillas of Colombia?

In Colombia, the Andes are split into three separate mountain ranges. They are the Cordilla Occidental (western range), the Cordilla Central (central range), and the Cordilla Oriental (eastern range). The city of Cali is located in the valley between the Occidental and Central ranges, while Bogotá is located between the Central and Oriental.

What is the **Atacama desert**?

One of the world's driest deserts, the Atacama is located in northern Chile. It is completely barren of plant life. The town of Calama, which is located in the Atacama, has never received rain. The Atacama is a source for nitrates and borax.

Who owns **Easter Island**?

Easter Island, which is located 2,237 miles west of Chile, is owned by Chile. On this island, there are over 100 large rocks carved into the shape of heads, complete with facial features. These large heads vary in size from 10 to 40 feet (3 to 12 meters) and were made out of a soft, volcanic rock.

Is the **Strait of Magellan** crooked?

Yes, it is! The Strait of Magellan is a winding waterway between South America and the islands of Tierra del Fuego at the southern tip of South America. This strait was discovered by the explorer Ferdinand Magellan in 1520 and has been used as a shortcut to avoid having to sail around Cape Horn, the southern tip of South America.

What is the world's largest **tropical rain forest**?

The Amazon forest is the world's largest tropical rain forest. It occupies one-third of Brazil's land area and receives over 80 inches (200 centimeters) of rain a year. The rain forest loses 15,000 square miles (38,850 square kilometers) of forest each year because of clear-cutting. The Amazon rain forest is home to about 90 percent of the Earth's animal and plant species and is a major producer of the world's oxygen.

Which country is the world's **leading copper producer**?

Chile produces an astounding 20 percent of the world's copper annually and contains approximately one-fourth of the world's copper reserves. The world's total copper production is 9.2 million metric tons; thus Chile's production is 1.8 million metric tons.

Which South American countries are members of **OPEC**?

Ecuador (500,000 barrels/day) and Venezuela (2.3 million barrels/day) are both members of OPEC (Organization of Petroleum Exporting Countries). Bolivia and Brazil may be members in the near future.

Which country is the world's **leading coffee producer**?

Brazil produces more than 32.6 million bags of coffee beans each year, followed by Colombia, which produces 11.5 million bags.

Where does **cocaine** come from?

Cocaine is produced from the coca plant, which was originally domesticated by the Incas. Coca paste from the coca plant is refined to make cocaine. Illegal cartels and individuals in Colombia and other South American countries are major exporters of cocaine, especially to the United States.

What percentage of **Americans** have used **cocaine**?

Approximately 14 percent of the U.S. population uses or has used cocaine.

HISTORY

Where did the **Inca civilization** develop?

In the fifteenth and sixteenth centuries, the civilization of the Incas developed in the altiplanos of the Andes mountains. Altiplanos are high plains located among the mountains that are suitable for habitation. The Bolivian capital of La Paz is also located on an altiplano. The Inca civilization lasted from approximately the eleventh through sixteenth centuries.

What is **Machu Picchu**?

Machu Picchu is an ancient Incan city constructed at an elevation of 8,000 feet (2,438 meters) above sea level, and located about 43 miles (69 kilometers) northwest of Cuzco in Peru. It was built by the Incan ruler Pachacuti Inca Yupanqui between 1460 and 1470. It is comprised of more than 200 buildings, which are visited by thousands of tourists each year. Tourists reach the city either by bus or by a ritualistic 20-mile hike to the summit. It was rediscovered by a Yale University team, headed by Hiram Bingham, in 1911. Some argue that other explorers may have discovered the site earlier, including a German businessman named Augusto Bern in the 1880s.

A view of Machu Picchu in the Andes Mountains (photo by Paul A. Tucci).

How was the **New World divided** between Spain and Portugal?

In 1493 Pope Alexander VI divided the New World into Spanish and Portuguese spheres of influence. A line was placed "100 leagues" (about 300 miles [480 kilometers]) west of the Azores islands, located several hundred miles west of Portugal in the Atlantic Ocean. Everything in the New World to the east of this Demarcation Line, which lay off the east coast of South America, belonged to Portugal, while the lands in the west belonged to Spain. Since this division provided little land for Portugal, the Portuguese were dissatisfied. The Treaty of Tordesillas established a new line about 800 miles (1,300 kilometers) to the west of the old line. Pope Julius II approved the line in 1506.

Which South American country was the **first to gain independence** from colonial rule?

In 1816, Argentina gained independence from Spain. International recognition of the independent country, then called the United Provinces of the Plate River, did not come until 1823, when the United States recognized the new state.

Who was **Che Guevara?**

Ernesto Che Guevara (1928–1967) was an Argentine Marxist revolutionary, physician, author, and guerrilla leader. He formed his political ideas after seeing the extreme poverty in Latin America, believing that the cause was economic inequality due to monopolistic capitalism, neo-colonialism, and imperialism. He was instrumental in helping Cuba's Fidel Castro overthrow the U.S.-backed dictator Fulgencio Batista.

Which **South American territory** has yet to gain its **independence**?

French Guyana, on the northeast coast of South America, has been a colony of France since 1817. It is officially a department (state) of France and is the launch site of the European Space Agency.

Most of the present-day country of Brazil was east of the line drawn in the Treaty of Tordesillas, so it became Portuguese territory in 1506. Brazil's official language is Portuguese, making it the only Portuguese-speaking country in South America.

Which place in **South America** is part of the **European Union**?

Although not a country, French Guyana, lying just north of Brazil and one of the 26 departments of France, is part of the European Union. It uses the euro as its national currency.

What is **Devil's Island**?

Devil's Island, located off the coast of French Guyana, became the overseas prison of France in the middle of the nineteenth century. France stopped using the island as a penal colony in 1938.

What was **Gran Colombia**?

After many wars against Spain for independence, Gran Colombia, led by Simon Bolivar, became an independent country in 1821. Gran Columbia consisted of the area that is now present-day Colombia, Panama, Ecuador, and Venezuela. In 1830, Gran Colombia was split into Colombia (which included Panama), Ecuador, and Venezuela.

Why did **Peru and Ecuador fight two wars** in the twentieth century?

When Ecuador split off from Gran Colombia in the nineteenth century, it signed a border agreement with Peru, defining its boundaries along the Maranon River. In 1941, Peru invaded Ecuador and occupied half the country for ten days. Afterwards, a peace treaty was brokered and guaranteed by the United States, Brazil, Argentina, and Chile. The United States mapped the border, leaving approximately 48 miles (78 kilometers) of a line in the Cordillera del Condor area unmarked. The area became a site for dispute in 1941, and again in 1995.

Who was **Simon Bolivar**?

In the early nineteenth century, Simon Bolivar led the fight in South America for independence from Spain. He is revered as a hero among South Americans for his role in the independence of Venezuela, Colombia, Ecuador, and Peru. Bolivia was named in honor of Bolivar.

PEOPLE, COUNTRIES, AND CITIES

Where is South America's **population clustered**?

The population of South America is approximately 371 million (23 percent higher than the United States). In South America, much of the population is found along the Atlantic Coast. Brazil's Atlantic coast is home to two of the world's largest urban areas—São Paulo (18.84 million people) and Rio de Janeiro (11.75 million people). Buenos Aires, Argentina (12.8 million people), is also a major urban area on the Atlantic coast.

What **percentage** of the world's **poor** live in **South America**?

Approximately 3.93 percent of all of the world's poor people live in South America. They are mostly concentrated in Brazil, Colombia, Venezuela, Peru, and Ecuador. Poor is defined as people having incomes of less than $1.00/day.

What are the most **heavily urbanized countries** in South America?

Argentina, Chile, Uruguay, and Venezuela all have an urbanization level (the percent of population who live in urban areas) of approximately 85 percent (the same percentage as the United States).

What is the **largest city** in the **Amazon River basin**?

Manaus, Brazil, is the largest city in the basin, with a population of just over 1.6 million. Manaus is the capital of Brazil's largest state, Amazonas, and is a major trading center for the region. When the Amazon basin was the only known source of rubber, Manaus experienced a boom, but subsequently declined in importance due to the planting of rubber in other regions of the world. Manaus has since recovered by becoming a duty-free trade area of 1.6 million people.

What is **MERCOSUR**?

The Southern Cone Common Market, also known as MERCOSUR, is a trade group that includes Brazil, Argentina, Paraguay, and Uruguay. It was established in 1995 to reduce trade barriers between those four countries and to promote economic unity.

How **large** is **Brazil**?

Brazil makes up just under 50 percent of the land area of the entire South American continent. It is the world's fifth-largest country, with 3.3 million square miles (8.5 million square kilometers) of territory.

When did **Brazil move its capital city**?

In 1960, Brazil moved its capital city from Rio de Janeiro to a brand new city in the center of the country, called Brasilia. Brasilia was designed and constructed on empty land near the center of the country in the 1950s. Brazil moved its capital from Rio de Janeiro to Brasilia to assert its independence, exchanging a colonial capital on the coast for a new interior capital. The interior, underdeveloped, location of the new capital allowed a fresh start as well as an opportunity to develop the region.

A beach near Rio de Janeiro, Brazil (photo by Paul A. Tucci).

Who designed and planned the capital of Brazil?

Brasilia was created in 1956 by two people: urban planner Lucio Costa and architect Oscar Niemeyer.

What **statue** overlooks **Rio de Janeiro**?

The 100-foot-high (30.5-meter-high) statue of "Christ the Redeemer" stands, arms outstretched, over the city of Rio de Janeiro. The statue of Jesus Christ, with its base on top of Corcovado Mountain at 2,340 feet (713 meters), was built in commemoration of the 100th anniversary of Brazilian independence.

What are the **capitals of Bolivia**?

Bolivia has two capitals—La Paz is the administrative capital, while Sucre is the constitutional and judicial capital. Several countries divide national functions between cities.

What **country is crossed** by both the **equator and a Tropic**?

Brazil is the only country crossed by the equator at 0° and the Tropic of Capricorn 23.5° South.

What South American city has more **Japanese residents** than any city outside of Japan?

São Paulo, Brazil, has more Japanese residents than any other city outside of Japan. Well over two million Japanese live in this urban area. The original settlers—791 farmers—traveled to Brazil from Kobe, Japan, in 1908.

What is **Mardi Gras**?

The Catholic festival Mardi Gras literally means "fat Tuesday" in French. Parades, dancing, and carnivals are all part of this pre-Ash Wednesday celebration. Known as Carnival, it is very popular in Rio de Janeiro, Brazil, and New Orleans, Louisiana. The Brazilian festival is a significant source of tourism-related income for the country.

What makes **Brazilian automobiles** run?

Over half of Brazilian automobiles use alternatives to petroleum known as gasohol and ethanol. Gasohol is made from sugarcane and ethanol is made from alcohol. The two fuels are much less expensive than petroleum-based gasoline.

An early twentieth-century Japanese poster advertising work opportunities in Brazil (image courtesy of Historical Museum of Japanese Immigration).

Who is **Alberto Fujimori** and what has he done for **Peru**?

Alberto Fujimori, who was president of Peru from 1990 to 2000, was credited with ending terrorism in Peru and turning around a devastated economy. Some believe, however, that he trampled on the rights of individuals and indigenous people during his oftentimes authoritarian rule. He later was convicted on charges of abuse of power in ordering the illegal search of the apartment of his security chief's spouse. He was sentenced to six years in prison.

Where has **one-third of the population of Suriname** emigrated to since 1975?

Suriname was a Dutch colony until it gained independence in 1975. Since 1975, approximately 200,000 of its residents have emigrated to the Netherlands.

What is the world's **highest capital city**?

La Paz, Bolivia, is the world's highest capital city. La Paz is located high in the Andes mountains at an elevation of 12,507 feet (3,700 meters). It was founded in 1548 by Spanish explorers and is now home to approximately 711,000 people.

What **port** does landlocked Bolivia use?

Having no access to the sea itself, Bolivia made an agreement in 1992 with Peru to use its port at Ilo.

What is the actual name of **Bogotá**?

Bogotá, Colombia, was originally called Santa Fe. More recently, the name became Santa Fe de Bogotá. Today, the Colombian capital is known as Bogotá for short, and about eight million people live in its metropolitan area.

What is a **cartel**?

A cartel is an organization made up of businesses that band together to eliminate competition, collude to fix prices, and control supply and production of a product or service. In South America, the word refers to the drug cartels of Colombia, most notably the Medellin and Cali cartels, both which were crushed by the Colombian government. Later, government members and lieutenants of former cartel operatives stepped in and created their own cartels. Today, they are still manufacturing and distributing cocaine and its derivatives into the United States, which is the biggest consumer of cocaine in the world.

How **long** is **Chile**?

Chile stretches approximately 2,700 miles (4,344 kilometers) along the western coast of South America. At its widest it is only 100 miles (161 kilometers) across. Chile is a classic example of an elongated country, which makes governing difficult.

Who's fighting over the **Falkland Islands**?

The Falkland Islands (also known as Islas Malvinas), located near the southern tip of South America, have long been a source of conflict between the United Kingdom and Argentina. Though the islands have been occupied by the British since 1833, Argentina has claimed the islands as its own since the eighteenth century. In 1982, Argentina invaded the islands, but the British regained possession within a matter of weeks. Argentina still claims the Islas Malvinas and is pursing its acquisition through diplomatic channels.

What is the world's **southernmost city**?

Ushuaia, in southern Argentina, is the world's southernmost city. Ushuaia sits on Tierra del Fuego Island, south of the Strait of Magellan.

What is the **Pan-American Highway**?

Begun in the 1930s, the Pan-American Highway is the result of an international effort to create a highway stretching from Fairbanks, Alaska, to Buenos Aires, Argentina. In 1962 a bridge, known as the Bridge of the Americas, was built over the Panama Canal to continue the highway over the canal. A 100-mile (161-kilometer) stretch of the highway in eastern Panama still remains unfinished.

What is the **primary religion** throughout Latin America?

Due to Spanish and Portuguese colonization, most Latin Americans are Catholic, about 83 percent. Protestants make up about seven percent of the region, and the rest are atheists, nonreligious, animists, or other religions.

What is a **plaza**?

Most Latin American cities have an open public square at the center of the downtown called the plaza. The plaza is used for festivals and ceremonies and is surrounded by a cathedral and shopping areas.

WESTERN EUROPE

PHYSICAL FEATURES
AND RESOURCES

What are the **Alps**?

The Alps are Europe's most famous mountain chain, running east-west for approximately 700 miles (1,125 kilometers). The Alps stretch from southeastern Spain to the Balkans, and include Mont Blanc, the highest point in western Europe.

What are the **Apennines**?

The Apennines are a mountain range extending from northern to southern Italy for approximately 600 miles (1,000 kilometers). The highest point is a place called Corno Grande, which reaches a summit at 9,500 feet (2,912 meters).

Where is the **Rock of Gibraltar**?

The Rock of Gibraltar is a limestone mountain located on the Gibraltar peninsula in southern Spain. The city of Gibraltar, located on this same peninsula, is actually a British colony, and is used as a naval air base. This is the perfect location from which to control the Strait of Gibraltar, the small waterway that connects the Mediterranean Sea with the Atlantic Ocean. Spain has continually advocated a claim for this area but has been consistently unable to retrieve this vital piece of land.

On the opposite side of the Strait of Gibraltar, at the northern tip of Morocco, Spain has its own autonomous community, consisting of Ceuta and Melilla, which is also strategically located to control the Strait of Gibraltar.

189

Lago di Barrea (Lake Barrea) and the city of Barrea, Abruzzo Province, in the Apennine Mountains of Italy (photo by Paul A. Tucci).

Where are the **Highlands**?

The island of Great Britain is divided into highlands and lowlands along the Tees-Exe line, which runs between Plymouth in the south and Middlesborough on the east coast. To the southeast of this line lie the flat plains of England, while to the northwest lie the Scottish Highlands.

How wide is the **Strait of Gibraltar**?

The strait, which connects the Mediterranean Sea to the Atlantic Ocean between Africa and Spain, is eight miles (13 kilometers) wide at its narrowest.

Where is the **southernmost glacier** in **Europe**?

The southernmost European glacier is also high atop the Apennines, near Corno Grande. Due to global warming it has lost a significant amount of its mass.

How many **volcanoes** are in **Europe**?

There are more than one hundred volcanoes in Europe.

Which **volcanoe** poses the **most risk** to people in Europe?

Mt. Vesuvius, on the western coast of Italy, lies very near a city of more than one million people: Naples. The volcano is still active, and geologists predict there is a very good chance it will erupt again in the future, potentially wiping out Naples and the surrounding area.

How many **volcanoes** does Iceland have?

Iceland, formed by volcanoes along the Mid-Atlantic Ridge, is home to more than 100 volcanoes. Of these, more than 20 have erupted over the past few centuries.

What is Iceland's **leading export**?

Over three-quarters of Iceland's exports are fish. The fish industry employs 12 percent of the nation's workforce, and the country is economically vulnerable to fluctuations in world fish prices.

A winged god, possibly Apollo, painted on the interior of a house in Herculaneum, another city that was completely buried in ash from the eruption of Mt. Vesuvius in 79 C.E. (photo by Paul A. Tucci).

How much **oil** is produced by **Europe**?

Europe produces 6,358,000 barrels of oil per day, which is approximately 7.9 percent of the world's total production of 80,247,000 barrels per day.

Who is the **biggest producer of oil** in Europe?

Norway wins as Europe's biggest oil producer, producing approximately 3.2 million barrels per day. The United Kingdom and Denmark are the second and third biggest oil producers in Europe.

What are the **seven hills of Rome**?

The city of Rome sits upon seven hills: Capitoline, Quirinal, Viminal, Esquiline, Caelian, Aventine, and Palatine. According to ancient legend, the first settlement in the area, the city of Romulus, was built upon Palatine.

Where is the **Jutland Peninsula**?

The Jutland Peninsula extends north from Germany and is home to the country of Denmark.

Where is the **Black Forest**?

Located in southwestern Germany, the Black Forest is a densely forested, mountainous region that is a popular location for vacationing, with its many health resorts and

191

wilderness trails. The Black Forest is the source of the Danube River and is renowned for its cuckoo clocks.

HISTORY

What was the **Potsdam Conference**?

At the end of World War II, the United States, United Kingdom, and U.S.S.R. met at Potsdam, Germany, from July 17 to August 2, 1945, for a conference to determine how to control Germany and other eastern territories. The Potsdam Conference divided Germany and Austria into Soviet, French, American, and British zones of control.

What was the **Berlin Wall**?

At the end of World War II, Germany was divided into four zones, each occupied separately by the United States, the United Kingdom, France, and the U.S.S.R. The city of Berlin, while located entirely within the Soviet-occupied zone, was itself divided into four zones. Soon thereafter, the Soviets stopped cooperating with the other Allied powers. The three zones occupied by the United States, United Kingdom, and France joined together to create West Germany, while the Soviet zone became East Germany. A similar split occurred in the city of Berlin.

The city of Berlin held the dichotomy of east versus west, Communist versus capitalist. Many people who lived in East Berlin could see that those in West Berlin generally had a higher standard of living. It is estimated that over two million East Germans fled to the West within Berlin. In August 1961, the Communist government, determined to stop this mass exodus, began to build the Berlin Wall, a wall that physically divided East and West Berlin. On the west side, the wall became the location of spray-painted messages that voiced free opinions; on the east side of the wall lay a deserted area of barbed wire and armed guards called "No Man's Land."

For decades, the Berlin Wall stood as the physical version of the psychological "iron curtain" that separated east from west. On November 8, 1989, the Berlin Wall came tumbling down, and soon thereafter the era of the Cold War also ended.

Where was **Checkpoint Charlie**?

Checkpoint Charlie was a famous crossing point on the Berlin Wall between East and West Berlin, used mainly by tourists and U.S. military personnel.

What is **Hadrian's Wall**?

Hadrian's Wall was built under the direction of the Roman Emperor Hadrian in 122 C.E. Located in northern Great Britain, it was intended to keep out the Caledonians of Scotland. Built of mud and stone, the Wall stretched nearly 75 miles (120 kilometers), from Solway Firth in the west to the Tyne River in the east (near Newcastle).

What was the **Maginot Line**?

The Maginot Line was a defensive zone that was built in the 1930s to defend France against the possibility of a German invasion. The zone consisted of underground tunnels, artillery, anti-tank obstacles, and many other defensive structures and stratagems to slow down invading Germans. The Maginot Line stretched for approximately 200 miles (322 kilometers) near the French-German border.

During World War II, when the Germans did invade France, the Germans bypassed the Maginot Line by storming through neutral Belgium. Thus, the Maginot Line had failed its one great test because it was too short. The Line was also rendered obsolete by the fact that it did not provide defense against the new, modern warfare that included aircraft.

What is **Benelux**?

Benelux stands for Belgium, the Netherlands, and Luxembourg, and represents an economic alliance between the three, that was formed in the 1940s. At the time Belgium was primarily industrial and the Netherlands was primarily agricultural, the two countries' economies complemented each other, a relationship strengthened by an economic union. Luxembourg, which has a varied economy and is extremely small, has long been closely affiliated with its two larger neighbors, and thus also benefited from the union. Today, the countries are referred to as Benelux whenever referred to in terms of market analyses, commerce, and trade.

How many Irish left during the **Great Starvation**?

In the mid-nineteenth century, Ireland suffered from the "Great Starvation." From 1845 to 1850, a fungus ravaged the potato crops of Ireland, destroying the primary food source of Irish peasants. Though many have called this tragic event the "Great Potato Famine," the mass starvation of the Irish people was caused more by the lack of assistance from the British government than by the famine

193

itself. It is estimated that over one million people died during these catastrophic times, and approximately twice that number left their homeland in an effort to find food and solace.

PEOPLE, COUNTRIES, AND CITIES

What is the **European Union**?

In 1951, six western European countries joined together in the European Coal and Steel Community. As more members joined, the organization grew in scope and soon became an organization that helped mend and meld the economies of Europe. In 1993, the European Community was renamed the European Union (EU). Today, there are 27 member states: Austria, Belgium, Bulgaria, Cyprus, Czech Republic, Denmark, Estonia, Finland, France, Germany, Greece, Hungary, Ireland, Italy, Latvia, Lithuania, Luxembourg, Malta, the Netherlands, Poland, Portugal, Romania, Slovakia, Slovenia, Spain, Sweden, and the United Kingdom. The European Union has a flag, an anthem, and in 1999 began using a single monetary unit (the "Euro") .

Where are the **low countries**?

Belgium, the Netherlands, and Luxembourg are known as the low countries because of their low elevation.

How do the **Netherlands** keep getting bigger?

For hundreds of years, the Dutch have been expanding the size of their country by building dikes and draining (and reclaiming) land. These lands, known as polders, have greatly expanded the size of the Netherlands and are now considered one of the seven wonders of the modern world.

A canal boat in Amsterdam, the Netherlands (photo by Paul A. Tucci).

What is **Randstad**?

The Randstad is a region of the Netherlands that includes the metropolitan areas of Amsterdam, The Hague, Rotterdam, and Utrech. The urban area of the Randstad holds nearly half the Netherlands' population.

What is **The Hague**?

The Hague is a city on the west coast of the Netherlands with an approximate population of 450,000. The Hague is the home of many international organizations, such as the International Court of Justice.

What are the two **cultural groups** that make up **Belgium**?

The Walloons in southern Belgium, called Walloonia, are descendants of the Celts and speak French. The Flemings in northern Belgium, called Flanders, are descendants of German Franks and speak Flemish, a language similar to Dutch. There is little unity within Belgium, for only 10 percent of Belgians are bilingual.

Who **settled Denmark**?

Surprisingly, Denmark was not settled by Europeans from the continent directly to its south, but was settled in the tenth century by Danes from Iceland and the Scandinavian Peninsula.

What's the difference between **England, Great Britain, and the United Kingdom**?

Northeast of France lie two large islands; Great Britain to the east and Ireland to the west. On the island of Great Britain there are three regions: England in the southeast, Wales in the southwest, and Scotland in the north. The other island, Ireland, is divided into two political divisions: the region called Northern Ireland in the north and the country of Ireland in the south. The United Kingdom is a country that includes all three regions on the island of Great Britain (England, Wales, and Scotland) and the one northern region on the island of Ireland (Northern Ireland).

Is **Scotland** a country?

While Scotland does have limited self-rule, it is still part of the country of the United Kingdom. Scotland occupies the northern portion of the island of Great Britain.

What are the **British Isles**?

The British Isles are composed of the two large islands of Great Britain and Ireland (separated by St. George's Channel) and the many small islands nearby. The British Isles include two countries: the United Kingdom and Ireland.

What is the **Commonwealth**?

The Commonwealth, also known as the British Commonwealth, consists of the United Kingdom and the now-independent former countries of the British Empire. The Commonwealth is not a policy-making body but is solely a loose voluntary association between countries that were formerly under British control.

What is **Land's End**?

Land's End has quite an appropriate name, as it is a cape at the southwestern tip of Great Britain that is the westernmost point of England; the "end of land" in the west.

What is a **moor**?

A moor is uncultivated pasture land. You'll find moors in the United Kingdom; in the United States most people call them fields or prairies.

Where is **Camelot**?

The legendary sixth-century castle of King Arthur has been said to be located either near Exeter or Winchester, England. Camelot was not only the home of King Arthur and Queen Guinevere but was also the location of the Round Table and its famous knights.

The city of Cannes on the scenic French Riviera.

Did people bathe in **Bath**?

The Roman baths found in Bath, England, were originally built on a Celtic holy site over a period of 300 years, beginning in 70 C.E. The structures include a hot bath, a tepid or medium bath, and a frigid bath complex. Though the ancient city lies buried beneath it, the modern city of Bath is also renowned for its hot springs, which once warmed the Romans and now offer a relaxing bath in this resort town.

Where is **Catalonia**?

Catalonia is an autonomous region in northeastern Spain. Because Catalonia is home to more than six million Spanish Catalans, who have their own language and culture, many in the region would like the territory to become an independent nation. Spain does not want Catalonia to secede, as Catalonia is responsible for a sizable portion of Spanish economic production. Catalonia's capital is Barcelona, the host city of the 1992 Summer Olympics.

Where is the **French Riviera**?

The French Riviera, also known as Côte d'Azur, is located in southeastern France, near the border with Italy, along the Mediterranean Sea. The French Riviera is a major

197

vacation spot for Europeans, with its mild Mediterranean climate and beautiful scenery. The tiny country of Monaco is located within the French Riviera, and adds to the Riviera's image of luxury with the multitude of casinos and hotels at Monte Carlo.

Where does the **Tour de France** begin and end?

The Tour de France changes its course each year, but the last leg is always along Paris's famous boulevard, the Champs-Élysées. The bicycle race is approximately 2,000 miles (3,200 kilometers) long and takes 25 to 30 days to complete.

What is the **Giro d'Italia**?

The Giro d'Italia is the second most important multi-stage bicycle race in Europe. It began in 1909, and instead of a yellow jacket for the fastest stage winner, recipients wear a pink-colored jersey reminiscent of the newspaper publisher and founder of the race's use of pink newsprint for his newspapers.

Where is **Gaul**?

Gaul was an ancient country that included most of modern-day France. It is thought to have been settled by Celtic people from the north, as well as Balkan people from the east. Other empires took control of Gaul, and it eventually became the kingdom of the Franks, the forerunners of the French.

What was the **French Community**?

Only a viable association during the 1950s and 1960s, the French Community, also known as Communaute française, was an organization tying together the former French colonies with France.

Who rules **Andorra**?

Since 1278, the tiny country of Andorra, nestled in the Pyrenees between France and Spain, has been jointly ruled by two people who live outside the country: the President of France and the Bishop of La Seu d'Urgell in northeastern Spain. France and Spain jointly take responsibility for the defense of Andorra.

What are the **largest cities** of Europe?

Western Europe's largest city is Paris, with nine million inhabitants. London (seven million), Milan (four million), Madrid (four million), and Athens (four million) are the next four largest cities.

What is the European country most visited by tourists?

Over 76 million people visit France each year. Spain is the next most popular with 56 million visitors.

Which country had the **world's first legislature**?

Though Iceland had been settled about 60 years earlier by the Norwegians, Iceland's parliament, the Althing, was created in 930 C.E.

What was the **first tunnel** through the Alps?

The Mont Cenis railroad tunnel was the first tunnel through the Alps and the first major railroad tunnel in the world. Opened in 1871, the tunnel spanned 8.5 miles (13.7 kilometers) and connected France and Italy.

Which **European country** produces the **most nuclear energy**?

France produces the second most nuclear energy in the world, generating 415 terawatt-hours per year. This is more than half of the 780 terawatts that the United States produces each year.

Where is the **highest life expectancy** on Earth?

People in Andorra live on average 83.52 years.

What are some of the **most expensive cities** in the world?

Half of the top ten most expensive cities in the world are in Europe: London, England; Copenhagen, Denmark; Geneva, Switzerland; Zurich, Switzerland; and Oslo, Norway.

What is Europe's **oldest independent state**?

San Marino claims to have been founded in the year 301 C.E. Its first constitution was established in 1600. San Marino is located on Mt. Titano in Italy and, at 24 square miles (62 square kilometers) in area, is one of the world's smallest independent countries.

Where is **Atlantis**?

Atlantis, the legendary underwater utopia supposedly located west of the Pillars of Hercules (the land on either side of the Strait of Gibraltar), was first described by Plato in the fourth century B.C.E. (Before the Common Era) as a magnificent civilization that was swallowed by the sea. Though Plato believed Atlantis to have been destroyed, the legend has grown over the centuries to describe this civilization as an underwater kingdom. Researchers now believe that the legend of Atlantis was based on the ancient Minoan civilization that lived on the Greek islands of Thira and Crete, which disappeared after a volcanic eruption in the sixteenth century B.C.E. Thus, the Minoan civilization in Thira and Crete fits the approximate date of Atlantis's destruction, but not its supposed location.

RUSSIA AND EASTERN EUROPE

RUSSIA AND THE FORMER SOVIET STATES

Where does **Asia** end and **Europe** begin?

Though Europe and Asia are actually part of one large landmass, tradition has split the region into two continents along the Ural Mountains in western Russia.

What was the **U.S.S.R.**?

The country called the Union of Soviet Socialist Republics (commonly known as the Soviet Union) was created in 1924, seven years after the Russian Revolution that overthrew the czarist monarchy. The Soviet Union consisted of Russia and its neighboring territories, such as Ukraine, Kazakhstan, and the Baltic States. By the end of 1991, communism had failed in the Soviet Union and many of its internal republics became independent states.

How many **republics or states comprised** the former **Soviet Union**?

The U.S.S.R. was comprised of the following 15 Socialist Republics: Armenia, Azerbaijan, Belarus, Estonia, Georgia, Kazakhstan, Kyrgyzstan, Latvia, Lithuania, Moldova, Russia, Tajikistan, Turkmenistan, Ukraine, and Uzbekistan.

What is the **Commonwealth of Independent States**?

The Commonwealth of Independent States (CIS), established by Russia just after the fall of the Soviet Union, is an organization that serves to keep the resources of the Soviet Union flowing between the now-independent countries. Ten of the 15 former

Ice melting along the Neva River in the Russian city of St. Petersburg, which in the past has also been called Leningrad and Petrograd (photo by Paul A. Tucci).

Soviet republics are members: Armenia, Azerbaijan, Belarus, Kazakhstan, Kyrgyzstan, Moldova, Russia, Tajikistan, Ukraine, and Uzbekistan. The Republic of Georgia withdrew in 2006, and Turkmenistan withdrew in 2005.

How large a part of the Soviet Union was **Russia**?

While the Soviet Union consisted of 15 Soviet Socialist Republics, the largest was the Russian Soviet Federated Socialist Republic (RSFSR). The RSFSR, also known as Russia, comprised three-quarters of the Soviet Union's territory and over half of its population.

How **big** is **Russia**?

With 6.6 million square miles (17 million square kilometers) and 145 million people, Russia is Europe's largest country and also its most populous. Additionally, Russia is the largest country in the world and the world's sixth-most populous. Within Europe, the second largest country is the Ukraine, with 233,100 square miles (603,729 square kilometers), and the second-most populous country is Germany, with 82 million people.

What is **Russia's official name**?

The official name of Russia is the Russian Federation. It consists of 83 federal subjects, including 46 provinces, 21 republics, 9 territories, 4 autonomous districts, 1 administrative sector, and 2 federal districts (Moscow and St. Petersburg).

Why is there a **tiny piece of Russia** in the middle of Eastern Europe?

The important seaport of Kaliningrad, wedged between Poland and Lithuania, was annexed by the U.S.S.R. at the end of World War II. Though once the capital of East Prussia and ethnically German, the Soviets quickly evicted the Germans and replaced them with ethnic Russians. In 1991 many of the autonomous republics within the U.S.S.R. gained independence. Though Kaliningrad lies west of these new countries, its inhabitants are ethnically Russian and thus remained part of the Russian state.

How **cold** is **Siberia**?

Siberia holds the record for the world's lowest temperature outside of Antarctica. On February 6, 1933, the temperature reached –90 degrees Fahrenheit in Oimyakon, Rus-

> ## What do the names St. Petersburg, Leningrad, and Petrograd have in common?
>
> St. Petersburg, Leningrad, and Petrograd were three names for the same city. Located in northwestern Russia along the Gulf of Finland, the city was originally founded as St. Petersburg in 1703 by Czar Peter the Great. Since "St. Petersburg" sounded too German to be the capital city of Russia, the city's name was changed to Petrograd in 1914. After the death of Communist leader Vladimir Lenin in 1924, the city's name was again changed, this time to Leningrad. After the change in the Soviet government in 1991, Leningrad once again became St. Petersburg.

sia. During the winter, almost all of Siberia has extremely cold temperatures, often reaching –50 degrees Fahrenheit.

Is it possible to **drive across Russia**?

It certainly is. Most of the urban areas of Russia have modern expressways. But in the vast expanse of the country east of Moscow, many of the roads are rough gravel or dirt and are only intermittently paved. Driving across Russia is therefore entirely dependent upon the season. During the winter, from November to May, the roads are frozen and can be driven upon; during the summer, many roads become quagmires and are unusable.

How do most people **travel across Russia**?

Most people, as well as goods, travel across Russia by airplanes or train. In 1891, Czar Alexander III launched the building of a railroad that would unify eastern and western Russia. Traveling from Moscow, through Siberia, to Vladivostok on the Pacific Coast, the Trans-Siberian Railroad was opened in 1904. The Trans-Siberian Railroad is the longest railroad line in the world.

How **long** does it take to **travel by train across Russia**?

Depending on the number of stops the train makes, it can take anywhere from five to eight days to travel across Russia. Ticket prices can be anywhere from $500 to more than $10,000, which allows for the most pampered class of service. Train lines end in Irkutsk; Beijing, China; and even Vladivostok, along the Eastern Pacific coast.

Which **Russian city** is among the **most expensive** in the world?

Moscow is ranked thirty-seventh in the world, in terms of expensive cities to live in. It is just ahead of Singapore; Athens, Greece; and Caracas, Venezuela.

How many **people live** in **Moscow**?

Approximately 10 million people call Moscow home. Nearly five million people live in St. Petersburg, Russia's second largest city.

How many **tourists visit Russia** each year?

Approximately 1.6 million people travel to Russia each year. The majority of these tourists are from Germany, followed by the United States, China, Great Britain, Italy, France, Japan, and Spain.

How **fast** is the **Russian economy expanding**?

The Russian economy grows at approximately 7.9 percent per year, with most of the growth coming from the energy sector.

How much market share does Russia's national airline, **Aeroflot**, have?

For domestic flights, Aeroflot has a 23 percent market share of all domestic travel in Russia. Nearly 10 million people fly Aeroflot each year. Siberian Airlines handles more than 3.5 million passengers each year.

What role did **Russia** play in **World War II**?

Russia was the target of Germany's war machine on its eastern front. More people were killed during Germany's war with Russia—30 million people (20 million of whom were civilians) lost their lives between 1941 and 1945—than in all other theaters of World War II combined. Without the active participation of the Russian military against the Nazis, the outcome of the war in Europe may have been drastically different.

What is the **longest river** in **Europe**?

The Volga River, which lies entirely within Russia, is Europe's longest river. The Volga River flows 2,290 miles (3,685 kilometers) from the Valdai Hills, near the city of Rzhev, into the Caspian Sea.

How big is **Siberia**?

Siberia makes up approximately three-fourths of Russia. Siberia is bounded on the west by the Ural Mountains, on the north by the Arctic Ocean, on the east by the Pacific Ocean, and on the south by China, Mongolia, and Kazakhstan. Russia conquered the area now known as Siberia in the late sixteenth century.

> ## What Russian city lost the most people during World War II?
>
> During the German 900-day siege of Leningrad (St. Petersburg) from 1941 to 1943, nearly 1.5 million people lost their lives.

What is the **largest city in Siberia**?

Located in southern Siberia—and on the Trans-Siberian Railroad route—Novosibirsk is the largest city in Siberia, with a population of 1.4 million.

How big is Russia's **Lake Baikal**?

Lake Baikal, located in southern Siberia, holds one-fifth of the world's non-frozen fresh water. Lake Baikal is also the world's deepest lake, with a maximum depth of just over one mile (5,371 feet [1,637 meters]). The crescent-shaped Lake Baikal is also famous for its crystal-clear water and bountiful plant and animal life.

How did **factories** in the **U.S.S.R.** end up on the **east side** of the **country**?

During World War II, the U.S.S.R. enacted their scorched-earth policy as Germany invaded from the west. The scorched-earth policy involved moving everything they could to the east and burning what they couldn't move. Factories were disassembled, shipped by train to the region near the Ural Mountains, and reassembled to keep Soviet industry working. The Ural Region is still a major manufacturing area for Russia.

What started the fighting in **Chechnya**?

Chechnya was once part of a Soviet republic called Chechen-Ingush. After the fall of the Soviet Union, Chechen-Ingush was divided into two internal republics, Chechnya in the east and Ingushetia in the west. Though the Chechens declared independence in 1992, Russia did not approve, and invaded Chechnya in 1994. The Russians crushed the rebellion, killing thousands of Chechens. However, the region is still in a state of unrest today.

Where was the **Pale of Settlement**?

During the eighteenth and nineteenth centuries, the Pale of Settlement was an area in which Russia attempted to restrict Jewish settlement. It extended from what is now eastern Poland to Ukraine and Belarus. Within the confines of the Pale, Jews were subjected to anti-Jewish regulations as well as mass killings.

Abandoned and crumbling buildings near the Chernobyl power plant are a stark reminder of the Soviet-era disaster in the Ukraine.

What was the world's **worst nuclear disaster**?

In April 1986, the Chernobyl nuclear power plant in the Ukraine, near the border with Belarus, had a major accident that released radiation into the atmosphere. The protective covering of the nuclear reactor exploded and deadly radiation escaped, immediately killing at least 31 people. The radiation exposure that initially occurred is still killing people through related diseases, and this will continue for many years. More than 100,000 people were evacuated from the region, and deaths due to radiation poisoning continue as radioactive isotopes spread across Europe.

EASTERN EUROPE

What are the **Baltic States**?

The three Baltic States of Estonia, Latvia, and Lithuania are so named because they lie on the Baltic Sea. These three countries became independent after the Soviet Union broke apart in 1991. Poland and Finland, which also lie on the sea, are sometimes also included as Baltic States.

What are the **Balkan States**?

The Balkan States and the Baltic States are two completely different regions. The countries lying on the Balkan Peninsula are commonly referred to as the Balkan

States (or the Balkans). The Balkan Peninsula itself lies between the Adriatic Sea (east of Italy) and the Black Sea. The countries on the peninsula include Albania, Bosnia and Herzegovina, Bulgaria, Croatia, Greece, Macedonia, Romania, Serbia and Montenegro, Slovenia, and the portion of Turkey that lies in Europe.

What is **balkanization**?

Balkanization is taken to mean the fragmentation of a country into ethnic, language, or cultural divisions by territory. It is what happened to Yugoslavia, a former country in the Balkans that disintegrated into different territories and ultimately countries. These new nations engaged in war with each other, practicing forced deportations of ethnic groups and mass killings of unwanted ethnic group members.

Does **Yugoslavia** still exist?

In 1991, the republics of Yugoslavia fell into disarray, and aligned themselves on ethnic and cultural fronts. The old geographic divisions of the country fell apart, and war broke out between various states, including Slovenia, Croatia, Bosnia and Herzegovina, Macedonia, and Serbia and Montenegro. Later, in 2006, Montenegro and Serbia also broke apart, and both declared their independence as separate countries in a referendum. Most recently, Kosovo unilaterally declared its independence from Serbia in 2008. Thus, over the course of just seventeen years, seven new countries were added to the map of the world.

What is **ethnic cleansing**?

Ethnic cleansing is the forced deportation or murder of people in certain ethnic groups within a country or region. Ethnic cleansing has been a part of world history for many millennia. Infamous examples of ethnic cleansing include the mass killings in the former Yugoslavia in the 1990s, the mass killings in Rwanda in 1994, and the Armenian genocide in Turkey in the 1930s.

How did the **Ukraine** help feed the Soviet Union?

Often called the "breadbasket of the Soviet Union," the Ukraine's rich wheat harvests were used to feed the U.S.S.R. Now, as an independent country, Ukraine produces four percent of the world's wheat and exports much of it to Russia.

Why does **Romania** have so many **orphans**?

The draconian population policies of Romania, which forbade birth control and abortion and required women to have five children, led to the birth of far more children than could be supported by the country. The population policies were implemented by communist dictator Nicolae Ceausescu, who ruled from 1965 until his capture and execution in 1989. The result of his policies has been thousands of Romanian children living in orphanages. Elections were held in 1990 and Romania is now struggling to improve conditions.

Why did **Macedonia's name** cause problems between that country and Greece?

When Macedonia declared independence in 1991, Greece felt indignant that a modern country would use what it felt was a historically Greek name. Greece blocked trade to Macedonia until 1995, when the two countries signed an agreement of understanding. The official name that the United Nations recognizes is the Former Yugoslav Republic of Macedonia, although most countries refer to it as the Republic of Macedonia, its constitutional name.

Is **Transylvania** a country?

The home of Count Dracula, Transylvania is a region located in central Romania. Transylvania is surrounded by the Transylvanian Alps and the Carpathian Mountains. A productive region in the western part of Romania, it contributes more than one-third of the total economic output of that country today.

Where is **Crimea**?

Protruding into the Black Sea, Crimea is a diamond-shaped peninsula attached—by the Isthmus of Perekop—to southern Ukraine. Though Crimea declared its independence from Ukraine in 1992, it later compromised and became an autonomous republic of Ukraine.

The Hagia Sophia in Istanbul now serves as a museum in Turkey. It was originally a church built by the Byzantine Emperor Justinian in the sixth century C.E.

Where is the **Putrid Sea**?

The Putrid Sea, also known as the Sivash Sea, lies to the east of the Isthmus of Perekop, between Crimea and Ukraine. It is a swampy area of salty lagoons.

What was the **Byzantine Empire**?

After the fall of Rome in 476 C.E., the eastern portion of the Roman Empire became the Byzantine Empire. The capital city of the Byzantine Empire was moved far east to Constantinople (now Istanbul). At one point, the empire included most of the eastern and southern coast of the Mediterranean Sea. The empire shrank in size until 1453, when Constantinople was conquered by the Ottoman Empire.

What **two countries** emerged from **Czechoslovakia**?

Czechoslovakia was created at the end of World War I (1918) by the Allies as a new country containing the area where the Czechs and Slovaks live. In 1967 and 1968, Czechoslovakia attempted to move away from Communism, but the Soviet Union and Warsaw Pact countries invaded, squelching such aspirations. This "Prague Spring," as it was known, was the first military action taken by the countries of the Warsaw Pact.

In 1992, the two republics of Czechoslovakia agreed to divide into two independent countries—the Czech Republic and Slovakia. The dissolution of Czechoslovakia was a peaceful one.

What two cities make up **Budapest**?

Budapest, Hungary, is actually two cities—Buda and Pest. The two cities are separated by the Danube River; Buda is on the west bank and Pest is on the east bank. The province in which the twin cities are located is also called Budapest.

What **playwright** and author became **president** of an **Eastern European country**?

Czech poet and author Vaclav Havel was both the last president of Czechoslovakia (1989–1992), and the first president of the Czech Republic (1992–2003). He has won numerous literary awards for his works, as well as the U.S. Presidential Medal of Freedom, the highest honor bestowed by the president of the United States to a civilian.

Who was **Lech Walesa**?

In 1970 Lech Walesa was one of the leaders of a shipyard workers strike in Gdansk, Poland. Later, he organized workers in non-communist labor unions, arguing for improved conditions for workers. In 1980, he led the Gdansk shipyard strike, which ended with the government agreeing with his provisions on behalf of workers, including the right to strike and the right to form independent unions. The Catholic Church, which is very influential in Poland, supported his activities, including his formation of the Solidarity Movement. Walesa was later elected president of Poland and served in office from 1990 to 1995.

Who are the **Maygar**?

Maygars are the predominant ethnic group of Hungary. The group originated in Asia, east of the Ural Mountains, and have a language much different from any other in Europe.

Do Caucasians come from the **Caucasus Mountains**?

In the latter half of the nineteenth century, scientists attempted to divide the world's peoples into "races." Each race was defined by the color of its skin, a process that was laced with stereotypes. These scientists used the term "Caucasian" for "white" people because they believed that the region of the Caucasus Mountains of Southwest Asia was their origin. Since that time, we have learned that there is but one human race and that it originated in Africa.

Which **river touches more countries** than any other?

The Danube River, which begins in Germany, passes through or borders 10 countries in Europe, more than any other river in the world. On its journey, the Danube River encounters Germany, Austria, Slovakia, Hungary, Croatia, Serbia and Montenegro, Romania, Bulgaria, Moldova, and Ukraine.

ASIA

Where does **Europe end** and **Asia begin**?

Though Europe and Asia are actually part of one large landmass, tradition has split the region into two continents along the Ural Mountains in western Russia.

CHINA AND MIDDLE ASIA

What is the **most populous country** in the world?

China is home to more than 1.32 billion people, which means that one in every five people on the planet is Chinese. India, the second-most populous country, has 1.13 billion people and is expected to surpass China's population sometime before 2050. Both countries have much larger populations than the third-largest country, the United States, which is home to 304 million people.

What is China's **one-child** rule?

In the late 1970s, the government of China decided that population control was needed because of a rapidly growing population that would soon outgrow the country's ability to feed itself. The policy mandated that every couple would only be allowed to have one child. The law exempts certain ethnic groups, rural families, and families where neither parent has no siblings. Punishment is strict and is primarily economic—China won't provide medical care or education funding for the second child. The one-child rule has been working and has slowed population growth, reducing the threat of overpopulation.

What is the **Forbidden City**?

The Forbidden City, located in the center of Beijing, China, was the home of the emperor and the entire imperial court for close to 500 years. Completed in 1420, the   **213**

With a population exceeding well over one billion people, China has put a law in place that limits most couples to only one child (photo by Paul A. Tucci).

Forbidden City (also known as the Purple Forbidden City or Gugong) was the home of 24 emperors from the Ming and Qing dynasties. For centuries, visitors were not allowed into this imperial city. In 1950, several decades after the last emperor was expelled, the city was made a museum and opened to the public.

How was **Taiwan created**?

In 1949, following the Communist revolution in China, the Chinese Nationalist government, led by Chiang Kai-shek, fled to the island of Taiwan and established a Chinese country there. When the Nationalists arrived, they were met by indigenous Taiwanese and Chinese who had been living there for millennia. Although the island has never been controlled by modern China, successive governments have been disputing this territorial issue since 1949. For about 25 years, most governments of the world recognized Taiwan, the Republic of China, as a sovereign country.

After President Nixon's policy of extending relations to the People's Republic of China in the 1970s, China has pressured all countries in the world to not recognize Taiwan, which has one of the most developed economies in the world. The United States extends *de facto* recognition to Taiwan through unofficial channels, and it still considers Taiwan to be a very strong ally. China has threatened the use of military action if the Taiwanese ever vote for independence from mainland China, and, in effect, officially form their own country. Proposals for degrees of independence by the Taiwan legislature have been met by threats from the Chinese government.

Is the **Great Wall of China** the only **man-made object** that can be **seen from space**?

No, it is not. Besides the Great Wall of China, there are many other man-made structures visible with the naked eye from space, such as urban areas and highways.

The immense Great Wall of China is the most famous man-made structure that is visible from outer space.

How **long** is China's **Great Wall**?

Located in northeastern China, the Great Wall stretches approximately 1,500 miles (2,400 kilometers). It averages 25 feet (7.6 meters) tall, is 15 to 30 feet (4.6 to 9.2 meters) wide at the base, and 10 to 15 feet (3 to 4.6 meters) wide at the top. The wall was originally erected to keep northern invaders out of China. Though part of the Wall was initially built in the third century B.C.E., the wall was expanded over the course of succeeding centuries.

What is the **Great Wall** of China **made of**?

Walking along sections of the Great Wall that have been restored, it is possible to see that it is made of stones and mortar. In other more remote stretches of the Wall, it was hastily put together with straw and mud, and in some places even rice and mud. The trick of the builders was to make it just high enough to prevent people or horses from breaching the walls. In some stretches, the Wall is only a few meters tall. In other places, it is several stories tall.

What is the **Terracotta Army**?

Located in Xian, China, in Shaangxi Province, which was once the capital of the Chinese Empire for 13 dynasties, the Terracotta Army is one of the great archaeological finds of

The Terracotta Army, created in the third century B.C.E., was discovered buried near Xian, China (image courtesy of Maros Mraz/GNU free documentation license).

the twentieth century. It was discovered by two farmers who were drilling for water in 1974. In the third century B.C.E., the very first emperor of China, Qin Shi Huang, united China, established the longest-running form of government, built the Great Wall, and built his own elaborate tomb. As a symbol of his rule and to guard himself in the afterlife, Qin Shi Huang had an entire army replicated—approximately 8,000 soldiers, 130 chariots with 520 horses, and 15 cavalry horses, all life size and each with distinct facial expressions—to guard the first emperor's mausoleum at Mt. Li in Xi'an, China.

What is the **Three Gorges Dam**?

The Yangtze River will soon be the site of world's largest electricity-generating facility, the Three Gorges Dam. Construction began in 1994 and should be completed around 2009. The construction of the dam requires the relocation of over one million Chinese living upstream of the dam, as the rising waters will flood towns and archaeological sites. The reservoir will be approximately 600 miles (965 kilometers) long, spanning the Yangtze River in Sandouping, Hubei Province, China. The dam will be about 600 feet (183 meters) high, and 1.5 miles (2.4 kilometers) wide.

When operational, the dam will produce 22,000 megawatts of power, more than eight times the power produced at Hoover Dam in the United States. The environmental impacts of the dam are now being felt, with increased drought, less rainfall, and mudslides that can endanger the millions of people who live nearby. More than 1.2 million people were relocated from the area, and 116 towns and cities were evacuated in order to complete the project.

How important is **China** to the **biodiversity** in the world?

China's territory covers 3,705,390 square miles (9,596,960 square kilometers) of land. This vast expanse includes a wide diversity of wildlife, as well as 10 percent of the world's plants with stems, roots, and leaves. Therefore, China's policies for managing its natural resources are very important for the health of the planet.

Where can you **see** the **Great Wall**?

There are several places where tourists may see the Great Wall that are within a one- to two-hour car ride from the capital city of Beijing. One of the most famous and most developed tourist areas is Badaling, where one can see both the Great Wall of China, in its restored and untouched splendor, and the nearby Ming Tombs.

What is the **oldest European settlement** in **eastern Asia**?

In 1557, the Portuguese established the trading colony of Macao on mainland China at the mouth of the Xi (Pearl) River. Macao was a territory of Portugal from 1849 until 1999, when it was returned to China.

Why isn't **Tibet** on the map?

Tibet is not on the map because it is no longer an independent country. Though Tibet was once a theocratic Buddhist kingdom, China annexed it in 1950. Tibet is now a mildly autonomous region in southwestern China with a puppet communist government. In addition to the destruction of the Tibetan Buddhist religion in the 1960s by China, China also moved Tibetans out of the area and moved ethnic Chinese into Tibet to help moderate Tibet's secessionist ideas.

How long have **Communists** been in **power** in **China**?

China's communist revolution took place in 1949, and Mao Zedong became the country's first "chairman." Communism has been the doctrine ever since.

What is the **highest railroad** in the world?

The Qinghai-Tibet railroad is the highest railroad in the world, climbing to more than 16,737 feet (5,072 meters). It connects Lhasa, Tibet, to the rest of China on 709 miles (1,142 kilometers) of track. Because of the high altitudes of the sections of the track, each rail car includes a supply of oxygen for its passengers. More than 1.5 million people used the train in its first year of operation in July 2006.

What is the world's most **commonly spoken language**?

Over one billion people around the world speak Mandarin, the official language of China. Other languages spoken in China include Yue (Cantonese), which is spoken by 71 million people; Wu (Shanghaiese), spoken by 70 million; and other minority languages such as Minbei (Fuzhou), Minnan (Hokkien-Taiwanese), Xiang, Gan, and Hakka dialects.

What is **pinyin**?

Pinyin is a new system for transliterating Chinese into the Roman alphabet. It replaced the Wade-Giles system in 1958, when the Chinese government started using

China is a leading rice producer, and the country is also famous for its remarkable rice farming terraces, such as these located in Yuanyang County.

pinyin for external press announcements. It has gradually gained acceptance, and is the reason why we now call the Chinese capital Beijing instead of Peking (a Wade-Giles transliteration).

How much **rice** does China produce?

China is the world's leading rice producer and is responsible for about one-third of total rice production in the world. The country produces about 188.5 million metric tons of rice each year. India, the next biggest producer, produces about 142 million metric tons. Ninety percent of the rice in the world is produced in Asia.

Who is the **biggest exporter of rice** in the world?

Thailand exports about 9.2 million metric tons of rice per year, about one-third of the total rice that is exported in the world. The Philippines imports the most rice in the world: approximately two million metric tons each year.

What **transition** took place within **Hong Kong** in 1997?

The sovereignty of Hong Kong, a former British territory, was transferred to China on July 1, 1997, following the expiration of a 99-year treaty signed by China and the United Kingdom. The treaty gave the United Kingdom the use of the trading port of Hong Kong, as well as a section of land on Mainland China known as the New Territories. Since the treaty for the New Territories was due to expire in the late 1990s, the United Kingdom and China entered into negotiations to hand over to China both Hong Kong (covered by an in-perpetuity agreement) and the New Territories. China agreed to the terms of the negotiation, which granted Hong Kong special administrative rights, limited sovereignty and self-rule, and allowed its economic structure to remain in place until 2047. Hong Kong was and continues to be one of the most developed territories in the world, as well as one of the great banking and financial centers.

What is the **least-densely populated country** in the world?

Mongolia (not to be confused with Inner Mongolia, which is a province in northern China), with its tiny population of 2.6 million people spread over 600,000 square miles (1,554,000 square kilometers) of territory, has a population density of about four people per square mile. Mongolia's density is limited because only one percent of the

> ## Why is the Aral Sea shrinking?
>
> The area of the Aral Sea, located on the border of Kazakhstan and Uzbekistan, has been reduced in size by one-half since 1960. Though it was once the world's fourth-largest lake, diversion of its feeder rivers for agricultural purposes has severely shrunken the lake. This shrinkage has exposed soil saturated with salt, which now destroys plants and vegetation across the nearby plains.

country can be used for agriculture; the remainder of the country is dry and used for nomadic herding. Mongolia was originally established in the thirteenth century when Genghis Khan overtook and unified much of mainland Asia.

Why did the **Soviet Union** invade **Afghanistan** in 1979?

The Soviet Union sent troops to Afghanistan in 1979 because it wanted to come to the aid of its ally, the Parcham faction, which was more moderate in its outlook toward moving the country towards communism. The Parcham faction had signed a treaty of Friendship and Cooperation with the U.S.S.R. the previous year. The United States began to fund the mujadeen resistance, who were opposed to both the Afghan government and the Soviet occupation. These same mujadeen would later form the ideological corps of the people who masterminded the attack on the World Trade Center in 2001.

Between 600,000 and 2 million Afghanis were killed during the war, which lasted until 1989. Of the 600,000 Soviet troops who served in Afghanistan, 14,453 lost their lives and more than 469,000 became sick or were wounded. More than five million Afghanis were displaced and fled to Pakistan. Another two million Afghanis were displaced within their own country, seeking shelter from the violence of the war and the factional fighting that ensued.

What were the **Buddhas of Bamiyan**?

The Buddhas of Bamiyan were two gigantic sculptures—one measuring 180 feet (55 meters) and the other 121 feet (37 meters) high—that were built in the sixth century C.E. Located 143 miles (230 kilometers) northwest of Kabul, Afghanistan, they were designated as a UNESCO World Heritage Site and were one of the great archaeological and religious sites in the world. In 2001, the Taliban regime in Afghanistan ordered their destruction, and they were dynamited. Many countries, including Japan and Switzerland, have pledged their support to rebuild these cultural treasures.

Why has **Afghanistan** been contested and **invaded** so many times?

Afghanistan sits at a crossroads linking Asia with the Middle East. For millennia, it was the major transit point along the Silk Road, which brought goods from Asia to and

from the Middle East. In recent times, it has been treated as a geographically strategic location because of its potential for influencing the policies of countries in both regions. It is an ethnically diverse region that is home to many tribes and cultures, all of whom have been vying for some form of control or voice in the way in which the country is governed. Therefore, in the past several hundred years, major geo-political players have sought to control, occupy, or colonize this country. Much of what we see happening in Afghanistan today has roots in conflicts dating back many centuries.

What is **Ulan Bator**?

Ulan Bator is the capital of the Republic of Mongolia.

One of the Buddhas of Bamiyan, a UNESCO World Heritage Site, which was destroyed by the Taliban government in Afghanistan in March 2001 (image courtesy UNESCO, A. Lezine).

How **big** was the **Mongol Empire**?

The Mongol Empire, under the reign of Genghis Khan (1162–1227 C.E.; also spelled Chingis Khan) and his son Ögedi Khan (1186–1241), became one of the greatest contiguous empires in world history. It would eventually stretch from what is now modern-day Korea and China in the east to Poland in the west, and to Vietnam and Oman in the south.

Who was **Kublai Khan**?

Kublai Khan (1215–1294 C.E.) was the grandson of Genghis Khan. Under him, the Mongol Empire reached its peak in 1279. He founded the Yuan dynasty, moved its capital to Beijing in the fourteenth century, and would later move it back to Mongolia, as his empire fell.

Who are the **Sherpa**?

The Sherpa are an indigenous ethnic group in Tibet and Nepal. They live among the mountains of the Himalayas and are often hired as guides for climbing expeditions to such peaks as Mt. Everest. In 1953, Norkey Tenzing (Sherpa) and Edmund Hillary (British) were the first two people to reach the 29,028-foot (8,848-meter) summit of Mt. Everest.

THE INDIAN SUBCONTINENT

What is in the **Taj Mahal**?

Located in Agra, India, the Taj Mahal is a mausoleum for the wife of the Mogul emperor Shah Jahan. After Arjuman Banu Bagam's death in 1631, her husband began construction of the mausoleum in 1632. Over 300 feet (91.4 meters) tall, the white marble mausoleum is a grandiose and striking memorial to her life and death.

An ancient drawing of the Kublai Khan (image courtesy of National Palace Museum, Taipei, Taiwan).

What makes **New Delhi** so new?

From 1773 to 1912, the capital of India was Calcutta, then was moved to Delhi in 1912. Because the British wanted to build a brand-new capital city, they began to construct a new city adjacent to Delhi (now known as Old Delhi). When construction was completed on this new city in 1931, it became the capital of India and was known as New Delhi. The metropolitan area of Old and New Delhi is one of the world's largest urban areas, containing a population of over 16 million people.

Where did **Bombay** go?

In 1996, India changed the name of the third largest metropolitan area (population 19 million) from Bombay to Mumbai.

What is **Bollywood**?

Known as "Bollywood," Mumbai, India (previously called Bombay), is the world's movie capital. The entertainment industry in India produces more than 1,000 films annually, which is much more than the United States.

Where is **Dum Dum** airport?

Each year, over two and a half million passengers pass through Dum Dum International Airport in Kolkatta (Calcutta), India.

Does **India** have a **population control program** similar to China's?

Though India does not limit births to one child per family, it does have one of the oldest population control programs in the world. The program, begun in the 1950s,

Does a cashmere sweater come from Kashmir?

The Kashmir goats, which make the fine wool known as cashmere, do come from the Kashmir region in India (a disputed area not controlled by any one nation). This region, which is located in northern India bordering on Pakistan, is plagued by violence and unrest. In 1947 the British decided to split the colony of India into two separate countries—one Hindu (India) and one Islamic (Pakistan). The state of Jammu-Kashmir has a mixed population of Hindus and Muslims, which has led to conflict within the region. India and Pakistan have waged three wars over this region, and sporadic violence continues today.

encourages the use of birth control and family planning, and the Indian government provides grants to people who undergo sterilization surgeries. Though India's population is growing more slowly (1.46%), the world's population growth rate is still lower, at 1.17 percent.

How does India's **caste system** work?

The caste system is the extremely rigid, hierarchical social class system of India. Based on the ancient Hindu text "Law of Manu," the caste system consists of four categories: Brahmans (priests), Kshatriyas (warriors), Vaisyas (merchants), and Sudras (servants). There are many people who are considered to be outside of the caste system, or who have no caste, called the Harijans or untouchables. The untouchables serve the most menial jobs and live literally outside of the social class system. The caste system in India dictates not only one's profession, but also whom one can marry, social contacts, and all other aspects of life.

Where is the world's **second-highest mountain**?

K2, at 28,250 feet (8,611 meters), is the world's second-tallest mountain. K2 sits in the disputed Kashmir region of northern Pakistan.

Where was **East Pakistan**?

When the British left South Asia, they divided the region into India and Pakistan. Muslim regions in Pakistan were located to the east and west of Hindu India. These two separate territories became East and West Pakistan, and were separated by approximately 1,000 miles (1,600 kilometers). Having been extremely geographically separated from West Pakistan for over three decades, East Pakistan declared independence and became Bangladesh in 1971.

Why does **Bangladesh flood** so often?

Given that most of Bangladesh's elevation is near sea level, and that it is located within the delta of the Ganges and Brahmaputra Rivers, it is not remarkable that the country is easily flooded by regular monsoons (periodic winds and accompanying rainfall) and hurricanes. Unfortunately, Bangladesh also suffers from a poor emergency warning system, so people in Bangladesh are not adequately warned of impending disaster.

JAPAN AND THE KOREAN PENINSULA

What are the **four main islands that make up Japan**?

The northernmost island, Hokkaido, is home to the city of Sapporo. The largest island, Honshu, is the Japanese core area that includes Tokyo and Osaka-Kyoto. The Japanese island of Honshu is the world's seventh largest island in the world, as well as the second most populous island in the world with 103 million people; it is the "mainland" island of Japan. Honshu covers 86,246 square miles (223,656 square kilometers), which makes it larger than the island of Great Britain. In the south are the islands of Shikoku and Kyushu. Kyushu is the southernmost island and was the first island where foreign traders were allowed into Japan. It has a population of over 13 million. Besides the four main islands, Japan includes 2,000 smaller islands. Japan's population is 128 million people and is declining at a rate of 0.02 percent per year.

What is the world's **most visited mountain**?

Japan's Mt. Fuji, a sacred and important volcano to the Japanese, is the country's most popular tourist spot and the world's most visited mountain. Mt. Fuji, which is shaped almost like a perfect cone, rises to 12,388 feet (3,776 meters) and last erupted in 1708.

Where is the **land of the rising sun**?

The Japanese name for Japan, Nippon, which means "origin of the sun," evolved into "land of the rising sun." The name probably derived from the fact that for

Japan's picturesque Mt. Fuji, near Tokyo (photo by Paul A. Tucci).

> ## How have the Kurile Islands kept World War II from ending?
>
> **B**efore World War II, Japan owned this chain of four islands located between Russia (south of the Kamchatka Peninsula) and Japan (north of Hokkaido). During the war, the then-Soviet Union took control of the islands and hasn't given them back. Japan has been requesting their return, but to no avail. Because of the Kurile Island controversy, Japan has yet to sign a peace treaty with Russia declaring the end of World War II.

centuries, Japan was the easternmost known land, and thus where the sun seemed to rise.

How **geologically active is Japan**?

Both volcanoes and earthquakes threaten Japan. Japan has 19 active volcanoes, several of which have erupted in the last decade. Earthquakes are also frequent occurrences, with many very destructive quakes in the last century. In 1923 a major earthquake (approximately 8.3 on the Richter scale) struck Yokohama and killed over 140,000 people. More recently, in 1995, an earthquake in Kobe killed 5,500 people.

How does **Japan get its oil**?

Having no oil itself, Japan must import all the oil it needs. To accommodate the amount of oil necessary, there is a constant stream of oil tankers—spaced approximately 300 miles (483 kilometers) apart—that bring oil to Japan 24 hours a day, 365 days a year. Behind only the United States and China, Japan is the third biggest importer of oil with more than 5.5 million barrels brought in every day.

What is the **life expectancy** in Japan?

Japan's life expectancy is now the third highest in the world, behind Andorra and Macau. People in Japan live, on average, upwards of 83 years.

Where is **Iwo Jima**?

Iwo Jima, one of the three islands that make up the Volcano Islands, is located southeast of Japan. One of the deadliest battles of World War II was fought on Iwo Jima, with approximately 20,000 Japanese and 6,000 American soldiers killed. The Japanese air base on Iwo Jima was captured by the United States on February 23, 1945. The island was returned to Japan in 1968.

How old is the Korean civilization?

The Korean civilization is more than 4,000 years old, with archaeological evidence dating back to 2333 B.C.E.

Where was the **first atomic bomb** used on a populated area?

The Japanese city of Hiroshima was leveled by an atomic bomb dropped by the United States on August 6, 1945. Three days later, the United States dropped a second bomb on Nagasaki, Japan. While these events may have hastened the Japanese surrender in World War II, over 115,000 people were killed immediately due to the blasts, and many more died later due to radiation-related diseases.

What is a **bullet train**?

Bullet trains, or *shinkansen*, as they are called in Japan, are similar to traditional passenger trains but have been enhanced to travel at speeds of up to 215 miles per hour (346 kilometers per hour). They have been used in Japan since 1965. In 2003 Japan set a world record, testing another generation of bullet trains capable of exceeding 361 miles per hour (581 kilometers per hour).

How did **North and South Korea** come to be?

From 1392 to 1910 the Korean peninsula was the home of the Choson Kingdom (as it is referred to by its inhabitants). In 1910 Japan took control of the peninsula but lost this territory after its defeat in World War II. North and South Korea are still divided by a line that lies near the latitude of 38 degrees north. This latitude marked the line dividing the Soviet occupation zone in the north and the American occupation zone in the south following World War II. From 1950 to 1953, the Communist North Koreans fought with the democratic South Koreans. With American forces involved in the Korean war, North Korea's army was eventually forced to retreat into China, though it later recaptured land to the 38th parallel, where the border between the two Koreas remains today.

What is **South Korea's national slogan**?

The national slogan of South Korea is the "Land of the Morning Calm."

How long did **Japan occupy** and colonize **Korea**?

Japan began its forced occupation of Korea in 1910 and left the country at the end of World War II in 1945. During the occupation, many Koreans were put into forced

> ## How many American troops are stationed in South Korea?
>
> Currently, 22,500 U.S. troops still occupy South Korea, but only 160 are actually stationed in the demilitarized zone.

labor, slavery, and military conscription. It is estimated that between 200,000 and 800,000 people died at the hands of the Japanese in Korea and the Manchuria region.

How does **South Korea rank** in the **world economy**?

South Korea ranks 13th in terms of its per capita GNP (gross national product) of $20,000. It is also the United States's seventh largest trading partner. Its economy is one of the fastest growing in the world, expanding at nearly seven percent per year.

What are some of the **largest South Korean companies**?

Samsung, Hyundai, LG (Lucky Goldstar), Hanjin Shipping, Daewoo, and Kia are South Korean companies with global brands. They are some of the largest companies in the world today.

What is **Truce Village**?

The Joint Security Area, or Truce Village, is the nickname given to Panmunjom, a former village destroyed during the war that is 53 kilometers from Seoul. It marks the spot along the 151-mile (243-kilometer) border between North and South Korea, and is the only area between the two countries that the United States still considers a combat zone. The border area strip, which is 2.5 miles (4 kilometers) wide, is patrolled by military forces on both sides. More than 1,000 North and South Koreans have been killed in this area since fighting "ended" in 1953.

How many times have **North and South Korea met** to **negotiate** a solution to the division of Korea in Panmunjom?

Both sides have met nearly 400 times to negotiate an end to the division.

Who was **Kim Il Sung**?

Sung was the leader of the Democratic People's Republic of Korea (North Korea) from its inception in 1948 until his death in 1994. His son, Kim Jong Il, took power after his death, and leads the republic today. Kim Il Sung was worshipped by his people under a state ideology called Juche, which was a complex movement to establish a unique North Korean political movement independent of Marxist-Leninist ideals and the de-Stalinization of the Soviet Union in the late 1950s.

What **catastrophe** happened in **North Korea in 1995**?

Because of their isolation from former trading partners in the Soviet Union, coupled with poor relations with neighboring China, failed economic policies that drove up the price of food, excessive military expenditures, and severe floods, the production of grain in North Korea fell below what was needed to sustain the population in 1995. This triggered food shortages and caused the deaths of between one and three million people. It is not clear how many people died during the three-year famine because the government does not report these types of figures.

The United States gave approximately $600 million to North Korea to help prevent mass starvation. The largest donor country in the world, America contributed more than 50 percent of the total aid given to North Korea to help stop the famine.

Who **provides food** to **North Korea** today?

The top exporters of food to North Korea are China, Japan, South Korea, and the United States.

What is the **Arirang Festival**?

The Arirang Festival or Mass Games, held every April 15 and lasting for two months, is a celebration including performances by thousands of students who create gigantic mosaic pictures by holding colored mosaic cards while standing in an enormous stadium. There are also dance performances during the ceremony. For this event, the North Korean government has even allowed some American tourists to attend.

SOUTHEAST ASIA

Where is **Indochina**?

Indochina is the peninsula in Southeast Asia composed of Myanmar, Thailand, Cambodia, Laos, Vietnam, and the mainland portion of Malaysia. During the colonial era, the eastern portion of Indochina was ruled by France and the west was ruled by Britain.

When did **Burma become Myanmar**?

In 1989, the name of Burma changed to Myanmar when the military took control of the country, following the president's resignation as a result of riots and national turmoil.

An early morning street scene in Hanoi, Vietnam (photo by Paul A. Tucci).

Why was the **Vietnam War** fought, and what were some of the consequences?

After World War II, Vietnamese nationalists and communists fought the French to gain their freedom from European colonization. Ultimately, the French left Vietnam in shambles, and the former colony divided into the primarily communist controlled North and the pro-West forces in the South. The United States initially was involved in the war to help its ally France regain control of the colony by sending in advisers and aiding in military supplies. Forces in the South saw the chance to gain American assistance, and they convinced various U.S. administrations to assist in ridding the country of their enemies. In an effort to stem the spread of communism, the United States continued to support South Vietnam until the fall of Saigon in 1975.

Ironically, during the height of the Vietnam War in 1970 while communist Vietnamese were being portrayed as a threat to the American way of life, President Richard Nixon and Secretary of State Henry Kissinger were courting the largest, most radical communist country, the People's Republic of China, to extend relations with the West. China had been funding and supporting militarily Vietnamese resistance against both French and American forces for decades.

The American occupation of Vietnam lasted from 1962 to 1975. In 1975, Saigon was overrun by communist forces, and the country was eventually unified under one communist regime. The aftermath and legacy of the war in Vietnam was the destabilization of and genocide in neighboring Laos and Cambodia, which cost millions of lives during the upheaval of the 1970s, as well as the displacement of millions of Vietnamese, some of whom emigrated to the U.S. as "boat people."

Which **Southeast Asian country** is one of the **world's richest**?

The tiny nation of Brunei (composed of just 2,226 square miles [5,765 square kilometers]), located on the island of Borneo, is one of the world's richest countries. Brunei's wealth is based on its oil and gas exports. Sultan and Prime Minister His Majesty Paduka Seri Baginda Sultan Haji Hassanal Bolkiah Mu'izzaddin Waddaulah, who is the ruler of Brunei, is one of the richest men in the world.

What is **Bandar Seri Begawan**?

The capital of Brunei Darussalam is Bandar Seri Begawan, or Bandar for short. It is said that during the Vietnam War the airport in Bandar frequently picked up the traces of U.S. and Vietnamese fighter airplanes engaged in battle nearby.

Where is **Brunei**?

Brunei lies on the northern tip of the island of Borneo, just south of the Philippines. It shares Borneo Island with four provinces of Indonesia (West, Central, South, and East Kalimantan) and two states of Malaysia (Sabah and Sarawak).

What are the **principal industries** in the **Maldives**?

Tourism and fisheries contribute the most to the Maldives's GDP, which is approximately $1.569 billion per year.

How many **islands** make up the Maldives?

There are approximately 1,192 islands and islets within the territory of the Maldives, of which only 250 are inhabited. Of the inhabited islands, some are actually hotel resorts only, as the islands have

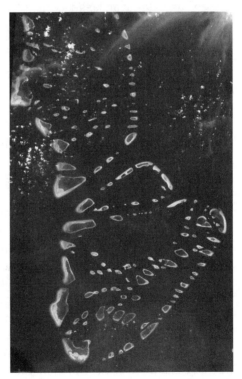

A satellite view of the Republic of Maldives, which may have its very existence threatened should sea levels continue to rise (image courtesy of NASA/GSFC/METI/ERSDAC/JAROS, and U.S./Japan ASTER Science Team).

229

only enough space for one or two hotels. The Maldives lie about 600 miles (965 kilometers) southwest of India.

Which country has the **lowest maximum elevation** in the world?

The Maldives has a natural maximum elevation of only 7.5 feet (2.3 meters) above sea level. The problem is that sea levels over the past 100 years are rising at a rate of approximately 9.75 inches (20 centimeters) per year, which may make the country, or parts of it, disappear entirely. The 2004 Asian tsunami caused the ocean to completely cover parts of the Maldives.

What is a **dhoni**?

A dhoni is a traditional wooden Maldivian fishing boat.

What **language** is spoken in the **Maldives**?

Dhivehi, an Indo-Aryan language, is spoken by the 300,000 inhabitants of the Maldives.

What is an **economic tiger**?

An economic tiger is a term applied to any rapidly developing Asian nation with the power and ability to become an influential, international economic powerhouse. South Korea, Taiwan, and Singapore are considered to be the three economic tigers. Hong Kong was once part of this Pacific Rim group, but, since its merger with China, it can no longer be considered an economic tiger.

What part of **Southeast Asia** is one of the most **contested geographic regions** in the world?

The Spratly Islands are a group of 100 small islands, islets, and reefs in the South China sea. They are located between Vietnam, the Philippines, and East Malaysia. Though only comprised of 2 square miles (5.2 square kilometers) and spread across 158,000 square miles (400,000 square kilometers) of the South China Sea, these islands are of strategic importance. Ownership is contested by Vietnam, China, the Philippines, Malaysia, and Taiwan. Oil fields that yield nearly 15 percent of the petroleum used by the Philippines have been the center of the dispute. Each country feels that it should be able to negotiate and profit from the lucrative contracts and benefits that oil exploration and production might bring.

What is the only Catholic country in Asia?

The Philippines is the only Catholic country in Asia— approximately 83 percent of its population is Catholic. Catholicism was firmly implanted in the Philippines when the land was under Spanish rule, from the sixteenth through nineteenth centuries.

THE PHILIPPINES AND INDONESIA

How many **islands** make up **the Philippines**?

The Philippines are composed of 7,100 islands. Only 1,000 of these are inhabited and 2,500 still remain unnamed. The islands are divided into three main groups: the Luzon region in the north, the Visayan region in the middle, and the Mindanao and Sulu region in the south.

When did the **United States control the Philippines**?

The Philippines were a Spanish colony until 1898, when the United States took control of the more than 7,100 islands at the close of the Spanish-American War. The islands remained under American control through 1946 (except for a two-year period of Japanese control during World War II). The Philippines became independent in 1946 and leased land to the United States for military bases until 1992, when the U.S. military presence in the Philippines ended.

How many **islands make** up **Indonesia**?

Indonesia is composed of over 13,500 islands. Of these, only 6,000 are inhabited. Indonesia is the world's largest archipelago and was formerly known as the Dutch East Indies. The area had been under the control of the Netherlands since around 1600, but declared its independence in 1945 (after being subjected to Japanese rule during much of World War II).

What is the world's most **densely populated island**?

The Indonesian island of Java is the world's most densely populated island. It has a total population of over 124 million, which, when placed within its 51,000 square miles (132,000 square kilometers) of land, gives it a density of over 2,431 people per square mile. Indonesia's capital city, Jakarta, is located on the island of Java.

A view of Singapore's prosperous financial district. The country is the smallest in southeast Asia, but has a high GDP per capita of $49,900.

What is Southeast Asia's largest **oil-producing** country?

Indonesia produces approximately 1,136,000 million barrels of oil per day, making it the twentieth biggest oil producer in the world.

What was **East Timor**?

East Timor—its former name—is now the Democratic Republic of Timor-Leste, a country adjacent to the most eastern province of Indonesia. After being under the control of Portugal for four centuries, the region came under Indonesian rule. Later, in 1975, East Timor declared independence. Indonesia consequently invaded, killing about 100,000 people. In 1991 the international community condemned Indonesia for this and the additional massacres and human rights violations that followed. Indonesia eventually relinquished control of East Timor, and Timor-Leste became an independent republic in 2002. With a per capita GDP of less than $800 per year, Timor-Leste is one of the poorest countries in the world.

What is the **smallest country in Southeast Asia**?

The Republic of Singapore is the smallest country in southeast Asia. It covers only 272 square miles (707 square kilometers) of land.

Is **Singapore** a city or a country?

Singapore, a city state, is both a city and a country, and one the wealthiest, most developed countries in the world. Born out of the extreme hardship and mass killings of World War II, the country was taken under United Kingdom control after the Japanese defeat. In 1963, Singapore merged with the states of Malaya, Sabah, and Sarawak to form what is Malaysia today. In 1965, Singapore split from Malaysia and formed its own country.

MIDDLE EAST

THE LAND AND HISTORY

What is the Middle East **in the middle of**?

At one time, it was common to refer to the Near, Middle, and Far East. Though two of the terms have fallen into disuse, the Middle East is still commonly used. The Near East, at its greatest extent in the sixteenth century, once referred to the Ottoman Empire, which included Eastern Europe, Western Asia, and Northern Africa. The Middle East referred to the area from Iran to India to Myanmar (formerly Burma). The Far East used to refer to Southeast Asia, China, Japan, and Korea.

Where is the **Middle East today**?

It is generally agreed that the Middle East includes Egypt, Israel, Syria, Lebanon, Jordan, the countries of the Arabian Peninsula (Saudi Arabia, Yemen, Oman, United Arab Emirates, Bahrain, and Qatar), Iraq, Kuwait, Turkey, and Iran. Most regional specialists also include the countries of northern Africa (Morocco, Algeria, Tunisia, and Libya). The new countries of Azerbaijan, Georgia, and Armenia (former Soviet republics) are also often included in the region.

What was the **Ottoman Empire**?

The Ottoman Empire began as a tiny state in the fourteenth century. It was centered in the city of Bursa in what is now northeast Turkey, and it rapidly expanded through the conquest of neighboring states. Its greatest expansion was in the sixteenth century, when it included southeastern Europe, the Middle East, and North Africa. Due to wars with other European countries in the seventeenth and eighteenth centuries, the

Why is the Dead Sea dead?

Water in the Dead Sea is extremely salty (nine times the saltiness of normal ocean water) and kills nearly all the plant and animal life that flows into the sea from the Jordan River. The Dead Sea lies on the border of Israel and Jordan and is the lowest land elevation on Earth, sinking to 1,378 feet (420 meters) below sea level.

Ottoman Empire began to decline and became known as the "sick man of Europe." The Empire's successor was Turkey, which became an independent country in 1922.

Is the Middle East a **desert**?

Actually, very little of the region is filled with sand dunes and sand storms. Coastal areas are temperate, and several areas boast very pleasant and moist Mediterranean climates. A lack of water is a problem for agriculture throughout the region, so countries are experimenting with technological solutions such as desalinization plants and drip-agriculture for water conservation.

Can you **ski** in the **Middle East**?

Yes, you can ski in the Middle East. Since many parts are mountainous, thousands of people ski in such places as Dizin and Shemshak, Iran, and in Faraya, Lebanon. People can even ski in Dubai, along the coast of the Persian Gulf, where each year thousands enjoy skiing in one of the largest indoor ski facilities in the world.

What is a **Mediterranean climate**?

A Mediterranean climate is a climate similar to the one found along the Mediterranean Sea: hot and dry in the summer and warm and wet in the winter. Areas that are renowned for having Mediterranean climates but are not near the Mediterranean Sea are California and Chile.

Is the **Empty Quarter** empty?

The Empty Quarter, also known as the Rub al Khali, is Saudi Arabia's vast, open desert. While the Empty Quarter is devoid of a large population, it is quite valuable, as it contains the world's largest petroleum reserves.

What is the Fertile Crescent?

The Fertile Crescent is an area located in a crescent-shaped region between the Persian Gulf in the east and Israel in the west. The development of this region, located along the Tigris and Euphrates Rivers, was begun thousands of years ago and was based on the availability of water from these rivers and the fertile soil they deposited. This area was a center of human civilization and was the location of the ancient Mesopotamian Empire.

What commodity makes the **Persian Gulf** so **strategically important**?

The Persian Gulf is strategically very important because a large share of the world's petroleum is transported through it in oil tankers. At the southeastern end of the Gulf is the Strait of Hormuz, a narrow chokepoint that can be (and has been) controlled to prevent ships from sailing in or out of the Gulf.

What was **Mesopotamia**?

This ancient region lay between the Tigris and Euphrates Rivers, from contemporary southern Turkey to the Persian Gulf. Mesopotamia was the home, at one time or another, to such civilizations as Babylonia, Assyria, and Sumer.

What was **Babylonia**?

Babylonia was an ancient country along the Euphrates River, in what is now Iraq. It began in the twenty-first century B.C.E. and was led by King Hammurabi in the eighteenth century B.C.E.

When did the **Suez Canal** begin to operate?

After a decade of construction, the Suez Canal, built by the French, opened on November 17, 1869. The 101-mile-long (162.5-kilometer-long) canal cuts through northeastern Egypt, making a passageway for ships to sail between the Mediterranean Sea and the Red Sea, which leads to the Indian Ocean.

Where are **Egypt's pyramids**?

The most famous group of pyramids in Egypt is located near the city of Giza, which is just outside of Cairo. This group includes the Great Pyramid, which is known as one of the Seven Wonders of the Ancient World. The pyramids in Egypt served as tombs for Pharaohs and were built from the twenty-seventh to the tenth century B.C.E. Other

The Giza Pyramids located just outside of Cairo, Egypt.

pyramids are located along the Nile River in southern Egypt and the northern Sudan. Approximately 70 pyramids remain in the region.

RELIGION

Which **religions began** in the **Middle East**?

Judaism, Zoroastrianism, Christianity, and Islam all have their roots in the Middle East. The region is consequently filled with sites that are considered holy by all four religions.

What is **Zoroastrianism**?

Zoroastrianism is a religion that began in present-day Iran about 1000 B.C.E. It is based upon the teachings of the prophet Zoroaster and is an ancient religion that some believe may have influenced the later religions of Christianity and Islam.

Which **religions** came to be **practiced first**?

Before the advent of Judaism, many earlier religions were practiced, including Animism, which is the worship of nature and natural phenomena like fire, wind, water, and earth. Judaism came into practice around 5000 B.C.E., followed by Zoroastrianism in 1000 B.C.E., Christianity in 30 C.E. and Islam in 600 C.E.

What is the difference between **Islam and a Muslim**?

Islam is a religion and a Muslim is a person who is a follower of Islam.

Who was **Muhammed**?

The prophet Muhammed was the founder of Islam. He was born in Mecca in 571 and fled to Medina later in life. According to the Islamic religion, he received prophecies from God, which were subsequently written in the Koran, the holy book of Islam. Muhammed's death in 632 C.E. led to the expansion of Islam around Eurasia.

The Zoroastrian religion once thrived in what is now Iran. Today, a small minority of worshippers remain, as well as some reminders of the past like this symbol found on ruins of Persepolis.

Why do people travel to **Mecca**?

The holy book of Islam, the Koran, requires that all Muslims make a journey, known as a pilgrimage, to the city of Mecca at some time in their lives. Mecca, located in Saudi Arabia, is the holiest city of Islam, as it was the birthplace of Muhammed, the founder of Islam in the sixth century, and contains many important religious sites. Non-Muslims are forbidden to enter Mecca.

What is the **holiest site** in **Mecca**?

The most important site in Mecca is the Great Mosque (called the Haram). It is located in the center of the city and houses the sacred Black Stone (Ka'abah). Muslims around the world face the Black Stone during prayer.

How many Muslims make a pilgrimage to Mecca each year?

Each year, approximately two million Muslims from around the world make a pilgrimage to Mecca during the last month of the Islamic calendar. While hundreds of years ago the trip took weeks or months to complete from the vast reaches of the Islamic Empire, today most people from great distances use the modern conveniences of travel and fly to Saudi Arabia.

The holy Islamic site of Mecca in Saudi Arabia swarms with faithful Muslims making their annual pilgrimage.

Who are the **Sunni and Shi'ite** Muslims?

The Sunni and Shi'ite are two sects of Islam. About 85 percent of Muslims are Sunnis, while Shi'ism is especially prevalent in Iran. After Muhammed's death, two relatives claimed to succeed him. Their followers developed into the two sects.

Who is an **ayatollah**?

The title of ayatollah was originally used for outstanding Shi'ite scholars. But in 1979 Iran went through a period of fundamentalism and replaced the secular shah (the former leader) with the ayatollah, a religious and political leader of the nation. Under the ayatollah's rule, traditional social values were mandated.

What is a **theocracy**?

A theocracy is a country ruled by religion. Iran has the world's largest theocracy, ruled by people who are national leaders as well as religious leaders. While there is a secular president, the division of power between mullahs (a Muslim trained in religious law) and the president is poorly defined.

How far did the **Islamic Empire** spread?

At its widest extent, the Islamic empire included most of northern Africa, Spain and Portugal, the Balkan Peninsula, India, Indonesia, Kazakhstan, and the southern reaches of Russia. Islamic influence also spread east into western China. The areas that were under Islamic rule from the seventh through the sixteenth century C.E. still keep the religion as a major aspect of their culture.

What is the world's **largest Islamic nation**?

Indonesia, the world's fourth-most populated country, is the world's largest Islamic nation. About 87 percent of Indonesians are Muslims. Islam spread to Indonesia during the Medieval era.

Are all **Israelis Jewish**?

No, they are not. About 80 percent of Israelis are Jewish, and the remaining 20 percent are predominantly Arabs.

CONFLICTS AND NATIONS

How did **Israel** become a **country**?

European Jews, who had been persecuted for centuries in Europe, began to emigrate to the area we know as Palestine in the late-nineteenth century. These new immigrants purchased land and set up communities in the vast desert and coastal areas. This area, which was under British administration, began to see a new influx of European Jews in the early part of the twentieth century. After World War I, the United Kingdom issued the Balfour Declaration, which accepted the establishment of a Jewish state in present-day Palestine. At the time, Palestine was not comprised of countries, but was merely a collection of territories ruled by colonial powers such as the United Kingdom, France, and the United States.

After World War II, hundreds of thousands of European Jews emigrated to the region, enough to cause the Arab community to revolt and riot against them. At this time, nearly 33 percent of all people in Palestine were Jewish. Jewish paramilitary groups increasingly fought against British occupying forces. The United Nations recognized Jerusalem as an international city for both Arabs and Jews, and it approved an interim plan to divide Palestine into Arab and Jewish states. The Jewish authorities accepted the plan, but new Arab countries that formed after World War I and World

War II did not. One day before the British mandate of control over the Palestine region expired in May 1948, the State of Israel was proclaimed.

What is the **Gaza Strip**?

This area of land along the Mediterranean Sea at the border of Israel and Egypt was part of Egypt until it was captured by Israel for a brief period in 1956 and 1957 and then permanently in the 1967 war. The Palestinian Liberation Organization (PLO) and Israel agreed to Palestinian self-rule in the Strip in 1994.

How has the **West Bank** caused conflict?

The West Bank, which refers to the western bank of the Jordan River, was supposed to become part of an independent Palestine at the same time Israel became a state. However, Arab attacks following the United Nations' 1947 proclamation of Israeli and Palestinian statehood led Israel to take over the West Bank when Israel became an independent state in 1948. Following a 1950 truce, Jordan occupied the West Bank, but Israel retook the land during the 1967 war against its Arab neighbors. Peace talks in the late 1980s led to an agreement between Israel and the Palestine Liberation Organization (PLO) for limited Palestinian self-rule in the West Bank.

Which member of the United Nations could not serve on the **Security Council**?

In order to sit on one of the four rotating spots on the nine-member Security Council, a member of the United Nations must also be a member of one of five regional groups. Israel was not eligble until 2000 because of its hostile relations with Arab nations.

Why is **Cyprus divided**?

The island of Cyprus, part of the European Union, is located in the eastern Mediterranean Sea. It became independent from the United Kingdom in 1960. In 1974, a coup occurred that overthrew the president. Turkey invaded the island and succeeded in taking control of its northern half. This became the Turkish Republic of Northern Cyprus, which is not internationally recognized as an independent country. Southern Cyprus remained an independent country—the Republic of Cyprus. Currently, a U.N. peacekeeping force of 938 soldiers monitors the cease-fire line between the two areas.

When did the **Iraq War begin?**

The war, which has been called the Second Persian Gulf War and Operation Iraqi Freedom, began on March 20, 2003. Although President George W. Bush proclaimed the end of the war aboard the aircraft carrier *U.S.S. Abraham Lincoln* on May 1, 2003, declaring the "mission accomplished," the war was still being waged five years later and with no end in sight.

The United States has been increasingly involved in military actions in the Middle East, including an ongoing invasion of Iraq. U.S. Army humvees are seen here at a military camp in Iraq.

Why did the **United States invade Iraq**?

The administration under President George W. Bush claimed that Iraq, led by Saddam Hussein, was a threat to the region because it believed that Iraq was developing weapons of mass destruction, including nuclear and biological weapons. This was later proven false by many investigators from both the U.S. and United Nations. The weapons programs had stopped at the time of the first Gulf War in 1991. The Bush administration also claimed that Iraq was somehow responsible for the 9/11 terrorist attacks, and the president repeated these claims in the news media for many years. These allegations have also been proven false; there was never a connection between the perpetrators of the 9/11 attacks and the country of Iraq.

How many **people** have been **killed** during the **U.S. war** with **Iraq**?

Numbers and methods used to compute the statistics of dead and wounded in Iraq vary considerably, from a high of over 1.2 million people (Opinion Research Business), which is more than the Rwanda genocide, to a low of 654,000 people (Lancet Survey). As of 2008, over 4,000 Americans have lost their lives in Iraq since the war began.

How much **money** have U.S. taxpayers spent on the **war in Iraq**?

The war in Iraq has cost the U.S. economy nearly $2 trillion.

What **nationalities** were the **9/11 terrorists**?

Out of the 19 terrorists who participated in the 9/11 attacks, 15 of them were from Saudi Arabia, two were from the United Arab Emirates, one was from Yemen, and one was from Morocco.

Who is **Osama bin Laden**?

Osama bin Laden is a Saudi Arabian militant and the leader of the terrorist organization Al Qaeda (the Base in Arabic). Al Qaeda is a Sunni Muslim organization that was created from veterans of the struggle against the Soviet occupation of Afghanistan. Bin Laden is the son of a wealthy Saudi Arabian family, who made their fortune in the Saudi civil construction business. He was expelled from Saudi Arabia, moved to Sudan, and now is alleged to live in the border region between Afghanistan and Pakistan. He is thought to be the mastermind behind the attacks on the World Trade Center in New York and the Pentagon in Virginia.

Bin Laden used his personal wealth to fund mujadeen insurgents in Afghanistan after the Soviet withdrawal in 1989, and he organized bombings against various American interests, including the American embassies in Tanzania and Kenya in 1998. He demands the end of foreign influence in Muslim countries and the creation of a new Muslim caliphate. He has been waging a jihad ("holy war," although many Muslims say this is a misuse of the term) to achieve his aims through violence and the killing of both civilians and military personnel. Bin Laden is on the FBI's most wanted list. A reward of $25 million for information leading to his arrest has been offered.

PEOPLE, COUNTRIES, AND CITIES

What are the **largest cities** in the Middle East?

Cairo, Egypt, is by far the largest city in the Middle East, with over 14.45 million people. Istanbul, Turkey, is an urban area with almost 11.3 million residents; and Tehran, Iran, is home to 7.7 million people.

Who lives in the **City of the Dead**?

There lies on the outskirts of Cairo, Egypt, a very old cemetery filled with mausoleums, memorials, and mosque-shaped tombs and shrines. Because of severe overcrowding in Cairo, squatters now live inside these memorials to the dead, in what is called by many the City of the Dead. Recently, the Egyptian government provided the area with electricity and water. Built as a home for the dead, this cemetery is now becoming a home for the living.

A view of the ancient section of Cairo, Egypt.

What does the suffix **"stan"** mean?

The suffix "stan" (as in Afghanistan) means nation or land. Therefore, Afghanistan literally means "land of the Afghans."

How many **countries end in the suffix "stan"**?

There are eight. Six are former republics of the Soviet Union: Kazakhstan, Turkestan, Turkmenistan, Kyrgyzstan, Tajikistan, and Uzbekistan. Afghanistan and Pakistan are the two others.

Where is **Asia Minor**?

Asia Minor is the term used for the larger part of Turkey that lies in Asia, east of the Strait of Bosporus. This part of Turkey is also known as Anatolia.

Where is the **Maghreb**?

From the Arabic word for "place of sunset" or "western," the Maghreb is a region of Northwest Africa composed of Morocco, Algeria, and Tunisia. Sometimes Libya is also included. The name "Maghreb" remains from the term used by the Islamic empire for this region.

245

Where in Egypt do **most Egyptians live**?

About 95 percent of Egypt's population lives within 12 miles (19 kilometers) of the Nile River. Since the rest of Egypt mainly consists of desert, the remaining residents live scattered across the country, primarily near oases or along the coast.

Who are the **Kurds**?

The Kurds are a Middle Eastern people who have no country of their own. They live in southeastern Turkey, northern Iran, northern Iraq, and other nearby countries. Since the Kurds have been persecuted by Iraq, the United Nations established a security zone for them north of 36 degrees north in Iraq. The Kurds hope that one day they will have their own nation-state in the region where they now live.

Which country is composed of **seven sheikdoms**?

The country of United Arab Emirates, located on the Arabian Peninsula, is composed of seven sheikdoms (also known as emirates). The seven emirates were defended by the United Kingdom in the nineteenth century, but merged to become an independent country in 1971. The seven emirates (Abu Dhabi, Ajman, Al Fujayrah, Ash Sharqah, Dubayy, Ra's a' Khaymah, and Umm al Qaywayn) are presently quite autonomous.

How many people **speak Arabic** in the world?

Approximately 200 million people are native speakers of Arabic, and an additional 200 million are non-native speakers.

Which Middle Eastern country has the highest percentage of **urban population**?

Ninety percent of Israel's population lives in urban areas, and its density is third only to those of Bahrain and Lebanon, at 702 people per square mile.

Which **Middle Eastern country** has the **largest population**?

Egypt is the most populous country in the Middle East, with over 75 million people. Just behind Egypt is Iran, with over 71 million people, followed by Turkey, with over 70 million people.

Which country has the most **Azeri-speaking people**?

Surprisingly, between 12 and 20 million Azerbaijanis (Azeri is their native language) live in Iran, considerably more than the eight million who live in Azerbaijan, Iran's neighbor to the north. It is difficult to know the exact number of Azeri speakers, since the government of Iran doesn't release such figures.

Which Middle Eastern country currently has the **longest-ruling leader**?

Hosni Mubarak has been the head of state of Egypt since 1981.

Who are the **Bedouin**?

The Bedouins are nomadic tribes that have inhabited areas of the Middle East and northern Africa for centuries. Grazing their goat and camel herds, the Bedouins travel wide distances and frequently cross international borders. Many Middle Eastern countries are attempting to halt the crossings. If these countries are successful, the Bedouin culture would be dramatically changed and perhaps even completely destroyed.

Which country has a **flag of only a single color** and no design or emblem?

The Libyan flag is green without a design. This North African country is located between Egypt and Algeria and has a population of over six million people.

Where is **Greater Syria**?

Some Syrians, Lebanese, Jordanians, and Palestinians see the boundaries of their countries as artificially imposed by colonial rulers, and hope to see a united Greater Syria composed of the four areas.

What was the **United Arab Republic**?

In 1958, non-neighbors Egypt and Syria united to form the country known as the United Arab Republic. This lasted only until 1961, when Syria decided to become independent. Even after the dissolution, Egypt kept the name United Arab Republic for a decade.

How do you **drive to the island** country of Bahrain?

Bahrain is located 15 miles (24 kilometers) off the coast of Saudi Arabia and has been connected by a four-lane causeway (bridge) to Saudi Arabia since 1986. Saudi Arabia paid for the construction of the causeway.

Where does the **name Saudi Arabia** come from?

The name comes from the ruling Al-Saud family, who have been in control of the country since 1932. Saudi Arabia is the world's only country named after its royal family.

Which **Middle Eastern country** has the most **tourism**?

Turkey has the most tourist arrivals, with more than 20 million people coming to visit each year. Because of the annual Haj, or Pilgrimage, to Saudi Arabia's religious sites of

Mecca and Medina, more than nine million people visit there each year. Over eight million people visit Egypt each year.

AFRICA

PHYSICAL FEATURES AND RESOURCES

Where is the **Kalahari Desert**?

The Kalahari Desert covers much of Botswana and Namibia. The Kalahari Desert is one of the largest deserts in the world, over 100,000 square miles (259,000 square kilometers), and lies on a high plateau at 3,000 feet (914 meters).

How big is the **Sahara Desert**?

The Sahara is the world's largest desert. It covers more than 3.5 million square miles (9 million square kilometers) in northern Africa. The Sahara receives less than 10 inches (25 centimeters) of rain each year but contains hundreds of individual oases. The elevation in the Sahara Desert ranges from 100 feet (30.5 meters) below, to more than 11,000 feet (3,353 meters) above sea level. There are people who live in the Sahara, mostly at or near oases.

Why are the **Blue Nile and White Nile** Rivers both called Niles?

The Nile River begins as two separate rivers—the White Nile and the Blue Nile. The White Nile begins its flow from Lake Victoria in Eastern Africa and the Blue Nile originates in the Ethiopian Highlands. The Blue Nile and the White Nile converge in Khartoum, the capital of the Sudan, and form the Nile River, which continues on to the Mediterranean.

Was the **flooding of the Nile** predictable before dams were built?

The summer floods of the Nile River were so predictable that the Egyptian calendar was based on their rise and fall. Flooding on the Nile occurred from late June until

A view of Amboseli National Park in Kenya, with Tanzania's Mt. Kilimanjaro in the background. The snow on top of the mountain is disappearing due to global warming.

late October. The floods brought nutrients and sediments beneficial to the nearby agricultural lands, making farming productive throughout the remainder of the year. Measuring scales called "nilometers" were placed along the river, and not only measured the river height but also served as a calendars.

Is there any **permanent ice** in **Africa**?

Though located within three degrees of the equator, there is ice year-round at the top of Mt. Kilimanjaro, 19,340 feet (5,895 meters) in the air.

Has **global warming** affected the ice on **Mt. Kilimanjaro**?

Global warming has indeed affected the glacial ice on top of Mt. Kilimanjaro in Tanzania and has caused as much as 80 percent of the ice to permanently disappear. Scientists believe that by the year 2020 there will be no more snow on top of the mountain.

How was **Mt. Kilimanjaro** formed?

Mt. Kilimanjaro, the tallest mountain in Africa at 19,340 feet (5,895 meters), was formed as a volcano, but is now dormant. The mountain is located in northeastern Tanzania and was first climbed in 1889 by German geographer Hans Meyer and Austrian climber Ludwig Purtscheller.

What is the Great Rift Valley?

In eastern Africa there lies a deep valley known as the Great Rift Valley. Over 3,000 miles (4,800 kilometers) in length and 20 to 60 miles (32 to 97 kilometers) wide, the rift spans the length of Africa and was formed by tectonic plates sliding apart. In 10 million years, eastern Africa will detach from the rest of Africa along this rift and form its own subcontinent.

What is the world's **longest freshwater lake**?

Lake Tanganyika is 420 miles (676 kilometers) long, the longest freshwater lake in the world, but it's only between 10 and 45 miles (16 to 72 kilometers) wide. It lies on the border between the Democratic Republic of Congo and Tanzania. It is also the world's second-deepest lake, with a maximum depth of 4,710 feet (1,436 meters).

What is Africa's **largest lake**?

Located in eastern Africa, Lake Victoria is Africa's largest lake, with a surface area of approximately 27,000 square miles (70,000 square kilometers). It is also the world's second-largest freshwater lake, after Lake Superior. Lake Victoria is bordered by Uganda, Kenya, and Tanzania. The lake was named by British explorer John Hanning Speke, the first European to see the lake (in 1858), in honor of the reigning British queen.

What is the **Bight of Bonny**?

Also called the Bight of Biafra, the Bight of Bonny is a bay located in the Gulf of Guinea, near Cameroon. The word "bight" is Old English for bay.

How did a **lake kill** more than 2,000 people?

In August 1986, Cameroon's Lake Nios, which sits upon a volcanic vent, produced an eruption of carbon dioxide and hydrogen sulfide gasses. The cloud of acidic gas blew into nearby villages, killing more than 2,000 people while they slept.

What country in Africa has the world's **highest minimum elevation**?

Lesotho, located in the mountains of South Africa, has an absolute minimum elevation of 4,530 feet (1,381 meters) above sea level in the valley of the Orange River, but almost all of the land in the country lies above 6,000 feet (1,829 meters).

251

Which **African country** is the world's **leading producer** of **cocoa beans**?

Côte d'Ivoire, also known as the Ivory Coast (named for the large amount of ivory collected from elephant herds that once roamed the country), produces 37.4 percent of the world's cocoa beans, the beans used to make chocolate. It produces more than 1.3 million tons each year. Ghana produces another 20 percent of the world's total, and both Cameroon and Nigeria contribute 10 percent to the total production of cocoa in the world. This means that nearly 70 percent of the world's cocoa beans are grown in Africa.

Which country produces the **most gold**?

South Africa's gold mines yield 28 percent of the world's gold annually. In 1886, gold was first discovered in South Africa at the mines near Witwatersrand, which is now South Africa's largest gold-producing area. In 2007, China overtook South Africa by producing 276 tons of gold, beating South Africa for the title of biggest gold producer by just four tons of gold. This is the first time that South Africa has not been first in gold production since 1905.

What **fish**, once thought to be **extinct**, suddenly appeared near Comoros?

In December 1938, a fisherman discovered a very strange-looking fish near the Comoros islands. Scientists discovered that this lobe-finned fish was a coelacanth, thought to have been extinct for over 70 million years. This species continues to live in the waters off of the Comoros islands, located between Madagascar and continental Africa.

HISTORY

How did the **Berlin Conference** of 1884 help expedite the colonization of Africa?

In 1884, a time when most of Africa had yet to be colonized, 13 European countries and the United States met in Berlin to divide Africa up among themselves. At the conference, geometric borders were drawn across Africa, completely ignoring the continent's cultural differences. Though the borders were established over 100 years ago, they still exist today and are responsible for much of the turmoil and trauma among the now-independent African countries.

How many African countries were **independent in 1950**?

In 1950 there were only four independent countries on the continent: Egypt, South Africa, Ethiopia, and Liberia. All other countries gained their independence in the decades that followed. Most recently, Eritrea became independent of Ethiopia in 1993.

What is significant about the country of Comoros?

The Union of Comoros, an island nation off the coast of Southeast Africa, consists of four main islands. In 1975, when a referendum was held to become independent from France, one of the four islands, Mahore (Mayotte), decided to remain a colony of France. To this day it is still administered by France.

Which countries **colonized Africa**?

Seven European countries colonized Africa: Belgium, France, Italy, Germany, Portugal, Spain, and the United Kingdom.

How many **African countries** were **colonized by Italy**?

Italy established colonies in what are now Libya, Eritrea, and Somalia.

Of all countries in Africa, which **countries** were **never colonized**?

Only three countries escaped the onslaught of European colonial occupation: Liberia, Ethiopia, and Sudan.

Who was **Haile Selassie**?

Haile Selassie was Emperor of Ethiopia, and was one of the great leaders of the twentieth century, as well as one of the most prominent figures in modern African history. He served as head of Ethiopia from 1930 to 1974, and was responsible for modernizing Ethiopia, forming the African Union, and uniting Africans both in Ethiopia and throughout the continent. He was a committed internationalist, and because of this the United Nations made Ethiopia a charter member when the U.N. was formed in 1945. Haile Selassie is revered by the Rastafarian religion as a living god.

Who is **Nelson Mandela**?

Nelson Mandela is a South African leader, who was the head of the African National Congress and its armed resistance group, Umkhonto we Sizwe. This group fought against the apartheid policies of South Africa. Mandela spent 27 years imprisoned on Robben Island for his political activities. Upon his release in 1990, Mandela asserted a policy of reconciliation and negotiation to end the racist policies of apartheid. He was the first fully representative, democratically elected president of South Africa, serving as head of state from 1994 to 1999. He was awarded the Nobel Peace Prize for his work in 1993.

The prison on Robben Island, Cape Town, South Africa, where Nelson Mandela was held for 27 years as a political prisoner.

What was **apartheid**?

South Africa's legalized form of racial discrimination was known as apartheid (separateness), and it classified individuals into one of four ethnic groups: white, black, Coloured (mixed race), and Asian. Apartheid laws limited where different groups could live and where they could work. Apartheid was repealed in 1990.

Why are there so many **starving people** in Ethiopia?

Droughts in the 1970s and 1980s, which were especially severe from 1984 to 1986, destroyed Ethiopian agriculture. Though relief food was shipped to Ethiopia, internal political corruption kept the food from reaching the starving victims. During the 1980s, approximately one million people died of starvation in Ethiopia.

What was the **Nazis' plan** for **Madagascar**?

In 1940, after the French defeat at the hands of the Germans, the Nazis developed a plan to forcibly relocate European Jews to the French colony of Madagascar. This plan never materialized, however, and the Nazis continued the extermination of Jews throughout Europe.

What is the **Organization of African Unity**?

Founded in 1963, the Organization of African Unity helped strengthen and defend African unity across the continent. The 53 member-countries sought to increase

How did American slavery help found Liberia?

In 1822 the American Colonization Society succeeded in founding a colony on the western coast of Africa for freed American slaves. The colony was named Liberia from the Latin word "liber," which means "free." This colony became independent in July 1847, and was Africa's first independent country.

development and economic unity within and between member-countries. It was disbanded in 2002 by South African President Thabo Mbeki when the new African Union was created with the same 53 member states. The African Union seeks to integrate political and socio-economic interests of the member states, speaking with a common voice on issues involving Africa. Its headquarters is in Addis Ababa, Ethiopia.

Which **country** in Africa is the only country that is **not a member** of the **African Union**?

Morocco withdrew from the previous Organization of African Unity in 1984 over a dispute with members on the recognition of Sahrawi by the organization. It did not become a member of the newly formed African Union.

PEOPLE, COUNTRIES, AND CITIES

How many **countries** are there in Africa?

Africa is home to more than a quarter of the world's countries. There are 47 independent countries on the continent itself and 53 if you include the nearby island countries of Comoros, Madagascar, Seychelles, Cape Verde, Mauritius, and São Tome and Principe.

How many African countries are **landlocked**?

Only 15 of Africa's 47 continental countries have no access to seas or oceans. These include Botswana, Burkina Faso, Burundi, Central African Republic, Chad, Ethiopia, Lesotho, Malawi, Mali, Niger, Rwanda, Swaziland, Uganda, Zambia, and Zimbabwe.

What is the **Sahrawi Arab Democratic Republic**?

Although not a fully recognized country because of an international border dispute, Sahrawi claims sovereignty over all of the territory of the former Spanish colony of

Western Sahara. A part of the territory of Western Sahara is considered by Morocco to be a province of Morocco. Sahrawi is a member of the African Union.

What is Africa's **most populous** country?

Nigeria, with a population of 148 million people, is Africa's most populous country. Nigeria is also the world's eighth-most populous country. If Nigeria's population continues to increase at the present rate, the population will increase to 214 million by the year 2022! The average Nigerian woman gives birth to 6.1 children in her lifetime.

How many **provinces** are there in **South Africa**?

In 1994, several new provinces (states) were created. South Africa is now divided into nine provinces: Eastern Cape, Free State, Gauteng, KwaZulu-Natal, Limpopo, Mpumalanga, Northern Cape, North-West, and Western Cape. The nine provinces incorporate homelands of indigenous peoples.

How many **capital cities** does **South Africa** have?

South Africa has three capitals: Pretoria is the administrative capital, Bloemfontein is the capital of the judiciary, and Cape Town is the legislative capital.

How many countries named **Congo** are there?

There are two countries named Congo, and just to make things a little more confusing, they're neighbors. The similarity between the names of the two—the Democratic Republic of the Congo and its western neighbor, the Republic of the Congo—makes distinguishing them a little difficult. In 1908 the Democratic Republic of the Congo was named Belgian Congo; in 1960 it was called Republic of the Congo; in 1964 it became People's Republic of the Congo; in 1966 it was called Democratic Republic of the Congo; in 1971 it was called Zaire; and finally, in 1997, again became the Democratic Republic of the Congo.

Where is **Timbuktu**?

Though the name is commonly used to refer to an extremely distant place, Timbuktu (or Tombouctou) is actually a town near the Niger River, in the African country of Mali. It has a population of about 30,000 and is a major salt-trading post for Saharan Desert camel caravan routes.

Who owns **Walvis Bay**?

When Namibia gained independence from South Africa in 1990, South Africa kept control of Walvis Bay, located approximately 400 miles (644 kilometers) north of the South African border. The excellent harbor and deep-water port of Walvis Bay was retained for four years, until it was returned to Namibia on March 1, 1994.

Where is the **Barbary Coast**?

The Barbary Coast refers to the countries in Northern Africa that are located along the Mediterranean Sea: Morocco, Algeria, Tunisia, Libya, and Egypt. Though the name Barbary Coast comes from the Berber people who inhabit the region, it is best known for its association with pirates from the sixteenth through the nineteenth centuries.

Why is **Cabinda** separate from **Angola**?

Cabinda, though a province of Angola, is separated from the bulk of Angola by approximately 25 miles (40 kilometers) of the Democratic Republic of Congo. In 1886, Belgium gave this tiny area to Angola. Cabindans have recently taken up arms against Angola in the hope of obtaining independence.

What was **Zimbabwe**'s **previous name**?

In April 1980, the British colony of Rhodesia was granted independence and renamed itself Zimbabwe. Rhodesia had been named for South African businessman Cecil Rhodes.

Where is **Ouagadougou**?

Ouagadougou (pronounced wah-gah-dah-goo) is the capital of the west African country of Burkina Faso. The city, with half a million residents, is home to the University of Ouagadougou.

Which country lies completely **within South Africa**?

The tiny country of Lesotho, which gained independence from the British in 1966, is completely surrounded by South Africa. Nearly 40 percent of the male workers migrate to South Africa for employment.

Is **Equatorial Guinea** on the **equator**?

No, Equatorial Guinea's southernmost point is still one degree north of the equator. Though Equatorial Guinea is close to the equator, its southern neighbor, Gabon, is truly equatorial.

Where is the **Horn of Africa**?

The Horn of Africa is the eastern protrusion of Africa that includes Somalia, Ethiopia, and Djibouti. The easternmost tip of the Horn is called Gees Gwardafuy.

How many people lived on the islands of **Seychelles** before 1770?

None. The country, which is composed of 115 islands northeast of Madagascar, was first inhabited by the colonizing French in the 1770s. The British later gained control of the area and brought Africans to the islands. The islands gained independence in 1976.

What is **Caprivi's Finger**?

Caprivi's Finger is the name of the narrow strip of land that protrudes from the northeast corner of the otherwise compact country of Namibia. The land was acquired by German Chancellor Georg Leo von Caprivi from the British in 1890 in order for Namibia, known then as German Southwest Africa, to have access to the Zambezi River. This strip of land is approximately 300 miles (480 kilometers) long but no more than 65 miles (105 kilometers) at its widest.

What language do people in **Madagascar speak**?

The people of Madagascar speak Malagasy. Malagasy is closely related to the languages spoken in Indonesia and Polynesia, rather than to any African language. The people of Madagascar are of Indonesian and Malaysian decent, having migrated there nearly 2,000 years ago.

Where was Kunta Kinte from?

Kunta Kinte, the protagonist of Alex Haley's novel *Roots,* was from Gambia. Though Gambia follows 200 miles (322 kilometers) of the Gambia River, it is a very thin country, averaging only 12 miles (19 kilometers) in width. Gambia lies entirely within Africa's smallest country, Senegal.

How many **official languages** are there in **South Africa**?

The country has 11 official languages: Afrikaans, English, Ndebele, Pedi, Sotho, Swazi, Tsonga, Tswana, Venda, Xhosa, and Zulu.

Which African country has **Spanish** as an official language?

Equatorial Guinea, which is composed of five islands in the Gulf of Guinea and a tiny sliver of land between Cameroon and Gabon, was a colony of Spain until 1968 and has kept Spanish as its official language. The capital city of Malabo is located on the island of Bioko, previously named Fernando Poo.

Where is the world's **largest church**?

The Basilica of Our Lady of Peace in Yamoussoukro, Ivory Coast, is the world's largest church, covering 100,000 square feet (9,290 square meters). Built in 1989 by President Felix Houphouet-Boigny, the church has seating for 18,000 people. In 1983, Houphouet-Boigny relocated the capital of the Ivory Coast from Abidjan to his hometown of Yamoussoukro.

How prevalent is **AIDS in Africa**?

Approximately 22.5 million Africans have AIDS, which makes up 64 percent of the world's AIDS cases. Almost all AIDS transmission in Africa is through heterosexual intercourse. Approximately 2.3 million have died of AIDS-related illnesses in Africa.

What is the name of the **currency of Botswana**?

The south African country of Botswana, consisting primarily of the Kalahari Desert, uses the pula, which means "rain," as their currency.

What is one of the only countries in the world to provide **constitutional protection to gays, lesbians, and bisexuals**?

South Africa's 1996 constitution protects gays, lesbians, and bisexuals against discrimination in both the public and private sectors. The "Equality Clause" in the Bill of

A building at the Murambi Technical School in Rwanda remains as a memorial to the 40,000 people who were killed there during the 1994 genocide.

Rights protects people from discrimination based on race, gender, pregnancy, marital status, ethnic or social origin, color, sexual orientation, age, disability, religion, conscience, belief, culture, language, and birth.

What was the **Rwanda genocide**?

The Rwanda genocide was the systematic killing of hundreds of thousands of people belonging to the Tutsi minority ethnic group by the majority Hutus. Most of the killing occurred within a span of 100 days in the summer of 1994. It was carried out by two Hutu extremist military groups. At least 500,000 people were immediately killed, and as many as one million people lost their lives by the end of the massacre. The genocide had its roots in the Rwanda Civil War, which pitted the majority Hutus against the Uganda-supported Tutsi minority.

OCEANIA AND ANTARCTICA

OCEANIA

What is **Oceania**?

Oceania is the region in the central and southern Pacific that includes Australia, New Zealand, Papua New Guinea, and the islands that compose Polynesia, Melanesia, and Micronesia.

Who **owns all of those islands** in the Pacific Ocean?

There are hundreds of islands in the Pacific Ocean. While many are part of independent countries, others remain colonies or vestiges of colonial empires.

What is **Micronesia**?

The region known as Micronesia consists of islands east of the Philippines, west of the International Date Line, north of the equator, and south of the Tropic of Cancer. Approximately 600 of the islands in this area united in 1986 to form an independent country, the Federated States of Micronesia. Though 600 islands joined the Federation, there are approximately 1,500 other islands in the region, including the independent countries of the Marshall Islands, Kiribati, and Palau.

What is **Polynesia**?

Polynesia consists of islands in the region bounded by Hawaii in the north and New Zealand in the southwest. The region includes the countries of Samoa, Tonga, and Tuvalu. Also located in this area is the colony of French Polynesia, which includes Tahiti and 117 other islands and atolls.

The Great Barrier Reef near Australia is the largest coral reef in the world, as well as a protected marine park that draws tourists from around the globe.

What is **Melanesia**?

Located northeast of Australia, Melanesia is a small region that lies south of the equator and west of 180 degrees longitude, but excludes New Zealand. Melanesia includes the countries of Vanuatu and Fiji as well as the Solomon Islands and New Caledonia.

What is a **coral reef**?

Coral reefs are formed by the accumulation of calcium carbonate that comes from the external skeletons of tiny animals called coral polyps. The polyps live in shallow, warm water, and thus congregate around islands in the tropics, where coral reefs are abundant.

Where is the **largest coral reef** in the world?

The Great Barrier Reef is the largest coral reef in the world. Located just off the northeastern coast of Australia, it extends for over 1,200 miles (1,930 kilometers). Much of the area is now protected as a marine park.

What is an **atoll**?

In addition to reefs, coral can also form atolls. Atolls are formed when a volcano, around which coral often grows, erodes away, leaving a circular wall of coral with a lagoon at the center.

How did the Guano Island Act help fertilize America?

In 1856, the United States Congress passed the Guano Island Act, which allowed the United States to take possession of any unclaimed island that contained guano. Guano, the excrement of sea birds, was mined for use as fertilizer before the widespread use of chemical fertilizers. Beginning in 1857, the Baker and Howland Islands, located southwest of Hawaii, were mined by the United States until their guano was depleted in 1891.

How did the **bikini** get its name?

In 1946, the United States began to test atomic weapons on the Bikini Atoll in the Marshall Islands. It was also in the late 1940s when the two-piece bathing suit made its debut and took its name from the intensely publicized Bikini Atoll.

How do the people on the **tiny islands** in the Southern Pacific Ocean **go shopping**?

Most basic goods can be bought on each populated island, and larger items can be shipped via airplane. When residents need to travel between islands, they often take to the air. Each island has an airstrip of adequate length for its own transportation needs. In the past, inhabitants used boats as their primary means of transportation between islands.

Where did **Gauguin** live?

The French painter Paul Gauguin moved to Tahiti in 1891 and later to other islands in French Polynesia to escape European civilization.

Which country has **more languages** than any other in the world?

Papua New Guinea is home to more than 700 different languages. The most common languages spoken there are Motu and pidgin English.

Where did the **mutineers of the Bounty** land?

In 1789, members of the crew of the HMS *Bounty* mutinied. After having dropped off 19 other members of the crew, including Captain William Bligh, the mutineers landed on the uninhabited island of Pitcairn. While the captain and his loyal crew successfully returned to England, the mutineers established a community composed of nine male mutineers, six male Polynesians, and 12 female Polynesians who had also been on board the *Bounty*. In 1856, approximately 200 of the mutineers' descendants voluntarily moved from Pitcairn Island to Norfolk Island because of overpopulation.

Where did **Charles Darwin** develop his theory of natural selection?

Charles Darwin graduated from Cambridge University in 1831 and spent the next five years of his life as a naturalist on board the HMS *Beagle*. The *Beagle* traveled around the world, including among its destinations the Galapagos Islands, west of South America, where Darwin spent six weeks collecting data from which he developed his theories of natural selection, published in *Origin of the Species* in 1859.

AUSTRALIA

How did **Australia** get its **name**?

Long throughout history there remained an assumed, yet completely undiscovered, land called Terra Australis Incognita, or unknown southern land. As early as the fourth century B.C.E. Aristotle believed that an extremely large continent, located in the Southern Hemisphere, lay undiscovered and would complete the symmetry of the land masses. For centuries, this unknown landmass remained a treasured legend and often appeared on maps in varied sizes and shapes. When the territory now known as Australia was discovered in the early seventeenth century, no one believed that this was the famed Terra Australis Incognita. During the early seventeenth century, the western coast of this territory was named New Holland and claimed for the Netherlands; in 1770 James Cook claimed the east coast of this territory for England and called it New South Wales. It wasn't until 1803 that Matthew Flinders circumnavigated this territory and proved that it was a continent and was the long sought-after Terra Australis Incognita. Finally, in the early nineteenth century, nearly two centuries after having been discovered, this land was finally named Australia.

Who are the **Aborigines**?

The Aborigines are the indigenous inhabitants of Australia, having migrated from Southeast Asia approximately 40,000 years ago. In the late eighteenth century, when European colonization began, there were over 300,000 Aborigines in Australia. Many were killed by European diseases and abuses, and by 1920 there were only 60,000 Aborigines remaining. Like the Maori of New Zealand, Australia's Aborigine population rebounded in the late twentieth century, and now stands at over 200,000. Most Aborigines now live in urban areas and are gaining political support and benefits.

Was Australia really used as a **penal colony**?

Yes, approximately two-thirds of Australia's initial settlers were convicts from Great Britain. From 1788 to 1850, when Australia was used as a penal colony, approximately 160,000 prisoners were sent to the continent. Though Britain stopped sending prisoners to Australia in 1850, free colonists began arriving with the first ship of convicts.

Where is the outback?

The outback is the general term used to describe the remote interior of Australia. Most of Australia's population is concentrated on the coast, since the interior is extremely dry and barren.

What is the **capital of Australia**?

The capital of Australia is Canberra, which is located in a federal territory (similar to Washington, D.C.) within the Australian state of New South Wales. When Australia was founded in 1901, both Sydney and Melbourne wanted to become the capital city. In 1908 it was ultimately decided that a brand-new capital city would be built and located away from the coast. Canberra is the largest city in Australia that is not on the coastline (population 334,000).

What's the **big red rock** in the middle of Australia?

The big, red, sandstone monolith in the center of Australia is called Uluru (its indigenous name; it was previously called Ayers Rock). It is the world's largest monolith, approximately 1.5 miles (2.4 kilometers) wide and one-fifth of a mile (1,100 feet [335 meters]) high.

Are **kangaroos** native to Australia?

Yes, kangaroos are native to Australia. Kangaroos range in size from giant kangaroos (five feet [1.5 meters] tall) to tiny, rat-sized kangaroos, called potoroos.

Is the **Tasmanian devil** a real animal?

The Tasmanian devil is a real animal, though it resembles few characteristics of its cartoon counterpart. The real Tasmanian devil is a carnivorous marsupial that lives on the island of Tasmania, just southeast of the Australian mainland.

Is Australia the **smallest continent**?

Despite being the sixth-largest *country* in the world, Australia is the smallest continent in the world. Australia is approximately three million square miles (7.8 million square kilometers) in area, just a little smaller than Brazil.

Which country is the world's leading **bauxite** producer?

Australia mines more bauxite than any other country in the world, producing 33 percent of the world's supply annually. An ore of aluminum, bauxite is especially prevalent in Australia's Darling Range, located in the southwestern part of the country.

Which country is the world's leading **lead** producer?

Australia produces approximately 23.5 percent of the world's lead annually. Most of the lead is mined near Mt. Isa in the northeast area of the country and near Broken Hill in the southeast.

How much **beef** does Australia export?

Australia is responsible for more than 26 percent of the world's beef exports—1,203,000 metric tons.

What is a **boomerang**?

The boomerang was developed as a hunting tool by the Aborigines of Australia. There are two types of boomerangs—those that return and those that do not return. The returning boomerang is used to kill small animals; the non-returning boomerang is used to kill large game or enemies.

NEW ZEALAND

Who are the **Maori**?

The Maori are the indigenous inhabitants of New Zealand. Around the ninth century, the Maori arrived in New Zealand from other Pacific Islands. In 1769 there were over 100,000 Maori, but their population decreased significantly (to 40,000) by the end of the nineteenth century, due to European colonization. In the twentieth century, the Maori population has expanded to 632,000 in New Zealand, with another 73,000 living in Australia. Maoris now comprise 15 percent of New Zealand's population.

A circa 1890, nineteenth-century painting of a Maori woman, the indigenous people of New Zealand (image courtesy of Alexander Turnbull Library).

What is **Aotearoa**?

Aotearoa is the Maori name for New Zealand. It means "The Land of the Long White Cloud."

What is the **Royal Flying Doctor Service**?

Created in 1928, the Royal Flying Doctor Service (RFDS) is a charity organization established to provide health care and

A view of the terrain in Wanaka on New Zealand's South Island (photo by Paul A. Tucci).

emergency services to the sparse population of Australia's outback. With 38 aircraft, the RFDS averages over 80 flights a day, helping more than 135,000 people per year.

Are there **more people or sheep** in New Zealand?

There are just over four million people in New Zealand, but nearly 56 million sheep. New Zealand has long been a leading exporter of wool.

Can you **ski** in **New Zealand**?

New Zealand's South Island is famous for having some of the best skiing in the southern hemisphere. The center of New Zealand skiing is Queenstown, which is a few hours by plane from capital city, Auckland. Ski season is July through September, when winter hits the countries of the southern latitudes.

Who are **Kiwis**?

One nickname for a New Zealander is "Kiwi," but kiwis are also a flightless bird and a type of fruit found in New Zealand. Kiwi birds, a national icon of New Zealand, have long thin beaks and lay eggs larger in proportion to their body size than any other bird. Most of the world's supply of kiwi fruit is also grown in New Zealand.

267

What is ANZUS?

In 1951, Australia, New Zealand, and the United States signed the Australia-New Zealand-United States (ANZUS) Treaty to protect each other militarily. In 1986, New Zealand banned nuclear weapons from its country and thereafter refused to allow U.S. nuclear-powered or nuclear-armed ships to dock in its harbors. New Zealand was summarily excluded from the Treaty.

What was the *Rainbow Warrior*?

In 1985, the Greenpeace ship *Rainbow Warrior* was in the Auckland, New Zealand, harbor when it exploded and sank, killing one Greenpeace staff member. It was later discovered that French secret agents planted bombs onboard the *Rainbow Warrior* in order to stop the organization from protesting French nuclear weapon tests in the Pacific. Following the incident, the French minister of defense and head of the secret service were forced to resign. New Zealand, a country very much opposed to nuclear weapons, maintained a poor relationship with France for many years following the bombing.

Which country was the world's **first welfare state**?

In 1936, New Zealand became the world's first welfare state by offering its citizens full social security and health benefits.

Which country first granted **women the right to vote**?

In 1893, New Zealand became the first country to give women the right to vote.

ANTARCTICA

How **thick is Antarctica's ice**?

Most of the ice in Antarctica is approximately one mile thick. Over 80 percent of the world's freshwater is stored in ice in Antarctica. Some have suggested that large chunks of ice be cut off from Antarctica and shipped to dry regions of the world, but this has yet to be done.

What **time** is it in **Antarctica**?

People based on the many research stations of Antarctica generally observe Greenwich Mean Time—that is, the same time zone as London, England.

Icebergs drift by a huge glacier in Antarctica.

Which continent has the **highest average elevation**?

The average elevation of Antarctica, approximately 8,000 feet (2,438 meters), is higher than that of any other continent. The highest point in Antarctica is Vinson Massif, with an elevation of 16,860 feet (5,139 meters).

How **dry** is Antarctica?

Though Antarctica is covered with ice, it is the driest continent on the planet. The ice in Antarctica has been there for thousands of years, and the continent receives less than two inches (5 centimeters) of precipitation annually—the Sahara Desert receives 10 inches (25 centimeters) each year.

Who **owns Antarctica**?

Though Antarctica is a cold, icy, barren territory, seven countries claimed portions of it in the early twentieth century. All of these claims were defined by lines of longitude, and problems arose as many of these claims overlapped. In 1959, the Antarctic Treaty was established, proclaiming that no additional claims could be made upon Antarctica and that the continent would be used solely for scientific purposes.

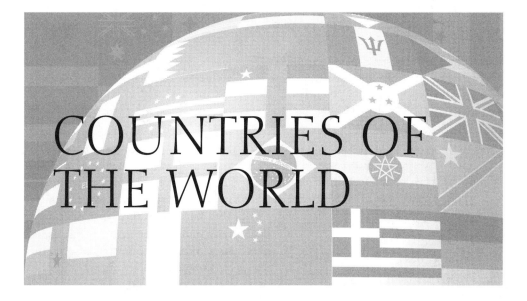

COUNTRIES OF THE WORLD

There are currently 195 countries in the world, and in this chapter you'll find key statistics on each one. You can use this chapter to find important geographic, political, and cultural information on countries mentioned in this book. The source of this information is the CIA's *World Factbook*.

Afghanistan

Long Name: Islamic Republic of Afghanistan

Location: Southern Asia, north and west of Pakistan, east of Iran

Area: 250,000 sq. mi. (647,500 sq. km)

Climate: arid to semiarid; cold winters and hot summers

Terrain: mostly rugged mountains; plains in north and southwest

Population: 32,738,376

Population Growth Rate: 2.626%

Birth Rate (per 1,000): 45.82

Death Rate (per 1,000): 19.56

Life Expectancy: 44.21 years

Ethnic Groups: Pashtun 42%, Tajik 27%, Hazara 9%, Uzbek 9%, Aimak 4%, Turkmen 3%, Baloch 2%, other 4%

Religion: Sunni Muslim 80%, Shi'a Muslim 19%, other 1%

Languages: Afghan Persian or Dari (official) 50%, Pashto (official) 35%, Turkic languages (primarily Uzbek and Turkmen) 11%, 30 minor languages (primarily Balochi and Pashai) 4%, much bilingualism

Literacy: 28.1%

Government Type: Islamic republic

Capital: Kabul

Independence: 19 August 1919 (from United Kingdom control over Afghan foreign affairs)

GDP Per Capita: $1,000

Occupations: agriculture 80%, industry 10%, services 10%

Currency: afghani (AFA)

Albania

Long Name: Republic of Albania

271

Location: Southeastern Europe, bordering the Adriatic Sea and Ionian Sea, between Greece in the south and Montenegro and Kosovo to the north

Area: 11,099 sq. mi. (28,748 sq. km)

Climate: mild temperate; cool, cloudy, wet winters; hot, clear, dry summers; interior is cooler and wetter

Terrain: mostly mountains and hills; small plains along coast

Population: 3,619,778

Population Growth Rate: 0.538%

Birth Rate (per 1,000): 15.22

Death Rate (per 1,000): 5.44

Life Expectancy: 77.78 years

Ethnic Groups: Albanian 95%, Greek 3%, other 2% (Vlach, Roma [Gypsy], Serb, Macedonian, Bulgarian) (1989 est.)

note: in 1989, other estimates of the Greek population ranged from 1% (official Albanian statistics) to 12% (from a Greek organization)

Religion: Muslim 70%, Albanian Orthodox 20%, Roman Catholic 10%

Languages: Albanian (official - derived from Tosk dialect), Greek, Vlach, Romani, Slavic dialects

Literacy: 98.7%

Government Type: emerging democracy

Capital: Tirana

Independence: 28 November 1912 (from the Ottoman Empire)

GDP Per Capita: $5,800

Occupations: agriculture 58%, industry 15%, services 27%

Currency: lek (ALL)

Algeria

Long Name: People's Democratic Republic of Algeria

Location: Northern Africa, bordering the Mediterranean Sea, between Morocco and Tunisia

Area: 919,590 sq. mi. (2,381,740 sq. km)

Climate: arid to semiarid; mild, wet winters with hot, dry summers along coast; drier with cold winters and hot summers on high plateau; sirocco is a hot, dust/sand-laden wind especially common in summer

Terrain: mostly high plateau and desert; some mountains; narrow, discontinuous coastal plain

Population: 33,769,668

Population Growth Rate: 1.209%

Birth Rate (per 1,000): 17.03

Death Rate (per 1,000): 4.62

Life Expectancy: 73.77 years

Ethnic Groups: Arab-Berber 99%, European less than 1%

Religion: Sunni Muslim (state religion) 99%, Christian and Jewish 1%

Languages: Arabic (official), French, Berber dialects

Literacy: 69.9%

Government Type: republic

Capital: Algiers

Independence: 5 July 1962 (from France)

GDP Per Capita: $6,700

Occupations: agriculture 14%, industry 13.4%, construction and public works 10%, trade 14.6%, government 32%, other 16%

Currency: Algerian dinar (DZD)

Andorra

Long Name: Principality of Andorra

Location: Southwestern Europe, between France and Spain

Area: 181 sq. mi. (468 sq. km)

Climate: temperate; snowy, cold winters and warm, dry summers

Terrain: rugged mountains dissected by narrow valleys

Population: 82,627

Population Growth Rate: 1.899%

Birth Rate (per 1,000): 10.59

Death Rate (per 1,000): 5.59

Life Expectancy: 82.67 years

Ethnic Groups: Spanish 43%, Andorran 33%, Portuguese 11%, French 7%, other 6%

Religion: Roman Catholic (predominant)

Languages: Catalan (official), French, Castilian, Portuguese

Literacy: 100%

Government Type: parliamentary democracy (since March 1993) that retains as its chiefs of state a coprincipality; the two princes are the president of France and bishop of Seo de Urgel, Spain, who are represented locally by coprinces' representatives

Capital: Andorra la Vella

Independence: 1278 (formed under the joint suzerainty of the French Count of Foix and the Spanish Bishop of Urgel)

GDP Per Capita: $38,800

Occupations: agriculture 0.3%, industry 20.3%, services 79.4%

Currency: euro (EUR)

Angola

Long Name: Republic of Angola

Location: Southern Africa, bordering the South Atlantic Ocean, between Namibia and Democratic Republic of the Congo

Area: 481,351.35 sq. mi. (1,246,700 sq. km)

Climate: semiarid in south and along coast to Luanda; north has cool, dry season (May to October) and hot, rainy season (November to April)

Terrain: narrow coastal plain rises abruptly to vast interior plateau

Population: 12,531,357

Population Growth Rate: 2.136%

Birth Rate (per 1,000): 44.09

Death Rate (per 1,000): 24.44

Life Expectancy: 37.92 years

Ethnic Groups: Ovimbundu 37%, Kimbundu 25%, Bakongo 13%, mestico (mixed European and native African) 2%, European 1%, other 22%

Religion: indigenous beliefs 47%, Roman Catholic 38%, Protestant 15%

Languages: Portuguese (official), Bantu and other African languages

Literacy: 67.4%

Government Type: republic; multiparty presidential regime

Capital: Luanda

Independence: 11 November 1975 (from Portugal)

GDP Per Capita: $7,800

Occupations: agriculture 85%, industry and services 15%

Currency: kwanza (AOA)

Antigua and Barbuda

Long Name: Antigua and Barbuda

Location: Caribbean, islands between the Caribbean Sea and the North Atlantic Ocean, east-southeast of Puerto Rico

Area: 171 sq. mi. (442.6 sq. km)

Climate: tropical maritime; little seasonal temperature variation

Terrain: mostly low-lying limestone and coral islands, with some higher volcanic areas

Population: 84,522

Population Growth Rate: 1.305%

Birth Rate (per 1,000): 16.78

Death Rate (per 1,000): 6.14

Life Expectancy: 74.25 years

Ethnic Groups: black 91%, mixed 4.4%, white 1.7%, other 2.9%

Religion: Anglican 25.7%, Seventh Day Adventist 12.3%, Pentecostal 10.6%, Moravian 10.5%, Roman Catholic 10.4%, Methodist 7.9%, Baptist 4.9%, Church of God 4.5%, other Christian 5.4%, other 2%, none or unspecified 5.8%

Languages: English (official), local dialects

Literacy: 85.8%

Government Type: constitutional monarchy with a parliamentary system of government

Capital: Saint John's

Independence: 1 November 1981 (from United Kingdom)

GDP Per Capita: $18,300

Occupations: agriculture 7%, industry 11%, services 82%

Currency: East Caribbean dollar (XCD)

Argentina

Long Name: Argentine Republic

Location: Southern South America, bordering the South Atlantic Ocean, between Chile and Uruguay

Area: 1,068,297 sq. mi. (2,766,890 sq. km)

Climate: mostly temperate; arid in southeast; subantarctic in southwest

Terrain: rich plains of the Pampas in northern half, flat to rolling plateau of Patagonia in south, rugged Andes along western border

Population: 40.482 million

Population Growth Rate: 1.068%

Birth Rate (per 1,000): 18.11

Death Rate (per 1,000): 7.43

Life Expectancy: 76.36 years

Ethnic Groups: white (mostly Spanish and Italian) 97%, mestizo (mixed white and Amerindian ancestry), Amerindian, or other non-white groups 3%

Religion: nominally Roman Catholic 92% (less than 20% practicing), Protestant 2%, Jewish 2%, other 4%

Languages: Spanish (official), Italian, English, German, French

Literacy: 97.2%

Government Type: republic

Capital: Buenos Aires

Independence: 9 July 1816 (from Spain)

GDP Per Capita: $13,100

Occupations: agriculture 1%, industry 23%, services 76%

Currency: Argentine peso (ARS)

Armenia

Long Name: Republic of Armenia

Location: Southwestern Asia, east of Turkey

Area: 11,484 sq. mi. (29,743 sq. km)

Climate: highland continental, hot summers, cold winters

Terrain: Armenian Highland with mountains; little forest land; fast flowing rivers; good soil in Aras River valley

Population: 2,968,586

Population Growth Rate: -0.077%

Birth Rate (per 1,000): 12.53

Death Rate (per 1,000): 8.34

Life Expectancy: 72.4 years

Ethnic Groups: Armenian 97.9%, Yezidi (Kurd) 1.3%, Russian 0.5%, other 0.3%

Religion: Armenian Apostolic 94.7%, other Christian 4%, Yezidi (monotheist with elements of nature worship) 1.3%

Languages: Armenian 97.7%, Yezidi 1%, Russian 0.9%, other 0.4%

Literacy: 99.4%

Government Type: republic

Capital: Yerevan

Independence: 21 September 1991 (from Soviet Union)

GDP Per Capita: $5,800

Occupations: agriculture 46.2%, industry 15.6%, services 38.2%

Currency: dram (AMD)

Australia

Long Name: Commonwealth of Australia

Location: Oceania, continent between the Indian Ocean and the South Pacific Ocean

Area: 2,967,896 sq. mi. (7,686,850 sq. km)

Climate: generally arid to semiarid; temperate in south and east; tropical in north

Terrain: mostly low plateau with deserts; fertile plain in southeast

Population: 21,007,310

Population Growth Rate: 1.221%

Birth Rate (per 1,000): 12.55

Death Rate (per 1,000): 6.68

Life Expectancy: 81.53 years

Ethnic Groups: white 92%, Asian 7%, aboriginal and other 1%

Religion: Catholic 26.4%, Anglican 20.5%, other Christian 20.5%, Buddhist 1.9%, Muslim 1.5%, other 1.2%, unspecified 12.7%, none 15.3%

Languages: English 79.1%, Chinese 2.1%, Italian 1.9%, other 11.1%, unspecified 5.8%

Literacy: 99%

Government Type: federal parliamentary democracy

Capital: Canberra

Independence: 1 January 1901 (federation of United Kingdom colonies)

GDP Per Capita: $37,300

Occupations: agriculture 3.6%, industry 21.2%, services 75.2%

Currency: Australian dollar (AUD)

Austria

Long Name: Republic of Austria

Location: Central Europe, north of Italy and Slovenia

Area: 32,382 sq. mi. (83,870 sq. km)

Climate: temperate; continental, cloudy; cold winters with frequent rain and some snow in lowlands and snow in mountains; moderate summers with occasional showers

Terrain: in the west and south mostly mountains (Alps); along the eastern and northern margins mostly flat or gently sloping

Population: 8,205,533

Population Growth Rate: 0.064%

Birth Rate (per 1,000): 8.66

Death Rate (per 1,000): 9.91

Life Expectancy: 79.36 years

Ethnic Groups: Austrians 91.1%, former Yugoslavs 4% (includes Croatians, Slovenes, Serbs, and Bosniaks), Turks 1.6%, German 0.9%, other or unspecified 2.4%

Religion: Roman Catholic 73.6%, Protestant 4.7%, Muslim 4.2%, other 3.5%, unspecified 2%, none 12%

Languages: German (official nationwide) 88.6%, Turkish 2.3%, Serbian 2.2%, Croatian (official in Burgenland) 1.6%, other (includes Slovene, official in Carinthia, and Hungarian, official in Burgenland) 5.3%

Literacy: 98%

Government Type: federal republic

Capital: Vienna

Independence: 976 (Margravate of Austria established); 17 September 1156 (Duchy of Austria founded); 11 August 1804 (Austrian Empire proclaimed); 12 November 1918 (republic proclaimed)

GDP Per Capita: $39,300

Occupations: agriculture 1.6%, industry 30.3%, services 68%

Currency: euro (EUR)

Azerbaijan

Long Name: Republic of Azerbaijan

Location: Southwestern Asia, bordering the Caspian Sea, between Iran and Russia, with a small European portion north of the Caucasus range

Area: 33,436 sq. mi. (86,600 sq. km)

Climate: dry, semiarid steppe

Terrain: large, flat Kur-Araz Ovaligi (Kura-Araks Lowland) (much of it below sea level) with Great Caucasus Mountains to the north, Qarabag Yaylasi (Karabakh Upland) in west; Baku lies on Abseron Yasaqligi (Apsheron Peninsula) that juts into Caspian Sea

Population: 8,177,717

Population Growth Rate: 0.723%

Birth Rate (per 1,000): 17.52

Death Rate (per 1,000): 8.32

Life Expectancy: 66.31 years

Ethnic Groups: Azeri 90.6%, Dagestani 2.2%, Russian 1.8%, Armenian 1.5%, other 3.9%

Religion: Muslim 93.4%, Russian Orthodox 2.5%, Armenian Orthodox 2.3%, other 1.8%

Languages: Azerbaijani (Azeri) 90.3%, Lezgi 2.2%, Russian 1.8%, Armenian 1.5%, other 3.3%, unspecified 1%

Literacy: 98.8%

Government Type: republic

Capital: Baku

Independence: 30 August 1991 (from Soviet Union)

GDP Per Capita: $8,000

Occupations: agriculture 41%, industry 7%, services 52%

Currency: Azerbaijani manat (AZN)

The Bahamas

Long Name: Commonwealth of The Bahamas

Location: Caribbean, chain of islands in the North Atlantic Ocean, southeast of Florida, northeast of Cuba

Area: 5,382 sq. mi. (13,940 sq. km)

Climate: tropical marine; moderated by warm waters of Gulf Stream

Terrain: long, flat coral formations with some low rounded hills

Population: 307,451

Population Growth Rate: 0.57%

Birth Rate (per 1,000): 17.06

Death Rate (per 1,000): 9.22

Life Expectancy: 65.72 years

Ethnic Groups: black 85%, white 12%, Asian and Hispanic 3%

Religion: Baptist 35.4%, Anglican 15.1%, Roman Catholic 13.5%, Pentecostal 8.1%, Church of God 4.8%, Methodist 4.2%, other Christian 15.2%, none or unspecified 2.9%, other 0.8%

Languages: English (official), Creole (among Haitian immigrants)

Literacy: 95.6%

Government Type: constitutional parliamentary democracy

Capital: Nassau

Independence: 10 July 1973 (from United Kingdom)

GDP Per Capita: $28,000

Occupations: agriculture 5%, industry 5%, tourism 50%, other services 40%

Currency: Bahamian dollar (BSD)

Bahrain

Long Name: Kingdom of Bahrain

Location: Middle East, archipelago in the Persian Gulf, east of Saudi Arabia

Area: 257 sq. mi. (665 sq. km)

Climate: arid; mild, pleasant winters; very hot, humid summers

Terrain: mostly low desert plain rising gently to low central escarpment

Population: 718,306

Population Growth Rate: 1.337%

Birth Rate (per 1,000): 17.26

Death Rate (per 1,000): 4.29

Life Expectancy: 74.92 years

Ethnic Groups: Bahraini 62.4%, non-Bahraini 37.6%

Religion: Muslim (Shi'a and Sunni) 81.2%, Christian 9%, other 9.8%

Languages: Arabic, English, Farsi, Urdu

Literacy: 86.5%

Government Type: constitutional monarchy

Capital: Manama

Independence: 15 August 1971 (from United Kingdom)

GDP Per Capita: $33,900

Occupations: agriculture 1%, industry 79%, services 20%

Currency: Bahraini dinar (BHD)

Bangladesh

Long Name: People's Republic of Bangladesh

Location: Southern Asia, bordering the Bay of Bengal, between Burma and India

Area: 55,598 sq. mi. (144,000 sq. km)

Climate: tropical; mild winter (October to March); hot, humid summer (March to June); humid, warm rainy monsoon (June to October)

Terrain: mostly flat alluvial plain; hilly in southeast

Population: 153,546,896

Population Growth Rate: 2.022%

Birth Rate (per 1,000): 28.86

Death Rate (per 1,000): 8

Life Expectancy: 63.21 years

Ethnic Groups: Bengali 98%, other 2% (includes tribal groups, non-Bengali Muslims)

Religion: Muslim 83%, Hindu 16%, other 1%

Languages: Bangla (official, also known as Bengali), English

Literacy: 43.1%

Government Type: parliamentary democracy

Capital: Dhaka

Independence: 16 December 1971 (from West Pakistan)

GDP Per Capita: $1,400

Occupations: agriculture 63%, industry 11%, services 26%

Currency: taka (BDT)

Barbados

Long Name: Barbados

Location: Caribbean, island in the North Atlantic Ocean, northeast of Venezuela

Area: 166 sq. mi. (431 sq. km)

Climate: tropical; rainy season (June to October)

Terrain: relatively flat; rises gently to central highland region

Population: 281,968

Population Growth Rate: 0.36%

Birth Rate (per 1,000): 12.48

Death Rate (per 1,000): 8.58

Life Expectancy: 73.21 years

Ethnic Groups: black 90%, white 4%, Asian and mixed 6%

Religion: Protestant 63.4% (Anglican 28.3%, Pentecostal 18.7%, Methodist 5.1%, other 11.3%), Roman Catholic 4.2%, other Christian 7%, other 4.8%, none or unspecified 20.6%

Languages: English

Literacy: 99.7%

Government Type: parliamentary democracy

Capital: Bridgetown

Independence: 30 November 1966 (from United Kingdom)

GDP Per Capita: $18,900

Occupations: agriculture 10%, industry 15%, services 75%

Currency: Barbadian dollar (BBD)

Belarus

Long Name: Republic of Belarus

Location: Eastern Europe, east of Poland

Area: 80,154 sq. mi. (207,600 sq. km)

Climate: cold winters, cool and moist summers; transitional between continental and maritime

Terrain: generally flat and contains much marshland

Population: 9,685,768

Population Growth Rate: -0.393%

Birth Rate (per 1,000): 9.62

Death Rate (per 1,000): 13.92

Life Expectancy: 70.34 years

Ethnic Groups: Belarusian 81.2%, Russian 11.4%, Polish 3.9%, Ukrainian 2.4%, other 1.1%

Religion: Eastern Orthodox 80%, other (including Roman Catholic, Protestant, Jewish, and Muslim) 20%

Languages: Belarusian, Russian, other

Literacy: 99.6%

Government Type: republic in name, although in fact a dictatorship

Capital: Minsk

Independence: 25 August 1991 (from Soviet Union)

GDP Per Capita: $10,600

Occupations: agriculture 14%, industry 34.7%, services 51.3%

Currency: Belarusian ruble (BYB/BYR)

Belgium

Long Name: Kingdom of Belgium

Location: Western Europe, bordering the North Sea, between France and the Netherlands

Area: 11,787 sq. mi. (30,528 sq. km)

Climate: temperate; mild winters, cool summers; rainy, humid, cloudy

Terrain: flat coastal plains in northwest, central rolling hills, rugged mountains of Ardennes Forest in southeast

Population: 10,403,951

Population Growth Rate: 0.106%

Birth Rate (per 1,000): 10.22

Death Rate (per 1,000): 10.38

Life Expectancy: 79.07 years

Ethnic Groups: Fleming 58%, Walloon 31%, mixed or other 11%

Religion: Roman Catholic 75%, other (includes Protestant) 25%

Languages: Dutch (official) 60%, French (official) 40%, German (official) less than 1%, legally bilingual (Dutch and French)

Literacy: 99%

Government Type: federal parliamentary democracy under a constitutional monarchy

Capital: Brussels

Independence: 4 October 1830 (a provisional government declared independence from the Netherlands); 21 July 1831 (King Leopold I ascended to the throne)

GDP Per Capita: $36,200

Occupations: agriculture 2%, industry 25%, services 73%

Currency: euro (EUR)

Belize

Long Name: Belize

Location: Central America, bordering the Caribbean Sea, between Guatemala and Mexico

Area: 8,867 sq. mi. (22,966 sq. km)

Climate: tropical; very hot and humid; rainy season (May to November); dry season (February to May)

Terrain: flat, swampy coastal plain; low mountains in south

Population: 301,270

Population Growth Rate: 2.207%

Birth Rate (per 1,000): 27.84

Death Rate (per 1,000): 5.77

Life Expectancy: 68.19 years

Ethnic Groups: mestizo 48.7%, Creole 24.9%, Maya 10.6%, Garifuna 6.1%, other 9.7%

Religion: Roman Catholic 49.6%, Protestant 27% (Pentecostal 7.4%, Anglican 5.3%, Seventh-Day Adventist 5.2%, Mennonite 4.1%, Methodist 3.5%, Jehovah's Witnesses 1.5%), other 14%, none 9.4%

Languages: Spanish 46%, Creole 32.9%, Mayan dialects 8.9%, English 3.9% (official), Garifuna 3.4% (Carib), German 3.3%, other 1.4%, unknown 0.2%

Literacy: 76.9%

Government Type: parliamentary democracy

Capital: Belmopan

Independence: 21 September 1981 (from United Kingdom)

GDP Per Capita: $7,900

Occupations: agriculture 22.5%, industry 15.2%, services 62.3%

Currency: Belizean dollar (BZD)

Bénin

Long Name: Republic of Benin

Location: Western Africa, bordering the Bight of Benin, between Nigeria and Togo

Area: 43,483 sq. mi. (112,620 sq. km)

Climate: tropical; hot, humid in south; semiarid in north

Terrain: mostly flat to undulating plain; some hills and low mountains

Population: 8,532,547

Population Growth Rate: 3.01%

Birth Rate (per 1,000): 39.8

Death Rate (per 1,000): 9.69

Life Expectancy: 58.56 years

Ethnic Groups: Fon and related 39.2%, Adja and related 15.2%, Yoruba and related 12.3%, Bariba and related 9.2%, Peulh and related 7%, Ottamari and related 6.1%, Yoa-Lokpa and related 4%, Dendi and related 2.5%, other 1.6% (includes Europeans), unspecified 2.9%

Religion: Christian 42.8% (Catholic 27.1%, Celestial 5%, Methodist 3.2%, other Protestant 2.2%, other 5.3%), Muslim 24.4%, Vodoun 17.3%, other 15.5%

Languages: French (official), Fon and Yoruba (most common vernaculars in south), tribal languages (at least six major ones in north)

Literacy: 34.7%

Government Type: republic

Capital: Porto-Novo

Independence: 1 August 1960 (from France)

GDP Per Capita: $1,400

Occupations: N/A

Currency: Communaute Financiere Africaine franc (XOF)

Bhutan

Long Name: Kingdom of Bhutan

Location: Southern Asia, between China and India

Area: 18,147 sq. mi. (47,000 sq. km)

Climate: varies; tropical in southern plains; cool winters and hot summers in central valleys; severe winters and cool summers in Himalayas

Terrain: mostly mountainous with some fertile valleys and savanna

Population: 682,321

Population Growth Rate: 1.301%

Birth Rate (per 1,000): 20.56

Death Rate (per 1,000): 7.54

Life Expectancy: 65.53 years

Ethnic Groups: Bhote 50%, ethnic Nepalese 35% (includes Lhotsampas - one of several Nepalese ethnic groups), indigenous or migrant tribes 15%

Religion: Lamaistic Buddhist 75%, Indian- and Nepalese-influenced Hinduism 25%

Languages: Dzongkha (official), Bhotes speak various Tibetan dialects, Nepalese speak various Nepalese dialects

Literacy: 47%

Government Type: in transition to constitutional monarchy; special treaty relationship with India

Capital: Thimphu

Independence: 1907 (became a unified kingdom under its first hereditary king)

GDP Per Capita: $5,200

Occupations: agriculture 63%, industry 6%, services 31%

Currency: ngultrum (BTN); Indian rupee (INR)

Bolivia

Long Name: Republic of Bolivia

Location: Central South America, southwest of Brazil

Area: 424,162 sq. mi. (1,098,580 sq. km)

Climate: varies with altitude; humid and tropical to cold and semiarid

Terrain: rugged Andes Mountains with a highland plateau (Altiplano), hills, lowland plains of the Amazon Basin

Population: 9,247,816

Population Growth Rate: 1.383%

Birth Rate (per 1,000): 22.31

Death Rate (per 1,000): 7.35

Life Expectancy: 66.53 years

Ethnic Groups: Quechua 30%, mestizo (mixed white and Amerindian ancestry) 30%, Aymara 25%, white 15%

Religion: Roman Catholic 95%, Protestant (Evangelical Methodist) 5%

Languages: Spanish 60.7% (official), Quechua 21.2% (official), Aymara 14.6% (official), foreign languages 2.4%, other 1.2%

Literacy: 86.7%

Government Type: republic

Capital: La Paz

Independence: 6 August 1825 (from Spain)

GDP Per Capita: $4,400

Occupations: agriculture 40%, industry 17%, services 43%

Currency: boliviano (BOB)

Bosnia and Herzegovina

Long Name: Bosnia and Herzegovina

Location: Southeastern Europe, bordering the Adriatic Sea and Croatia

Area: 19,772 sq. mi. (51,209 sq. km)

Climate: hot summers and cold winters; areas of high elevation have short, cool summers and long, severe winters; mild, rainy winters along coast

Terrain: mountains and valleys

Population: 4,590,310

Population Growth Rate: 0.666%

Birth Rate (per 1,000): 8.82

Death Rate (per 1,000): 8.54

Life Expectancy: 78.33 years

Ethnic Groups: Bosniak 48%, Serb 37.1%, Croat 14.3%, other 0.6%

Religion: Muslim 40%, Orthodox 31%, Roman Catholic 15%, other 14%

Languages: Bosnian, Croatian, Serbian

Literacy: 96.7%

Government Type: emerging federal democratic republic

Capital: Sarajevo

Independence: 1 March 1992 (from Yugoslavia; referendum for independence completed 1 March 1992; independence declared 3 March 1992)

GDP Per Capita: $6,100

Occupations: N/A

Currency: konvertibilna marka (convertible mark) (BAM)

Botswana

Long Name: Republic of Botswana

Location: Southern Africa, north of South Africa

Area: 231,803 sq. mi. (600,370 sq. km)

Climate: semiarid; warm winters and hot summers

Terrain: predominantly flat to gently rolling tableland; Kalahari Desert in southwest

Population: 1,842,323

Population Growth Rate: 1.434%

Birth Rate (per 1,000): 22.96

Death Rate (per 1,000): 14.02

Life Expectancy: 50.16 years

Ethnic Groups: Tswana (or Setswana) 79%, Kalanga 11%, Basarwa 3%, other, including Kgalagadi and white 7%

Religion: Christian 71.6%, Badimo 6%, other 1.4%, unspecified 0.4%, none 20.6%

Languages: Setswana 78.2%, Kalanga 7.9%, Sekgalagadi 2.8%, English 2.1% (official), other 8.6%, unspecified 0.4%

Literacy: 81.2%

Government Type: parliamentary republic

Capital: Gaborone

Independence: 30 September 1966 (from United Kingdom)

GDP Per Capita: $14,300

Occupations: N/A

Currency: pula (BWP)

Brazil

Long Name: Federative Republic of Brazil

Location: Eastern South America, bordering the Atlantic Ocean

Area: 3,286,473 sq. mi. (8,511,965 sq. km)

Climate: mostly tropical, but temperate in south

Terrain: mostly flat to rolling lowlands in north; some plains, hills, mountains, and narrow coastal belt

Population: 196,342,592

Population Growth Rate: 1.228%

Birth Rate (per 1,000): 18.72

Death Rate (per 1,000): 6.35

Life Expectancy: 71.71 years

Ethnic Groups: white 53.7%, mulatto (mixed white and black) 38.5%, black 6.2%, other (includes Japanese, Arab, Amerindian) 0.9%, unspecified 0.7%

Religion: Roman Catholic (nominal) 73.6%, Protestant 15.4%, Spiritualist 1.3%, Bantu/voodoo 0.3%, other 1.8%, unspecified 0.2%, none 7.4%

Languages: Portuguese (official and most widely spoken language); note - less common languages include Spanish (border areas and schools), German, Italian, Japanese, English, and a large number of minor Amerindian languages

Literacy: 88.6%

Government Type: federal republic

Capital: Brasilia

Independence: 7 September 1822 (from Portugal)

GDP Per Capita: $9,500

Occupations: agriculture 20%, industry 14%, services 66%

Currency: real (BRL)

Brunei

Long Name: Brunei Darussalam

Location: Southeastern Asia, bordering the South China Sea and Malaysia

Area: 2,228 sq. mi. (5,770 sq. km)

Climate: tropical; hot, humid, rainy

Terrain: flat coastal plain rises to mountains in east; hilly lowland in west

Population: 381,371

Population Growth Rate: 1.785%

Birth Rate (per 1,000): 18.39

Death Rate (per 1,000): 3.28

Life Expectancy: 75.52 years

Ethnic Groups: Malay 66.3%, Chinese 11.2%, indigenous 3.4%, other 19.1%

Religion: Muslim (official) 67%, Buddhist 13%, Christian 10%, other (includes indigenous beliefs) 10%

Languages: Malay (official), English, Chinese

Literacy: 92.7%

Government Type: constitutional sultanate

Capital: Bandar Seri Begawan

Independence: 1 January 1984 (from United Kingdom)

GDP Per Capita: $51,000

Occupations: agriculture 2.9%, industry 61.1%, services 36%

Currency: Bruneian dollar (BND)

Bulgaria

Long Name: Republic of Bulgaria

Location: Southeastern Europe, bordering the Black Sea, between Romania and Turkey

Area: 42,822 sq. mi. (110,910 sq. km)

Climate: temperate; cold, damp winters; hot, dry summers

Terrain: mostly mountains with lowlands in north and southeast

Population: 7,262,675

Population Growth Rate: -0.813%

Birth Rate (per 1,000): 9.58

Death Rate (per 1,000): 14.3

Life Expectancy: 72.83 years

Ethnic Groups: Bulgarian 83.9%, Turk 9.4%, Roma 4.7%, other 2% (including Macedonian, Armenian, Tatar, Circassian)

Religion: Bulgarian Orthodox 82.6%, Muslim 12.2%, other Christian 1.2%, other 4%

Languages: Bulgarian 84.5%, Turkish 9.6%, Roma 4.1%, other and unspecified 1.8%

Literacy: 98.2%

Government Type: parliamentary democracy

Capital: Sofia

Independence: 3 March 1878 (as an autonomous principality within the Ottoman Empire); 22 September 1908 (complete independence from the Ottoman Empire)

GDP Per Capita: $11,800

Occupations: agriculture 8.5%, industry 33.6%, services 57.9%

Currency: lev (BGN)

Burkina Faso

Long Name: Burkina Faso

Location: Western Africa, north of Ghana

Area: 105,869 sq. mi. (274,200 sq. km)

Climate: tropical; warm, dry winters; hot, wet summers

Terrain: mostly flat to dissected, undulating plains; hills in west and southeast

Population: 15,264,735

Population Growth Rate: 3.109%

Birth Rate (per 1,000): 44.68

Death Rate (per 1,000): 13.59

Life Expectancy: 52.55 years

Ethnic Groups: Mossi over 40%, other approximately 60% (includes Gurunsi, Senufo, Lobi, Bobo, Mande, and Fulani)

Religion: Muslim 50%, indigenous beliefs 40%, Christian (mainly Roman Catholic) 10%

283

Languages: French (official), native African languages belonging to Sudanic family spoken by 90% of the population

Literacy: 21.8%

Government Type: parliamentary republic

Capital: Ouagadougou

Independence: 5 August 1960 (from France)

GDP Per Capita: $1,200

Occupations: agriculture 90%, industry and services 10%

Currency: Communaute Financiere Africaine franc (XOF)

Burundi

Long Name: Republic of Burundi

Location: Central Africa, east of Democratic Republic of the Congo

Area: 10,745 sq. mi. (27,830 sq. km)

Climate: equatorial; high plateau with considerable altitude variation (772 m to 2,670 m above sea level); average annual temperature varies with altitude from 23 to 17 degrees centigrade but is generally moderate as the average altitude is about 1,700 m; average annual rainfall is about 150 cm; two wet seasons (February to May and September to November), and two dry seasons (June to August and December to January)

Terrain: hilly and mountainous, dropping to a plateau in east, some plains

Population: 8,691,005

Population Growth Rate: 3.443%

Birth Rate (per 1,000): 41.72

Death Rate (per 1,000): 12.91

Life Expectancy: 51.71 years

Ethnic Groups: Hutu (Bantu) 85%, Tutsi (Hamitic) 14%, Twa (Pygmy) 1%, Europeans 3,000, South Asians 2,000

Religion: Christian 67% (Roman Catholic 62%, Protestant 5%), indigenous beliefs 23%, Muslim 10%

Languages: Kirundi (official), French (official), Swahili (along Lake Tanganyika and in the Bujumbura area)

Literacy: 59.3%

Government Type: republic

Capital: Bujumbura

Independence: 1 July 1962 (from UN trusteeship under Belgian administration)

GDP Per Capita: $300

Occupations: agriculture 93.6%, industry 2.3%, services 4.1%

Currency: Burundi franc (BIF)

Cambodia

Long Name: Kingdom of Cambodia

Location: Southeastern Asia, bordering the Gulf of Thailand, between Thailand, Vietnam, and Laos

Area: 69,900 sq. mi. (181,040 sq. km)

Climate: tropical; rainy, monsoon season (May to November); dry season (December to April); little seasonal temperature variation

Terrain: mostly low, flat plains; mountains in southwest and north

Population: 14,241,640

Population Growth Rate: 1.752%

Birth Rate (per 1,000): 25.68

Death Rate (per 1,000): 8.16

Life Expectancy: 61.69 years

Ethnic Groups: Khmer 90%, Vietnamese 5%, Chinese 1%, other 4%

Religion: Theravada Buddhist 95%, other 5%

Languages: Khmer (official) 95%, French, English

Literacy: 73.6%

Government Type: multiparty democracy under a constitutional monarchy

Capital: Phnom Penh

Independence: 9 November 1953 (from France)

GDP Per Capita: $1,900

Occupations: agriculture 75%, others N/A

Currency: riel (KHR)

Cameroon

Long Name: Republic of Cameroon

Location: Western Africa, bordering the Bight of Biafra, between Equatorial Guinea and Nigeria

Area: 183,568 sq. mi. (475,440 sq. km)

Climate: varies with terrain, from tropical along coast to semiarid and hot in north

Terrain: diverse, with coastal plain in southwest, dissected plateau in center, mountains in west, plains in north

Population: 18,467,692

Population Growth Rate: 2.218%

Birth Rate (per 1,000): 34.59

Death Rate (per 1,000): 12.41

Life Expectancy: 53.3 years

Ethnic Groups: Cameroon Highlanders 31%, Equatorial Bantu 19%, Kirdi 11%, Fulani 10%, Northwestern Bantu 8%, Eastern Nigritic 7%, other African 13%, non-African less than 1%

Religion: indigenous beliefs 40%, Christian 40%, Muslim 20%

Languages: 24 major African language groups, English (official), French (official)

Literacy: 67.9%

Government Type: republic; multiparty presidential regime

Capital: Yaounde

Independence: 1 January 1960 (from French-administered UN trusteeship)

GDP Per Capita: $2,200

Occupations: agriculture 70%, industry 13%, services 17%

Currency: Communaute Financiere Africaine franc (XAF)

Canada

Long Name: Canada

Location: Northern North America, bordering the North Atlantic Ocean on the east, North Pacific Ocean on the west, and the Arctic Ocean on the north, north of the conterminous US

Area: 3,855,085 sq. mi. (9,984,670 sq km)

Climate: varies from temperate in south to subarctic and arctic in north

Terrain: mostly plains with mountains in west and lowlands in southeast

Population: 33,212,696

Population Growth Rate: 0.83%

Birth Rate (per 1,000): 10.29

Death Rate (per 1,000): 7.61

Life Expectancy: 81.16 years

Ethnic Groups: British Isles origin 28%, French origin 23%, other European 15%, Amerindian 2%, other, mostly Asian, African, Arab 6%, mixed background 26%

Religion: Roman Catholic 42.6%, Protestant 23.3% (including United

285

Church 9.5%, Anglican 6.8%, Baptist 2.4%, Lutheran 2%), other Christian 4.4%, Muslim 1.9%, other and unspecified 11.8%, none 16%

Languages: English (official) 59.3%, French (official) 23.2%, other 17.5%

Literacy: 99%

Government Type: constitutional monarchy that is also a parliamentary democracy and a federation

Capital: Ottawa

Independence: 1 July 1867 (union of British North American colonies); 11 December 1931 (recognized by United Kingdom)

GDP Per Capita: $38,600

Occupations: agriculture 2%, manufacturing 13%, construction 6%, services 76%, other 3%

Currency: Canadian dollar (CAD)

Cape Verde

Long Name: Republic of Cape Verde

Location: Western Africa, group of islands in the North Atlantic Ocean, west of Senegal

Area: 1,557 sq. mi. (4,033 sq. km)

Climate: temperate; warm, dry summer; precipitation meager and very erratic

Terrain: steep, rugged, rocky, volcanic

Population: 426,998

Population Growth Rate: 0.595%

Birth Rate (per 1,000): 23.95

Death Rate (per 1,000): 6.26

Life Expectancy: 71.33 years

Ethnic Groups: Creole (mulatto) 71%, African 28%, European 1%

Religion: Roman Catholic (infused with indigenous beliefs), Protestant (mostly Church of the Nazarene)

Languages: Portuguese, Crioulo (a blend of Portuguese and West African words)

Literacy: 76.6%

Government Type: republic

Capital: Praia

Independence: 5 July 1975 (from Portugal)

GDP Per Capita: $3,200

Occupations: N/A

Currency: Cape Verdean escudo (CVE)

Central African Republic

Long Name: Central African Republic

Location: Central Africa, north of Democratic Republic of the Congo

Area: 240,534 sq. mi. (622,984) sq. km

Climate: tropical; hot, dry winters; mild to hot, wet summers

Terrain: vast, flat to rolling, monotonous plateau; scattered hills in northeast and southwest

Population: 4,444,330

Population Growth Rate: 1.509%

Birth Rate (per 1,000): 33.13

Death Rate (per 1,000): 18.04

Life Expectancy: 44.22 years

Ethnic Groups: Baya 33%, Banda 27%, Mandjia 13%, Sara 10%, Mboum 7%, M'Baka 4%, Yakoma 4%, other 2%

Religion: indigenous beliefs 35%, Protestant 25%, Roman Catholic 25%, Muslim 15%

Languages: French (official), Sangho (lingua franca and national language), tribal languages

Literacy: 48.6%

Government Type: republic

Capital: Bangui

Independence: 13 August 1960 (from France)

GDP Per Capita: $700

Occupations: N/A

Currency: Communaute Financiere Africaine franc (XAF)

Chad

Long Name: Republic of Chad

Location: Central Africa, south of Libya

Area: 495,753 sq. mi. (1,284,000 sq. km)

Climate: tropical in south, desert in north

Terrain: broad, arid plains in center, desert in north, mountains in northwest, lowlands in south

Population: 10,111,337

Population Growth Rate: 2.195%

Birth Rate (per 1,000): 41.61

Death Rate (per 1,000): 16.39

Life Expectancy: 47.43 years

Ethnic Groups: Sara 27.7%, Arab 12.3%, Mayo-Kebbi 11.5%, Kanem-Bornou 9%, Ouaddai 8.7%, Hadjarai 6.7%, Tandjile 6.5%, Gorane 6.3%, Fitri-Batha 4.7%, other 6.4%, unknown 0.3%

Religion: Muslim 53.1%, Catholic 20.1%, Protestant 14.2%, animist 7.3%, other 0.5%, unknown 1.7%, atheist 3.1%

Languages: French (official), Arabic (official), Sara (in south), more than 120 different languages and dialects

Literacy: 25.7%

Government Type: republic

Capital: N'Djamena

Independence: 11 August 1960 (from France)

GDP Per Capita: $1,500

Occupations: agriculture 80% (subsistence farming, herding, and fishing), industry and services 20%

Currency: Communaute Financiere Africaine franc (XAF)

Chile

Long Name: Republic of Chile

Location: Southern South America, bordering the South Pacific Ocean, between Argentina and Peru

Area: 292,259 sq. mi. (756,950 sq. km)

Climate: temperate; desert in north; Mediterranean in central region; cool and damp in south

Terrain: low coastal mountains; fertile central valley; rugged Andes in east

Population: 16,454,143

Population Growth Rate: 0.905%

Birth Rate (per 1,000): 14.82

Death Rate (per 1,000): 5.77

Life Expectancy: 77.15 years

Ethnic Groups: white and white-Amerindian 95.4%, Mapuche 4%, other indigenous groups 0.6%

Religion: Roman Catholic 70%, Evangelical 15.1%, Jehovah's Witness 1.1%, other Christian 1%, other 4.6%, none 8.3%

Languages: Spanish (official), Mapudungun, German, English

Literacy: 95.7%

Government Type: republic

Capital: Santiago

Independence: 18 September 1810 (from Spain)

GDP Per Capita: $14,300

Occupations: agriculture 13.6%, industry 23.4%, services 63%

Currency: Chilean peso (CLP)

China

Long Name: People's Republic of China

Location: Eastern Asia, bordering the East China Sea, Korea Bay, Yellow Sea, and South China Sea, between North Korea and Vietnam

Area: 3,705,390 sq. mi. (9,596,960 sq. km)

Climate: extremely diverse; tropical in south to subarctic in north

Terrain: mostly mountains, high plateaus, deserts in west; plains, deltas, and hills in east

Population: 1,330,044,544

Population Growth Rate: 0.629%

Birth Rate (per 1,000): 13.71

Death Rate (per 1,000): 7.03

Life Expectancy: 73.18 years

Ethnic Groups: Han Chinese 91.5%, Zhuang, Manchu, Hui, Miao, Uyghur, Tujia, Yi, Mongol, Tibetan, Buyi, Dong, Yao, Korean, and other nationalities 8.5%

Religion: Daoist (Taoist), Buddhist, Christian 3%-4%, Muslim 1%-2%

Languages: Standard Chinese or Mandarin (Putonghua, based on the Beijing dialect), Yue (Cantonese), Wu (Shanghainese), Minbei (Fuzhou), Minnan (Hokkien-Taiwanese), Xiang, Gan, Hakka dialects, minority languages

Literacy: 90.9%

Government Type: Communist state

Capital: Beijing

Independence: 221 BC (unification under the Qin or Ch'in Dynasty); 1 January 1912 (Manchu Dynasty replaced by a Republic); 1 October 1949 (People's Republic established)

GDP Per Capita: $5,400

Occupations: agriculture 43%, industry 25%, services 32%

Currency: Renminbi (RMB), also referred by the unit yuan (CNY)

Colombia

Long Name: Republic of Colombia

Location: Northern South America, bordering the Caribbean Sea, between Panama and Venezuela, and bordering the North Pacific Ocean, between Ecuador and Panama

Area: 439,734 sq. mi. (1,138,910 sq. km)

Climate: tropical along coast and eastern plains; cooler in highlands

Terrain: flat coastal lowlands, central highlands, high Andes Mountains, eastern lowland plains

Population: 45,013,672

Population Growth Rate: 1.405%

Birth Rate (per 1,000): 19.86

Death Rate (per 1,000): 5.54

Life Expectancy: 72.54 years

Ethnic Groups: mestizo 58%, white 20%, mulatto 14%, black 4%, mixed black-Amerindian 3%, Amerindian 1%

Religion: Roman Catholic 90%, other 10%

Languages: Spanish

Literacy: 92.8%

Government Type: republic; executive branch dominates government structure

Capital: Bogotá

Independence: 20 July 1810 (from Spain)

GDP Per Capita: $7,400

Occupations: agriculture 22.7%, industry 18.7%, services 58.5%

Currency: Colombian peso (COP)

Comoros

Long Name: Union of the Comoros

Location: Southern Africa, group of islands at the northern mouth of the Mozambique Channel, about two-thirds of the way between northern Madagascar and northern Mozambique

Area: 838 sq. mi. (2,170 sq. km)

Climate: tropical marine; rainy season (November to May)

Terrain: volcanic islands, interiors vary from steep mountains to low hills

Population: 731,775

Population Growth Rate: 2.803%

Birth Rate (per 1,000): 35.78

Death Rate (per 1,000): 7.76

Life Expectancy: 63.1 years

Ethnic Groups: Antalote, Cafre, Makoa, Oimatsaha, Sakalava

Religion: Sunni Muslim 98%, Roman Catholic 2%

Languages: Arabic (official), French (official), Shikomoro (a blend of Swahili and Arabic)

Literacy: 56.5%

Government Type: republic

Capital: Moroni

Independence: 6 July 1975 (from France)

GDP Per Capita: $1,100

Occupations: agriculture 80%, industry and services 20%

Currency: Comoran franc (KMF)

Democratic Republic of the Congo

Long Name: Democratic Republic of the Congo

Location: Central Africa, northeast of Angola

Area: 905,564 sq. mi. (2,345,410 sq. km)

Climate: tropical; hot and humid in equatorial river basin; cooler and drier in southern highlands; cooler and wetter in eastern highlands; north of Equator - wet season (April to October), dry season (December to February); south of Equator - wet season (November to March), dry season (April to October)

Terrain: vast central basin is a low-lying plateau; mountains in east

Population: 66,514,504

Population Growth Rate: 3.236%

Birth Rate (per 1,000): 43

Death Rate (per 1,000): 1.24

Life Expectancy: 53.98 years

Ethnic Groups: over 200 African ethnic groups of which the majority are Bantu; the four largest tribes - Mongo, Luba, Kongo (all Bantu), and the Mangbetu-Azande (Hamitic) make up about 45% of the population

Religion: Roman Catholic 50%, Protestant 20%, Kimbanguist 10%, Muslim 10%, other (includes syncretic sects and indigenous beliefs) 10%

Languages: French (official), Lingala (a lingua franca trade language), Kingwana (a dialect of Kiswahili or Swahili), Kikongo, Tshiluba

Literacy: 67.2%

Government Type: republic

Capital: Kinshasa

Independence: 30 June 1960 (from Belgium)

GDP Per Capita: $300

Occupations: N/A

Currency: Congolese franc (CDF)

Republic of the Congo

Long Name: Republic of the Congo

Location: Western Africa, bordering the South Atlantic Ocean, between Angola and Gabon

Area: 132,046 sq. mi. (342,000 sq. km)

Climate: tropical; rainy season (March to June); dry season (June to October); persistent high temperatures and humidity; particularly enervating climate astride the Equator

Terrain: coastal plain, southern basin, central plateau, northern basin

Population: 3,903,318

Population Growth Rate: 2.696%

Birth Rate (per 1,000): 41.76

Death Rate (per 1,000): 12.28

Life Expectancy: 53.74 years

Ethnic Groups: Kongo 48%, Sangha 20%, M'Bochi 12%, Teke 17%, Europeans and other 3%

Religion: Christian 50%, animist 48%, Muslim 2%

Languages: French (official), Lingala and Monokutuba (lingua franca trade languages), many local languages and dialects (of which Kikongo is the most widespread)

Literacy: 83.8%

Government Type: republic

Capital: Brazzaville

Independence: 15 August 1960 (from France)

GDP Per Capita: $3,400

Occupations: N/A

Currency: Communaute Financiere Africaine franc (XAF)

Costa Rica

Long Name: Republic of Costa Rica

Location: Central America, bordering both the Caribbean Sea and the North Pacific Ocean, between Nicaragua and Panama

Area: 19,730 sq. mi. (51,100 sq. km)

Climate: tropical and subtropical; dry season (December to April); rainy season (May to November); cooler in highlands

Terrain: coastal plains separated by rugged mountains including over 100 volcanic cones, of which several are major volcanoes

Population: 4,195,914

Population Growth Rate: 1.388%

Birth Rate (per 1,000): 17.71

Death Rate (per 1,000): 4.31

Life Expectancy: 77.4 years

Ethnic Groups: white (including mestizo) 94%, black 3%, Amerindian 1%, Chinese 1%, other 1%

Religion: Roman Catholic 76.3%, Evangelical 13.7%, Jehovah's Witnesses 1.3%, other Protestant 0.7%, other 4.8%, none 3.2%

Languages: Spanish (official), English

Literacy: 94.9%

Government Type: democratic republic

Capital: San Jose

Independence: 15 September 1821 (from Spain)

GDP Per Capita: $11,100

Occupations: agriculture 14%, industry 22%, services 64%

Currency: Costa Rican colon (CRC)

Côte d'Ivoire

Long Name: Republic of Cote d'Ivoire

Location: Western Africa, bordering the North Atlantic Ocean, between Ghana and Liberia

Area: 124,502 sq. mi. (322,460 sq. km)

Climate: tropical along coast, semiarid in far north; three seasons - warm and dry (November to March), hot and dry (March to May), hot and wet (June to October)

Terrain: mostly flat to undulating plains; mountains in northwest

Population: 20,179,602

Population Growth Rate: 2.156%

Birth Rate (per 1,000): 32.73

Death Rate (per 1,000): 11.17

Life Expectancy: 54.64 years

Ethnic Groups: Akan 42.1%, Voltaiques or Gur 17.6%, Northern Mandes 16.5%, Krous 11%, Southern Mandes 10%, other 2.8% (includes 130,000 Lebanese and 14,000 French)

Religion: Muslim 38.6%, Christian 32.8%, indigenous 11.9%, none 16.7%

Languages: French (official), 60 native dialects with Dioula the most widely spoken

Literacy: 48.7%

Government Type: republic; multiparty presidential regime established 1960

Capital: Yamoussoukro

Independence: 7 August 1960 (from France)

GDP Per Capita: $1,700

Occupations: N/A

Currency: Communaute Financiere Africaine franc (XOF)

Croatia

Long Name: Republic of Croatia

Location: Southeastern Europe, bordering the Adriatic Sea, between Bosnia and Herzegovina and Slovenia

Area: 21,831 sq. mi. (56,542 sq. km)

Climate: Mediterranean and continental; continental climate predominant with hot summers and cold winters; mild winters, dry summers along coast

Terrain: geographically diverse; flat plains along Hungarian border, low mountains and highlands near Adriatic coastline and islands

Population: 4,491,543

Population Growth Rate: -0.043%

Birth Rate (per 1,000): 9.64

Death Rate (per 1,000): 11.66

Life Expectancy: 75.13 years

Ethnic Groups: Croat 89.6%, Serb 4.5%, other 5.9% (including Bosniak, Hungarian, Slovene, Czech, and Roma)

Religion: Roman Catholic 87.8%, Orthodox 4.4%, other Christian 0.4%, Muslim 1.3%, other and unspecified 0.9%, none 5.2%

Languages: Croatian 96.1%, Serbian 1%, other and undesignated 2.9% (including Italian, Hungarian, Czech, Slovak, and German)

Literacy: 98.1%

Government Type: presidential/parliamentary democracy

Capital: Zagreb

Independence: 25 June 1991 (from Yugoslavia)

GDP Per Capita: $15,500

Occupations: agriculture 2.7%, industry 32.8%, services 64.5%

Currency: kuna (HRK)

Cuba

Long Name: Republic of Cuba

Location: Caribbean, island between the Caribbean Sea and the North Atlantic Ocean, 150 km south of Key West, Florida

Area: 42,803 sq. mi. (110,860 sq. km)

Climate: tropical; moderated by trade winds; dry season (November to April); rainy season (May to October)

Terrain: mostly flat to rolling plains, with rugged hills and mountains in the southeast

Population: 11,423,952

Population Growth Rate: 0.251%

Birth Rate (per 1,000): 11.27

Death Rate (per 1,000): 7.19

Life Expectancy: 77.27 years

Ethnic Groups: white 65.1%, mulatto and mestizo 24.8%, black 10.1%

Religion: nominally 85% Roman Catholic prior to Castro assuming power; Protestants, Jehovah's Witnesses, Jews, and Santeria are also represented

Languages: Spanish

Literacy: 99.8%

Government Type: Communist state

Capital: Havana

Independence: 20 May 1902 (from Spain 10 December 1898; administered by the US from 1898 to 1902)

GDP Per Capita: $11,000

Occupations: agriculture 20%, industry 19.4%, services 60.6%

Currency: Cuban peso (CUP) and Convertible peso (CUC)

Cyprus

Long Name: Republic of Cyprus

Location: Middle East, island in the Mediterranean Sea, south of Turkey

Area: 3,571 sq. mi. (9,250 sq. km)

Climate: temperate; Mediterranean with hot, dry summers and cool winters

Terrain: central plain with mountains to north and south; scattered but significant plains along southern coast

Population: 792,604

Population Growth Rate: 0.522%

Birth Rate (per 1,000): 12.56

Death Rate (per 1,000): 7.76

Life Expectancy: 78.15 years

Ethnic Groups: Greek 77%, Turkish 18%, other 5%

Religion: Greek Orthodox 78%, Muslim 18%, other (includes Maronite and Armenian Apostolic) 4%

Languages: Greek, Turkish, English

Literacy: 97.6%

Government Type: republic

Capital: Nicosia

Independence: 16 August 1960 (from United Kingdom); note - Turkish Cypriots proclaimed self-rule on 13 February 1975 and independence in 1983, but these proclamations are only recognized by Turkey

GDP Per Capita: $27,100

Occupations: agriculture 8.5%, industry 20.5%, services 71%

Currency: Cypriot pound (CYP); euro (EUR) after 1 January 2008

Czech Republic

Long Name: Czech Republic

Location: Central Europe, southeast of Germany

Area: 30,450 sq. mi. (78,866 sq. km)

Climate: temperate; cool summers; cold, cloudy, humid winters

Terrain: Bohemia in the west consists of rolling plains, hills, and plateaus surrounded by low mountains; Moravia in the east consists of very hilly country

Population: 10,220,911

Population Growth Rate: -0.082%

Birth Rate (per 1,000): 8.89

Death Rate (per 1,000): 10.69

Life Expectancy: 76.62 years

Ethnic Groups: Czech 90.4%, Moravian 3.7%, Slovak 1.9%, other 4%

Religion: Roman Catholic 26.8%, Protestant 2.1%, other 3.3%, unspecified 8.8%, unaffiliated 59%

Languages: Czech 94.9%, Slovak 2%, other 2.3%, unidentified 0.8%

Literacy: 99%

Government Type: parliamentary democracy

Capital: Prague

Independence: 1 January 1993 (Czechoslovakia split into the Czech Republic and Slovakia)

GDP Per Capita: $24,500

Occupations: agriculture 4.1%, industry 37.6%, services 58.3%

Currency: Czech koruna (CZK)

Denmark

Long Name: Kingdom of Denmark

Location: Northern Europe, bordering the Baltic Sea and the North Sea, on a peninsula north of Germany (Jutland); also includes two major islands (Sjaelland and Fyn)

Area: 16,639 sq. mi. (43,094 sq. km)

Climate: temperate; humid and overcast; mild, windy winters and cool summers

Terrain: low and flat to gently rolling plains

Population: 5,484,723

Population Growth Rate: 0.295%

Birth Rate (per 1,000): 10.71

Death Rate (per 1,000): 10.25

Life Expectancy: 78.13 years

Ethnic Groups: Scandinavian, Inuit, Faroese, German, Turkish, Iranian, Somali

Religion: Evangelical Lutheran 95%, other Christian (includes Protestant and Roman Catholic) 3%, Muslim 2%

Languages: Danish, Faroese, Greenlandic (an Inuit dialect), German (small minority), English (as the predominant second language)

Literacy: 99%

Government Type: constitutional monarchy

Capital: Copenhagen

Independence: first organized as a unified state in 10th century; in 1849 became a constitutional monarchy

GDP Per Capita: $37,200

Occupations: agriculture 3%, industry 21%, services 76%

Currency: Danish krone (DKK)

Djibouti

Long Name: Republic of Djibouti

Location: Eastern Africa, bordering the Gulf of Aden and the Red Sea, between Eritrea and Somalia

Area: 8,880 sq. mi. (23,000 sq. km)

Climate: desert; torrid, dry

Terrain: coastal plain and plateau separated by central mountains

Population: 506,221

Population Growth Rate: 1.945%

Birth Rate (per 1,000): 38.61

Death Rate (per 1,000): 19.16

Life Expectancy: 43.31 years

Ethnic Groups: Somali 60%, Afar 35%, other 5% (includes French, Arab, Ethiopian, and Italian)

Religion: Muslim 94%, Christian 6%

Languages: French (official), Arabic (official), Somali, Afar

Literacy: 67.9%

Government Type: republic

Capital: Djibouti

Independence: 27 June 1977 (from France)

GDP Per Capita: $2,300

Occupations: N/A

Currency: Djiboutian franc (DJF)

Dominica

Long Name: Commonwealth of Dominica

Location: Caribbean, island between the Caribbean Sea and the North Atlantic Ocean, about half way between Puerto Rico and Trinidad and Tobago

Area: 291 sq. mi. (754 sq. km)

Climate: tropical; moderated by northeast trade winds; heavy rainfall

Terrain: rugged mountains of volcanic origin

Population: 72,514

Population Growth Rate: 0.196%

Birth Rate (per 1,000): 15.73

Death Rate (per 1,000): 8.32

Life Expectancy: 75.33 years

Ethnic Groups: black 86.8%, mixed 8.9%, Carib Amerindian 2.9%, white 0.8%, other 0.7%

Religion: Roman Catholic 61.4%, Seventh Day Adventist 6%, Pentecostal 5.6%, Baptist 4.1%, Methodist 3.7%, Church of God 1.2%, Jehovah's Witnesses 1.2%, other Christian 7.7%,

Rastafarian 1.3%, other or unspecified 1.6%, none 6.1%

Languages: English (official), French patois

Literacy: 94%

Government Type: parliamentary democracy

Capital: Roseau

Independence: 3 November 1978 (from United Kingdom)

GDP Per Capita: $9,000

Occupations: agriculture 40%, industry 32%, services 28%

Currency: East Caribbean dollar (XCD)

Dominican Republic

Long Name: Dominican Republic

Location: Caribbean, eastern two-thirds of the island of Hispaniola, between the Caribbean Sea and the North Atlantic Ocean, east of Haiti

Area: 18,815 sq. mi. (48,730 sq. km)

Climate: tropical maritime; little seasonal temperature variation; seasonal variation in rainfall

Terrain: rugged highlands and mountains with fertile valleys interspersed

Population: 9,507,133

Population Growth Rate: 1.495%

Birth Rate (per 1,000): 22.65

Death Rate (per 1,000): 5.3

Life Expectancy: 73.39 years

Ethnic Groups: mixed 73%, white 16%, black 11%

Religion: Roman Catholic 95%, other 5%

Languages: Spanish

Literacy: 87%

Government Type: democratic republic

Capital: Santo Domingo

Independence: 27 February 1844 (from Haiti)

GDP Per Capita: $6,600

Occupations: agriculture 17%, industry 24.3%, services 58.7%

Currency: Dominican peso (DOP)

Ecuador

Long Name: Republic of Ecuador

Location: Western South America, bordering the Pacific Ocean at the Equator, between Colombia and Peru

Area: 109,483 sq. mi. (283,560 sq. km)

Climate: tropical along coast, becoming cooler inland at higher elevations; tropical in Amazonian jungle lowlands

Terrain: coastal plain (costa), inter-Andean central highlands (sierra), and flat to rolling eastern jungle (oriente)

Population: 13,927,650

Population Growth Rate: 0.935%

Birth Rate (per 1,000): 21.54

Death Rate (per 1,000): 4.21

Life Expectancy: 76.81 years

Ethnic Groups: mestizo (mixed Amerindian and white) 65%, Amerindian 25%, Spanish and others 7%, black 3%

Religion: Roman Catholic 95%, other 5%

Languages: Spanish (official), Amerindian languages (especially Quechua)

Literacy: 91%

Government Type: republic

Capital: Quito

Independence: 24 May 1822 (from Spain)

GDP Per Capita: $7,200

Occupations: agriculture 8%, industry 24%, services 68%

Currency: US dollar (USD)

Egypt

Long Name: Arab Republic of Egypt

Location: Northern Africa, bordering the Mediterranean Sea, between Libya and the Gaza Strip, and the Red Sea north of Sudan, and includes the Asian Sinai Peninsula

Area: 386,660 sq. mi. (1,001,450 sq. km)

Climate: desert; hot, dry summers with moderate winters

Terrain: vast desert plateau interrupted by Nile valley and delta

Population: 81,713,520

Population Growth Rate: 1.682%

Birth Rate (per 1,000): 22.12

Death Rate (per 1,000): 5.09

Life Expectancy: 71.85 years

Ethnic Groups: Egyptian 99.6%, other 0.4% (2006 census)

Religion: Muslim (mostly Sunni) 90%, Coptic 9%, other Christian 1%

Languages: Arabic (official), English and French widely understood by educated classes

Literacy: 71.4%

Government Type: republic

Capital: Cairo

Independence: 28 February 1922 (from United Kingdom)

GDP Per Capita: $5,000

Occupations: agriculture 32%, industry 17%, services 51%

Currency: Egyptian pound (EGP)

El Salvador

Long Name: Republic of El Salvador

Location: Central America, bordering the North Pacific Ocean, between Guatemala and Honduras

Area: 8,124 sq. mi. (21,040 sq. km)

Climate: tropical; rainy season (May to October); dry season (November to April); tropical on coast; temperate in uplands

Terrain: mostly mountains with narrow coastal belt and central plateau

Population: 7,066,403

Population Growth Rate: 1.679%

Birth Rate (per 1,000): 25.72

Death Rate (per 1,000): 5.53

Life Expectancy: 72.06 years

Ethnic Groups: mestizo 90%, white 9%, Amerindian 1%

Religion: Roman Catholic 57.1%, Protestant 21.2%, Jehovah's Witnesses 1.9%, Mormon 0.7%, other religions 2.3%, none 16.8%

Languages: Spanish, Nahua (among some Amerindians)

Literacy: 80.2%

Government Type: republic

Capital: San Salvador

Independence: 15 September 1821 (from Spain)

GDP Per Capita: $6,000

Occupations: agriculture, 19%, industry 23%, services 58%

Currency: US dollar (USD)

Equatorial Guinea

Long Name: Republic of Equatorial Guinea

Location: Western Africa, bordering the Bight of Biafra, between Cameroon and Gabon

Area: 10,831 sq. mi. (28,051 sq. km)

Climate: tropical; always hot, humid

Terrain: coastal plains rise to interior hills; islands are volcanic

Population: 616,459

Population Growth Rate: 2.732%

Birth Rate (per 1,000): 37.04

Death Rate (per 1,000): 9.72

Life Expectancy: 61.23 years

Ethnic Groups: Fang 85.7%, Bubi 6.5%, Mdowe 3.6%, Annobon 1.6%, Bujeba 1.1%, other 1.4%

Religion: nominally Christian and predominantly Roman Catholic, pagan practices

Languages: Spanish 67.6% (official), other 32.4% (includes French [official], Fang, Bubi)

Literacy: 87%

Government Type: republic

Capital: Malabo

Independence: 12 October 1968 (from Spain)

GDP Per Capita: $28,200

Occupations: N/A

Currency: Communaute Financiere Africaine franc (XAF)

Eritrea

Long Name: State of Eritrea

Location: Eastern Africa, bordering the Red Sea, between Djibouti and Sudan

Area: 46,842 sq. mi. (121,320 sq. km)

Climate: hot, dry desert strip along Red Sea coast; cooler and wetter in the central highlands (up to 61 cm of rainfall annually, heaviest June to September); semiarid in western hills and lowlands

Terrain: dominated by extension of Ethiopian north-south trending highlands, descending on the east to a coastal desert plain, on the northwest to hilly terrain and on the southwest to flat-to-rolling plains

Population: 5,502,026

Population Growth Rate: 2.631%

Birth Rate (per 1,000): 34.94

Death Rate (per 1,000): 8.63

Life Expectancy: 61.38 years

Ethnic Groups: Tigrinya 50%, Tigre and Kunama 40%, Afar 4%, Saho (Red Sea coast dwellers) 3%, other 3%

Religion: Muslim, Coptic Christian, Roman Catholic, Protestant

Languages: Afar, Arabic, Tigre and Kunama, Tigrinya, other Cushitic languages

Literacy: 58.6%

Government Type: transitional government

Capital: Asmara

Independence: 24 May 1993 (from Ethiopia)

GDP Per Capita: $800

Occupations: agriculture 80%, industry and services 20%

Currency: nakfa (ERN)

Estonia

Long Name: Republic of Estonia

Location: Eastern Europe, bordering the Baltic Sea and Gulf of Finland, between Latvia and Russia

Area: 17,462 sq. mi. (45,226 sq. km)

Climate: maritime, wet, moderate winters, cool summers

Terrain: marshy, lowlands; flat in the north, hilly in the south

Population: 1,307,605

Population Growth Rate: -0.632%

Birth Rate (per 1,000): 10.28

Death Rate (per 1,000): 13.35

Life Expectancy: 72.56 years

Ethnic Groups: Estonian 67.9%, Russian 25.6%, Ukrainian 2.1%, Belarusian 1.3%, Finn 0.9%, other 2.2%

Religion: Evangelical Lutheran 13.6%, Orthodox 12.8%, other Christian (including Methodist, Seventh-Day Adventist, Roman Catholic, Pentecostal) 1.4%, unaffiliated 34.1%, other and unspecified 32%, none 6.1%

Languages: Estonian (official) 67.3%, Russian 29.7%, other 2.3%, unknown 0.7%

Literacy: 99.8%

Government Type: parliamentary republic

Capital: Tallinn

Independence: 20 August 1991 (from Soviet Union)

GDP Per Capita: $21,800

Occupations: agriculture 11%, industry 20%, services 69%

Currency: Estonian kroon (EEK)

Ethiopia

Long Name: Federal Democratic Republic of Ethiopia

Location: Eastern Africa, west of Somalia

Area: 435,184 sq. mi. (1,127,127 sq. km)

Climate: tropical monsoon with wide topographic-induced variation

Terrain: high plateau with central mountain range divided by Great Rift Valley

Population: 82,544,840

Population Growth Rate: 3.212%

Birth Rate (per 1,000): 43.97

Death Rate (per 1,000): 11.83

Life Expectancy: 54.99 years

Ethnic Groups: Oromo 32.1%, Amara 30.1%, Tigraway 6.2%, Somalie 5.9%, Guragie 4.3%, Sidama 3.5%, Welaita 2.4%, other 15.4%

Religion: Christian 60.8% (Orthodox 50.6%, Protestant 10.2%), Muslim 32.8%, traditional 4.6%, other 1.8%

Languages: Amarigna 32.7%, Oromigna 31.6%, Tigrigna 6.1%, Somaligna 6%, Guaragigna 3.5%, Sidamigna 3.5%, Hadiyigna 1.7%, other 14.8%, English (major foreign language taught in schools)

Literacy: 42.7%

Government Type: federal republic

Capital: Addis Ababa

Independence: oldest independent country in Africa and one of the oldest in the world - at least 2,000 years

GDP Per Capita: $700

Occupations: agriculture 80%, industry 8%, services 12%

Currency: birr (ETB)

Fiji

Long Name: Republic of the Fiji Islands

Location: Oceania, island group in the South Pacific Ocean, about two-thirds of the way from Hawaii to New Zealand

Area: 7,054 sq. mi. (18,270 sq. km)

Climate: tropical marine; only slight seasonal temperature variation

Terrain: mostly mountains of volcanic origin

Population: 931,741

Population Growth Rate: 1.388%

Birth Rate (per 1,000): 22.15

Death Rate (per 1,000): 5.66

Life Expectancy: 70.44 years

Ethnic Groups: Fijian 57.3% (predominantly Melanesian with a Polynesian admixture), Indian 37.6%, Rotuman 1.2%, other 3.9% (European, other Pacific Islanders, Chinese)

Religion: Christian 53% (Methodist 34.5%, Roman Catholic 7.2%, Assembly of God 3.8%, Seventh Day Adventist 2.6%, other 4.9%), Hindu 34% (Sanatan 25%, Arya Samaj 1.2%, other 7.8%), Muslim 7% (Sunni 4.2%. other 2.8%), other or unspecified 5.6%, none 0.3%

Languages: English (official), Fijian (official), Hindustani

Literacy: 93.7%

Government Type: republic

Capital: Suva

Independence: 10 October 1970 (from United Kingdom)

GDP Per Capita: $3,900

Occupations: agriculture 70%, industry and services 30%

Currency: Fijian dollar (FJD)

Finland

Long Name: Republic of Finland

Location: Northern Europe, bordering the Baltic Sea, Gulf of Bothnia, and Gulf of Finland, between Sweden and Russia

Area: 130,558 sq. mi. (338,145 sq. km)

Climate: cold temperate; potentially subarctic but comparatively mild because of moderating influence of the North Atlantic Current, Baltic Sea, and more than 60,000 lakes

Terrain: mostly low, flat to rolling plains interspersed with lakes and low hills

Population: 5,244,749

Population Growth Rate: 0.112%

Birth Rate (per 1,000): 10.39

Death Rate (per 1,000): 10

Life Expectancy: 78.82 years

Ethnic Groups: Finn 93.4%, Swede 5.6%, Russian 0.5%, Estonian 0.3%, Roma (Gypsy) 0.1%, Sami 0.1%

Religion: Lutheran Church of Finland 82.5%, Orthodox Church 1.1%, other Christian 1.1%, other 0.1%, none 15.1%

Languages: Finnish 91.5% (official), Swedish 5.5% (official), other 3% (small Sami- and Russian-speaking minorities)

Literacy: 100%

Government Type: republic

Capital: Helsinki

Independence: 6 December 1917 (from Russia)

GDP Per Capita: $36,000

Occupations: agriculture and forestry 4.4%, industry 18.6%, construction 6%, commerce 16.3%, finance, insurance, and business services 13.9%, transport and communications 7.6%, public services 33.2%

Currency: euro (EUR)

France

Long Name: French Republic

Location: Western Europe, bordering the Bay of Biscay and English Channel, between Belgium and Spain, southeast of the United Kingdom; bordering the Mediterranean Sea, between Italy and Spain

Area: 211,208 sq. mi. (547,030 sq. km)

Climate: generally cool winters and mild summers, but mild winters and hot summers along the Mediterranean; occasional strong, cold, dry, north-to-northwesterly wind known as mistral

Terrain: mostly flat plains or gently rolling hills in north and west; remainder is mountainous, especially Pyrenees in south, Alps in east

Population: 62,150,775

Population Growth Rate: 0.574%

Birth Rate (per 1,000): 12.73

Death Rate (per 1,000): 8.48

Life Expectancy: 80.87 years

Ethnic Groups: Celtic and Latin with Teutonic, Slavic, North African, Indochinese, Basque minorities

Religion: Roman Catholic 83%-88%, Protestant 2%, Jewish 1%, Muslim 5%-10%, unaffiliated 4%

Languages: French 100%, rapidly declining regional dialects and languages (Provencal, Breton, Alsatian, Corsican, Catalan, Basque, Flemish)

Literacy: 99%

Government Type: republic

Capital: Paris

Independence: 486 (Frankish tribes unified); 843 (Western Francia established from the division of the Carolingian Empire)

GDP Per Capita: $32,600

Occupations: agriculture 4.1%, industry 24.4%, services 71.5%

Currency: euro (EUR)

Gabon

Long Name: Gabonese Republic

Location: Western Africa, bordering the Atlantic Ocean at the Equator, between Republic of the Congo and Equatorial Guinea

Area: 103,346 sq. mi. (267,667 sq. km)

Climate: tropical; always hot, humid

Terrain: narrow coastal plain; hilly interior; savanna in east and south

Population: 1,485,832

Population Growth Rate: 1.954%

Birth Rate (per 1,000): 35.75

Death Rate (per 1,000): 12.59

Life Expectancy: 53.52 years

Ethnic Groups: Bantu tribes, including four major tribal groupings (Fang, Bapounou, Nzebi, Obamba); other Africans and Europeans, 154,000, including 10,700 French and 11,000 persons of dual nationality

Religion: Christian 55%-75%, animist, Muslim less than 1%

Languages: French (official), Fang, Myene, Nzebi, Bapounou/Eschira, Bandjabi

Literacy: 63.2%

Government Type: republic; multiparty presidential regime

Capital: Libreville

Independence: 17 August 1960 (from France)

GDP Per Capita: $14,000

Occupations: agriculture 60%, industry 15%, services 25%

Currency: Communaute Financiere Africaine franc (XAF)

The Gambia

Long Name: Republic of The Gambia

Location: Western Africa, bordering the North Atlantic Ocean and Senegal

Area: 4,363 sq. mi. (11,300 sq. km)

Climate: tropical; hot, rainy season (June to November); cooler, dry season (November to May)

Terrain: flood plain of the Gambia River flanked by some low hills

Population: 1,735,464

Population Growth Rate: 2.724%

Birth Rate (per 1,000): 38.36

Death Rate (per 1,000): 11.74

Life Expectancy: 54.95 years

Ethnic Groups: African 99% (Mandinka 42%, Fula 18%, Wolof 16%, Jola 10%, Serahuli 9%, other 4%), non-African 1%

Religion: Muslim 90%, Christian 8%, indigenous beliefs 2%

Languages: English (official), Mandinka, Wolof, Fula, other indigenous vernaculars

Literacy: 40.1%

Government Type: republic

Capital: Banjul

Independence: 18 February 1965 (from United Kingdom)

GDP Per Capita: $1,200

Occupations: agriculture 75%, industry 19%, services 6%

Currency: dalasi (GMD)

Georgia

Long Name: Georgia

Location: Southwestern Asia, bordering the Black Sea, between Turkey and Russia

Area: 26,911 sq. mi. (69,700 sq. km)

Climate: warm and pleasant; Mediterranean-like on Black Sea coast

Terrain: largely mountainous with Great Caucasus Mountains in the north and Lesser Caucasus Mountains in the south; Kolkhet'is Dablobi (Kolkhida Lowland) opens to the Black Sea in the west; Mtkvari River Basin in the east; good soils in river valley flood plains, foothills of Kolkhida Lowland

Population: 4,630,841

Population Growth Rate: -0.325%

Birth Rate (per 1,000): 10.62

Death Rate (per 1,000): 9.51

Life Expectancy: 76.51 years

Ethnic Groups: Georgian 83.8%, Azeri 6.5%, Armenian 5.7%, Russian 1.5%, other 2.5%

Religion: Orthodox Christian 83.9%, Muslim 9.9%, Armenian-Gregorian 3.9%, Catholic 0.8%, other 0.8%, none 0.7%

Languages: Georgian 71% (official), Russian 9%, Armenian 7%, Azeri 6%, other 7%

Literacy: 100%

Government Type: republic

Capital: T'bilisi

Independence: 9 April 1991 (from Soviet Union)

GDP Per Capita: $4,400

Occupations: agriculture 55.6%, industry 8.9%, services 35.5%

Currency: lari (GEL)

Germany

Long Name: Federal Republic of Germany

Location: Central Europe, bordering the Baltic Sea and the North Sea, between the Netherlands and Poland, south of Denmark

Area: 137,846 sq. mi. (357,021 sq. km)

Climate: temperate and marine; cool, cloudy, wet winters and summers; occasional warm mountain (foehn) wind

Terrain: lowlands in north, uplands in center, Bavarian Alps in south

Population: 82,369,552

Population Growth Rate: -0.044%

Birth Rate (per 1,000): 8.18

Death Rate (per 1,000): 10.8

Life Expectancy: 79.1 years

Ethnic Groups: German 91.5%, Turkish 2.4%, other 6.1% (made up largely of Greek, Italian, Polish, Russian, Serbo-Croatian, Spanish)

Religion: Protestant 34%, Roman Catholic 34%, Muslim 3.7%, unaffiliated or other 28.3%

Languages: German

Literacy: 99%

Government Type: federal republic

Capital: Berlin

Independence: 18 January 1871 (German Empire unification); divided into four zones of occupation (United Kingdom, US, USSR, and later, France) in 1945 following World War II; Federal Republic of Germany (FRG or West Germany) proclaimed 23 May 1949 and included the former United Kingdom, US, and French zones; German Democratic Republic (GDR or East Germany) proclaimed 7 October 1949 and included the former USSR zone; unification of West Germany and East Germany took place 3 October 1990; all four powers formally relinquished rights 15 March 1991

GDP Per Capita: $34,100

Occupations: agriculture 2.8%, industry 33.4%, services 63.8%

Currency: euro (EUR)

Ghana

Long Name: Republic of Ghana

Location: Western Africa, bordering the Gulf of Guinea, between Cote d'Ivoire and Togo

Area: 92,456 sq. mi. (239,460 sq. km)

Climate: tropical; warm and comparatively dry along southeast coast; hot and humid in southwest; hot and dry in north

Terrain: mostly low plains with dissected plateau in south-central area

Population: 23,382,848

Population Growth Rate: 1.928%

Birth Rate (per 1,000): 29.22

Death Rate (per 1,000): 9.39

Life Expectancy: 59.49 years

Ethnic Groups: Akan 45.3%, Mole-Dagbon 15.2%, Ewe 11.7%, Ga-Dangme 7.3%, Guan 4%, Gurma 3.6%, Grusi 2.6%, Mande-Busanga 1%, other tribes 1.4%, other 7.8%

Religion: Christian 68.8% (Pentecostal/Charismatic 24.1%, Protestant 18.6%, Catholic 15.1%, other 11%), Muslim 15.9%, traditional 8.5%, other 0.7%, none 6.1%

Languages: Asante 14.8%, Ewe 12.7%, Fante 9.9%, Boron (Brong) 4.6%, Dagomba 4.3%, Dangme 4.3%, Dagarte (Dagaba) 3.7%, Akyem 3.4%, Ga 3.4%, Akuapem 2.9%, other 36.1% (includes English [official])

Literacy: 57.9%

Government Type: constitutional democracy

Capital: Accra

Independence: 6 March 1957 (from United Kingdom)

GDP Per Capita: $1,400

Occupations: agriculture 56%, industry 15%, services 29%

Currency: Ghana cedi (GHC)

Greece

Long Name: Hellenic Republic

Location: Southern Europe, bordering the Aegean Sea, Ionian Sea, and the Mediterranean Sea, between Albania and Turkey

Area: 50,942 sq. mi. (131,940 sq. km)

Climate: temperate; mild, wet winters; hot, dry summers

Terrain: mostly mountains with ranges extending into the sea as peninsulas or chains of islands

Population: 10,722,816

Population Growth Rate: 0.146%

Birth Rate (per 1,000): 9.54

Death Rate (per 1,000): 10.42

Life Expectancy: 79.52 years

Ethnic Groups: Greek 93%, other (foreign citizens) 7%

Religion: Greek Orthodox 98%, Muslim 1.3%, other 0.7%

Languages: Greek 99% (official), other 1% (includes English and French)

Literacy: 96%

Government Type: parliamentary republic

Capital: Athens

Independence: 1829 (from the Ottoman Empire)

GDP Per Capita: $30,600

Occupations: agriculture 12%, industry 20%, services 68%

Currency: euro (EUR)

Grenada

Long Name: Grenada

Location: Caribbean, island between the Caribbean Sea and Atlantic Ocean, north of Trinidad and Tobago

Area: 133 sq. mi. (344 sq. km)

Climate: tropical; tempered by northeast trade winds

Terrain: volcanic in origin with central mountains

Population: 90,343

Population Growth Rate: 0.406%

Birth Rate (per 1,000): 21.61

Death Rate (per 1,000): 6.31

Life Expectancy: 65.6 years

Ethnic Groups: black 82%, mixed black and European 13%, European and East Indian 5%, and trace of Arawak/Carib Amerindian

Religion: Roman Catholic 53%, Anglican 13.8%, other Protestant 33.2%

Languages: English (official), French patois

Literacy: 96%

Government Type: parliamentary democracy

Capital: Saint George's

Independence: 7 February 1974 (from United Kingdom)

GDP Per Capita: $10,500

Occupations: agriculture 24%, industry 14%, services 62%

Currency: East Caribbean dollar (XCD)

Guatemala

Long Name: Republic of Guatemala

Location: Central America, bordering the North Pacific Ocean, between El Salvador and Mexico, and bordering the Gulf of Honduras (Caribbean Sea) between Honduras and Belize

Area: 42,042 sq. mi. (108,890 sq. km)

Climate: tropical; hot, humid in lowlands; cooler in highlands

Terrain: mostly mountains with narrow coastal plains and rolling limestone plateau

Population: 13,002,206

Population Growth Rate: 2.11%

Birth Rate (per 1,000): 28.55

Death Rate (per 1,000): 5.19

Life Expectancy: 69.99 years

Ethnic Groups: Mestizo (mixed Amerindian-Spanish - in local Spanish called Ladino) and European 59.4%, K'iche 9.1%, Kaqchikel 8.4%, Mam 7.9%, Q'eqchi 6.3%, other Mayan 8.6%, indigenous non-Mayan 0.2%, other 0.1%

Religion: Roman Catholic, Protestant, indigenous Mayan beliefs

Languages: Spanish 60%, Amerindian languages 40% (23 officially recognized Amerindian languages, including Quiche, Cakchiquel, Kekchi, Mam, Garifuna, and Xinca)

Literacy: 69.1%

Government Type: constitutional democratic republic

Capital: Guatemala

Independence: 15 September 1821 (from Spain)

GDP Per Capita: $5,100

Occupations: agriculture 50%, industry 15%, services 35%

Currency: quetzal (GTQ), US dollar (USD), others allowed

Guinea

Long Name: Republic of Guinea

Location: Western Africa, bordering the North Atlantic Ocean, between Guinea-Bissau and Sierra Leone

Area: 94,925 sq. mi. (245,857 sq. km)

Climate: generally hot and humid; monsoonal-type rainy season (June to November) with southwesterly winds; dry season (December to May) with northeasterly harmattan winds

Terrain: generally flat coastal plain, hilly to mountainous interior

Population: 9,806,509

Population Growth Rate: 2.492%

Birth Rate (per 1,000): 37.84

Death Rate (per 1,000): 11.29

Life Expectancy: 56.58 years

Ethnic Groups: Peuhl 40%, Malinke 30%, Soussou 20%, smaller ethnic groups 10%

Religion: Muslim 85%, Christian 8%, indigenous beliefs 7%

Languages: French (official); note - each ethnic group has its own language

Literacy: 29.5%

Government Type: republic

Capital: Conakry

Independence: 2 October 1958 (from France)

GDP Per Capita: $1,100

Occupations: agriculture 76%, industry and services 24%

Currency: Guinean franc (GNF)

Guinea-Bissau

Long Name: Republic of Guinea-Bissau

Location: Western Africa, bordering the North Atlantic Ocean, between Guinea and Senegal

Area: 13,946 sq. mi. (36,120 sq. km)

Climate: tropical; generally hot and humid; monsoonal-type rainy season (June to November) with southwesterly winds; dry season (December to May) with northeasterly harmattan winds

Terrain: mostly low coastal plain rising to savanna in east

Population: 1,503,182

Population Growth Rate: 2.035%

Birth Rate (per 1,000): 36.4

Death Rate (per 1,000): 16.05

Life Expectancy: 47.52 years

Ethnic Groups: African 99% (includes Balanta 30%, Fula 20%, Manjaca 14%, Mandinga 13%, Papel 7%), European and mulatto less than 1%

Religion: Muslim 50%, indigenous beliefs 40%, Christian 10%

Languages: Portuguese (official), Crioulo, African languages

Literacy: 42.4%

Government Type: republic

Capital: Bissau

Independence: 24 September 1973 (declared); 10 September 1974 (from Portugal)

GDP Per Capita: $600

Occupations: agriculture 82%, industry and services 18%

Currency: Communaute Financiere Africaine franc (XOF)

Guyana

Long Name: Cooperative Republic of Guyana

Location: Northern South America, bordering the North Atlantic Ocean, between Suriname and Venezuela

Area: 83,000 sq. mi. (214,970 sq. km)

Climate: tropical; hot, humid, moderated by northeast trade winds; two rainy seasons (May to August, November to January)

Terrain: mostly rolling highlands; low coastal plain; savanna in south

Population: 770,794

Population Growth Rate: 0.211%

Birth Rate (per 1,000): 17.85

Death Rate (per 1,000): 8.29

Life Expectancy: 66.43 years

Ethnic Groups: East Indian 43.5%, black (African) 30.2%, mixed 16.7%, Amerindian 9.1%, other 0.5%

Religion: Hindu 28.4%, Pentecostal 16.9%, Roman Catholic 8.1%, Anglican 6.9%, Seventh Day Adventist 5%, Methodist 1.7%, Jehovah Witness 1.1%, other Christian 17.7%, Muslim 7.2%, other 4.3%, none 4.3%

Languages: English, Amerindian dialects, Creole, Caribbean Hindustani (a dialect of Hindi), Urd

Literacy: 98.8%

Government Type: republic

Capital: Georgetown

Independence: 26 May 1966 (from United Kingdom)

GDP Per Capita: $3,700

Occupations: N/A

Currency: Guyanese dollar (GYD)

Haiti

Long Name: Republic of Haiti

Location: Caribbean, western one-third of the island of Hispaniola, between the Caribbean Sea and the North Atlantic Ocean, west of the Dominican Republic

Area: 10,714 sq. mi. (27,750 sq. km)

Climate: tropical; semiarid where mountains in east cut off trade winds

Terrain: mostly rough and mountainous

Population: 8,924,553

Population Growth Rate: 2.493%

Birth Rate (per 1,000): 35.69

Death Rate (per 1,000): 10.15

Life Expectancy: 57.56 years

Ethnic Groups: black 95%, mulatto and white 5%

Religion: Roman Catholic 80%, Protestant 16% (Baptist 10%, Pentecostal 4%, Adventist 1%, other 1%), none 1%, other 3% (voodoo practiced by about half the population)

Languages: French (official), Creole (official)

Literacy: 52.9%

Government Type: republic

Capital: Port-au-Prince

Independence: 1 January 1804 (from France)

GDP Per Capita: $1,300

Occupations: agriculture 66%, industry 9%, services 25%

Currency: gourde (HTG)

Honduras

Long Name: Republic of Honduras

Location: Central America, bordering the Caribbean Sea, between Guatemala and Nicaragua and bordering the Gulf of Fonseca (North Pacific Ocean), between El Salvador and Nicaragua

Area: 43,278 sq. mi. (112,090 sq. km)

Climate: subtropical in lowlands, temperate in mountains

Terrain: mostly mountains in interior, narrow coastal plains

Population: 7,639,327

Population Growth Rate: 2.024%

Birth Rate (per 1,000): 26.93

Death Rate (per 1,000): 5.36

Life Expectancy: 69.37 years

Ethnic Groups: mestizo (mixed Amerindian and European) 90%, Amerindian 7%, black 2%, white 1%

Religion: Roman Catholic 97%, Protestant 3%

305

Languages: Spanish, Amerindian dialects

Literacy: 80%

Government Type: democratic constitutional republic

Capital: Tegucigalpa

Independence: 15 September 1821 (from Spain)

GDP Per Capita: $4,300

Occupations: agriculture 34%, industry 23%, services 43%

Currency: lempira (HNL)

Hungary

Long Name: Republic of Hungary

Location: Central Europe, northwest of Romania

Area: 35,919 sq. mi. (93,030 sq. km)

Climate: temperate; cold, cloudy, humid winters; warm summers

Terrain: mostly flat to rolling plains; hills and low mountains on the Slovakian border

Population: 9,930,915

Population Growth Rate: -0.254%

Birth Rate (per 1,000): 9.59

Death Rate (per 1,000): 12.99

Life Expectancy: 73.18 years

Ethnic Groups: Hungarian 92.3%, Roma 1.9%, other or unknown 5.8%

Religion: Roman Catholic 51.9%, Calvinist 15.9%, Lutheran 3%, Greek Catholic 2.6%, other Christian 1%, other or unspecified 11.1%, unaffiliated 14.5%

Languages: Hungarian 93.6%, other or unspecified 6.4%

Literacy: 99.4%

Government Type: parliamentary democracy

Capital: Budapest

Independence: 25 December 1000 (crowning of King Stephen I, traditional founding date)

GDP Per Capita: $19,300

Occupations: agriculture 5.5%, industry 33.3%, services 61.2%

Currency: forint (HUF)

Iceland

Long Name: Republic of Iceland

Location: Northern Europe, island between the Greenland Sea and the North Atlantic Ocean, northwest of the United Kingdom

Area: 39,768 sq. mi. (103,000 sq. km)

Climate: temperate; moderated by North Atlantic Current; mild, windy winters; damp, cool summers

Terrain: mostly plateau interspersed with mountain peaks, icefields; coast deeply indented by bays and fiords

Population: 304,367

Population Growth Rate: 0.783%

Birth Rate (per 1,000): 13.5

Death Rate (per 1,000): 6.81

Life Expectancy: 80.55 years

Ethnic Groups: homogeneous mixture of descendants of Norse and Celts 94%, population of foreign origin 6%

Religion: Lutheran Church of Iceland 82.1%, Roman Catholic Church 2.4%, Reykjavik Free Church 2.3%, Hafnarfjorour Free Church 1.6%, other Christian 2.8%, other religions 0.9%, unaffiliated 2.6%, other or unspecified 5.5%

Languages: Icelandic, English, Nordic languages, German widely spoken

Literacy: 99%

Government Type: constitutional republic

Capital: Reykjavik

Independence: 1 December 1918 (became a sovereign state under the Danish Crown); 17 June 1944 (from Denmark)

GDP Per Capita: $40,400

Occupations: agriculture 5.1%, industry 23%, services 71.8%

Currency: Icelandic krona (ISK)

India

Long Name: Republic of India

Location: Southern Asia, bordering the Arabian Sea and the Bay of Bengal, between Burma and Pakistan

Area: 1,269,340 sq. mi. (3,287,590 sq. km)

Climate: varies from tropical monsoon in south to temperate in north

Terrain: upland plain (Deccan Plateau) in south, flat to rolling plain along the Ganges, deserts in west, Himalayas in north

Population: 1,147,995,904

Population Growth Rate: 1.578%

Birth Rate (per 1,000): 22.22

Death Rate (per 1,000): 6.4

Life Expectancy: 69.25 years

Ethnic Groups: Indo-Aryan 72%, Dravidian 25%, Mongoloid and other 3%

Religion: Hindu 80.5%, Muslim 13.4%, Christian 2.3%, Sikh 1.9%, other 1.8%, unspecified 0.1%

Languages: Hindi 41%, Bengali 8.1%, Telugu 7.2%, Marathi 7%, Tamil 5.9%, Urdu 5%, Gujarati 4.5%, Kannada 3.7%, Malayalam 3.2%, Oriya 3.2%, Punjabi 2.8%, Assamese 1.3%, Maithili 1.2%, other 5.9% (English

commonly spoken by the well-educated)

Literacy: 61%

Government Type: federal republic

Capital: New Delhi

Independence: 15 August 1947 (from United Kingdom)

GDP Per Capita: $2,600

Occupations: agriculture 60%, industry 12%, services 28%

Currency: Indian rupee (INR)

Indonesia

Long Name: Republic of Indonesia

Location: Southeastern Asia, archipelago between the Indian Ocean and the Pacific Ocean

Area: 741,097 sq. mi. (1,919,440 sq. km)

Climate: tropical; hot, humid; more moderate in highlands

Terrain: mostly coastal lowlands; larger islands have interior mountains

Population: 237,512,352

Population Growth Rate: 1.175%

Birth Rate (per 1,000): 19.24

Death Rate (per 1,000): 6.24

Life Expectancy: 70.46 years

Ethnic Groups: Javanese 40.6%, Sundanese 15%, Madurese 3.3%, Minangkabau 2.7%, Betawi 2.4%, Bugis 2.4%, Banten 2%, Banjar 1.7%, other or unspecified 29.9%

Religion: Muslim 86.1%, Protestant 5.7%, Roman Catholic 3%, Hindu 1.8%, other or unspecified 3.4%

Languages: Bahasa Indonesia (official, modified form of Malay), English, Dutch, local dialects (the most widely spoken of which is Javanese)

Literacy: 90.4%

307

Government Type: republic

Capital: Jakarta

Independence: 17 August 1945

GDP Per Capita: $3,600

Occupations: agriculture 43.3%, industry 18%, services 38.7%

Currency: Indonesian rupiah (IDR)

Iran

Long Name: Islamic Republic of Iran

Location: Middle East, bordering the Gulf of Oman, the Persian Gulf, and the Caspian Sea, between Iraq and Pakistan

Area: 636,293 sq. mi. (1,648,000 sq. km)

Climate: mostly arid or semiarid, subtropical along Caspian coast

Terrain: rugged, mountainous rim; high, central basin with deserts, mountains; small, discontinuous plains along both coasts

Population: 65,875,224

Population Growth Rate: 0.792%

Birth Rate (per 1,000): 16.89

Death Rate (per 1,000): 5.69

Life Expectancy: 70.86 years

Ethnic Groups: Persian 51%, Azeri 24%, Gilaki and Mazandarani 8%, Kurd 7%, Arab 3%, Lur 2%, Baloch 2%, Turkmen 2%, other 1%

Religion: Muslim 98% (Shi'a 89%, Sunni 9%), other (includes Zoroastrian, Jewish, Christian, and Baha'i) 2%

Languages: Persian and Persian dialects 58%, Turkic and Turkic dialects 26%, Kurdish 9%, Luri 2%, Balochi 1%, Arabic 1%, Turkish 1%, other 2%

Literacy: 77%

Government Type: theocratic republic

Capital: Tehran

Independence: 1 April 1979 (Islamic Republic of Iran proclaimed)

GDP Per Capita: $11,700

Occupations: agriculture 25%, industry 31%, services 45%

Currency: Iranian rial (IRR)

Iraq

Long Name: Republic of Iraq

Location: Middle East, bordering the Persian Gulf, between Iran and Kuwait

Area: 168,754 sq. mi. (437,072 sq. km)

Climate: mostly desert; mild to cool winters with dry, hot, cloudless summers; northern mountainous regions along Iranian and Turkish borders experience cold winters with occasionally heavy snows that melt in early spring, sometimes causing extensive flooding in central and southern Iraq

Terrain: mostly broad plains; reedy marshes along Iranian border in south with large flooded areas; mountains along borders with Iran and Turkey

Population: 28,221,180

Population Growth Rate: 2.562%

Birth Rate (per 1,000): 30.77

Death Rate (per 1,000): 5.14

Life Expectancy: 69.62 years

Ethnic Groups: Arab 75%-80%, Kurdish 15%-20%, Turkoman, Assyrian, or other 5%

Religion: Muslim 97% (Shi'a 60%-65%, Sunni 32%-37%), Christian or other 3%

Languages: Arabic, Kurdish (official in Kurdish regions), Turkoman (a Turkish dialect), Assyrian (Neo-Aramaic), Armenian

Literacy: 74.1%

Government Type: parliamentary democracy

Capital: Baghdad

Independence: 3 October 1932 (from League of Nations mandate under British administration)

GDP Per Capita: $3,700

Occupations: N/A

Currency: New Iraqi dinar (NID)

Ireland

Long Name: Ireland

Location: Western Europe, occupying five-sixths of the island of Ireland in the North Atlantic Ocean, west of Great Britain

Area: 27,135 sq. mi. (70,280 sq. km)

Climate: temperate maritime; modified by North Atlantic Current; mild winters, cool summers; consistently humid; overcast about half the time

Terrain: mostly level to rolling interior plain surrounded by rugged hills and low mountains; sea cliffs on west coast

Population: 4,156,119

Population Growth Rate: 1.133%

Birth Rate (per 1,000): 14.33

Death Rate (per 1,000): 7.77

Life Expectancy: 78.07 years

Ethnic Groups: Irish 87.4%, other white 7.5%, Asian 1.3%, black 1.1%, mixed 1.1%, unspecified 1.6%

Religion: Roman Catholic 87.4%, Church of Ireland 2.9%, other Christian 1.9%, other 2.1%, unspecified 1.5%, none 4.2%

Languages: English (official) is the language generally used, Irish (Gaelic or Gaeilge) (official) spoken mainly in areas located along the western seaboard

Literacy: 99%

Government Type: republic, parliamentary democracy

Capital: Dublin

Independence: 6 December 1921 (from United Kingdom by treaty)

GDP Per Capita: $46,600

Occupations: agriculture 6%, industry 27%, services 67%

Currency: euro (EUR)

Israel

Long Name: State of Israel

Location: Middle East, bordering the Mediterranean Sea, between Egypt and Lebanon

Area: 8,019 sq. mi. (20,770 sq. km)

Climate: temperate; hot and dry in southern and eastern desert areas

Terrain: Negev desert in the south; low coastal plain; central mountains; Jordan Rift Valley

Population: 7,112,359

Population Growth Rate: 1.713%

Birth Rate (per 1,000): 20.02

Death Rate (per 1,000): 5.41

Life Expectancy: 80.61 years

Ethnic Groups: Jewish 76.4% (of which Israel-born 67.1%, Europe/America-born 22.6%, Africa-born 5.9%, Asia-born 4.2%), non-Jewish 23.6% (mostly Arab)

Religion: Jewish 76.4%, Muslim 16%, Arab Christians 1.7%, other Christian 0.4%, Druze 1.6%, unspecified 3.9%

Languages: Hebrew (official), Arabic used officially for Arab minority, English most commonly used foreign language

Literacy: 97.1%

Government Type: parliamentary democracy

Capital: Jerusalem

Independence: 14 May 1948 (from League of Nations mandate under British administration)

GDP Per Capita: $26,600

Occupations: agriculture 18.5%, industry 23.7%, services 50%, other 7.8%

Currency: new Israeli shekel (ILS)

Italy

Long Name: Italian Republic

Location: Southern Europe, a peninsula extending into the central Mediterranean Sea, northeast of Tunisia

Area: 116,305 sq. mi. (301,230 sq. km)

Climate: predominantly Mediterranean; Alpine in far north; hot, dry in south

Terrain: mostly rugged and mountainous; some plains, coastal lowlands

Population: 58,145,320

Population Growth Rate: -0.019%

Birth Rate (per 1,000): 8.36

Death Rate (per 1,000): 10.61

Life Expectancy: 80.07 years

Ethnic Groups: Italian (includes small clusters of German-, French-, and Slovene-Italians in the north and Albanian-Italians and Greek-Italians in the south)

Religion: Roman Catholic 90% (approximately; about one-third practicing), other 10% (includes mature Protestant and Jewish communities and a growing Muslim immigrant community)

Languages: Italian (official), German (parts of Trentino-Alto Adige region are predominantly German speaking), French (small French-speaking minority in Valle d'Aosta region), Slovene (Slovene-speaking minority in the Trieste-Gorizia area)

Literacy: 98.4%

Government Type: republic

Capital: Rome

Independence: 17 March 1861 (Kingdom of Italy proclaimed; Italy was not finally unified until 1870)

GDP Per Capita: $30,900

Occupations: agriculture 5%, industry 32%, services 63%

Currency: euro (EUR)

Jamaica

Long Name: Jamaica

Location: Caribbean, island in the Caribbean Sea, south of Cuba

Area: 4,244 sq. mi. (10,991 sq. km)

Climate: tropical; hot, humid; temperate interior

Terrain: mostly mountains, with narrow, discontinuous coastal plain

Population: 2,804,332

Population Growth Rate: 0.779%

Birth Rate (per 1,000): 20.04

Death Rate (per 1,000): 6.37

Life Expectancy: 73.59 years

Ethnic Groups: black 91.2%, mixed 6.2%, other or unknown 2.6%

Religion: Protestant 62.5% (Seventh-Day Adventist 10.8%, Pentecostal 9.5%, Other Church of God 8.3%, Baptist 7.2%, New Testament Church of God 6.3%, Church of God in Jamaica 4.8%, Church of God of

Prophecy 4.3%, Anglican 3.6%, other Christian 7.7%), Roman Catholic 2.6%, other or unspecified 14.2%, none 20.9%

Languages: English, English patois

Literacy: 87.9%

Government Type: constitutional parliamentary democracy

Capital: Kingston

Independence: 6 August 1962 (from United Kingdom)

GDP Per Capita: $7,400

Occupations: agriculture 17%, industry 19%, services 64%

Currency: Jamaican dollar (JMD)

Japan

Long Name: Japan

Location: Eastern Asia, island chain between the North Pacific Ocean and the Sea of Japan, east of the Korean Peninsula

Area: 145,882 sq. mi. (377,835 sq. km)

Climate: varies from tropical in south to cool temperate in north

Terrain: mostly rugged and mountainous

Population: 127,288,416

Population Growth Rate: -0.139%

Birth Rate (per 1,000): 7.87

Death Rate (per 1,000): 9.26

Life Expectancy: 82.07 years

Ethnic Groups: Japanese 98.5%, Koreans 0.5%, Chinese 0.4%, other 0.6%

Religion: observe both Shinto and Buddhist 84%, other 16% (including Christian 0.7%)

Languages: Japanese

Literacy: 99%

Government Type: constitutional monarchy with a parliamentary government

Capital: Tokyo

Independence: 660 B.C.E. (traditional founding by Emperor Jimmu)

GDP Per Capita: $33,500

Occupations: agriculture 4.6%, industry 27.8%, services 67.7%

Currency: yen (JPY)

Jordan

Long Name: Hashemite Kingdom of Jordan

Location: Middle East, northwest of Saudi Arabia

Area: 35,637 sq. mi. (92,300 sq. km)

Climate: mostly arid desert; rainy season in west (November to April)

Terrain: mostly desert plateau in east, highland area in west; Great Rift Valley separates East and West Banks of the Jordan River

Population: 6,198,677

Population Growth Rate: 2.338%

Birth Rate (per 1,000): 20.13

Death Rate (per 1,000): 2.72

Life Expectancy: 78.71 years

Ethnic Groups: Arab 98%, Circassian 1%, Armenian 1%

Religion: Sunni Muslim 92%, Christian 6% (majority Greek Orthodox, but some Greek and Roman Catholics, Syrian Orthodox, Coptic Orthodox, Armenian Orthodox, and Protestant denominations), other 2% (several small Shi'a Muslim and Druze populations)

Languages: Arabic (official), English widely understood among upper and middle classes

Literacy: 89.9%

Government Type: constitutional monarchy

Capital: Amman

Independence: 25 May 1946 (from League of Nations mandate under British administration)

GDP Per Capita: $4,700

Occupations: agriculture 5%, industry 12.5%, services 82.5%

Currency: Jordanian dinar (JOD)

Kazakhstan

Long Name: Republic of Kazakhstan

Location: Central Asia, northwest of China; a small portion west of the Ural River in eastern-most Europe

Area: 1,049,151 sq. mi. (2,717,300 sq. km)

Climate: continental, cold winters and hot summers, arid and semiarid

Terrain: extends from the Volga to the Altai Mountains and from the plains in western Siberia to oases and desert in Central Asia

Population: 15,340,533

Population Growth Rate: 0.374%

Birth Rate (per 1,000): 16.44

Death Rate (per 1,000): 9.39

Life Expectancy: 67.55 years

Ethnic Groups: Kazakh (Qazaq) 53.4%, Russian 30%, Ukrainian 3.7%, Uzbek 2.5%, German 2.4%, Tatar 1.7%, Uygur 1.4%, other 4.9%

Religion: Muslim 47%, Russian Orthodox 44%, Protestant 2%, other 7%

Languages: Kazakh (Qazaq, state language) 64.4%, Russian (official, used in everyday business, designated the "language of interethnic communication") 95%

Literacy: 99.5%

Government Type: republic; authoritarian presidential rule, with little power outside the executive branch

Capital: Astana

Independence: 16 December 1991 (from Soviet Union)

GDP Per Capita: $11,000

Occupations: agriculture 32.2%, industry 18%, services 49.8%

Currency: tenge (KZT)

Kenya

Long Name: Republic of Kenya

Location: Eastern Africa, bordering the Indian Ocean, between Somalia and Tanzania

Area: 224,961 sq. mi. (582,650 sq. km)

Climate: varies from tropical along coast to arid in interior

Terrain: low plains rise to central highlands bisected by Great Rift Valley; fertile plateau in west

Population: 37,953,840

Population Growth Rate: 2.758%

Birth Rate (per 1,000): 37.89

Death Rate (per 1,000): 10.3

Life Expectancy: 56.64 years

Ethnic Groups: Kikuyu 22%, Luhya 14%, Luo 13%, Kalenjin 12%, Kamba 11%, Kisii 6%, Meru 6%, other African 15%, non-African (Asian, European, and Arab) 1%

Religion: Protestant 45%, Roman Catholic 33%, Muslim 10%, indigenous beliefs 10%, other 2%

Languages: English (official), Kiswahili (official), numerous indigenous languages

Literacy: 85.1%

Government Type: republic

Capital: Nairobi

Independence: 12 December 1963 (from United Kingdom)

GDP Per Capita: $1,700

Occupations: agriculture 75%, industry and services 25%

Currency: Kenyan shilling (KES)

Kiribati

Long Name: Republic of Kiribati

Location: Oceania, group of 33 coral atolls in the Pacific Ocean, straddling the Equator; the capital Tarawa is about half way between Hawaii and Australia; note - on 1 January 1995, Kiribati proclaimed that all of its territory lies in the same time zone as its Gilbert Islands group (UTC +12) even though the Phoenix Islands and the Line Islands under its jurisdiction lie on the other side of the International Date Line

Area: 313 sq. mi. (811 sq. km)

Climate: tropical; marine, hot and humid, moderated by trade winds

Terrain: mostly low-lying coral atolls surrounded by extensive reefs

Population: 110,356

Population Growth Rate: 2.235%

Birth Rate (per 1,000): 30.31

Death Rate (per 1,000): 7.97

Life Expectancy: 62.85 years

Ethnic Groups: Micronesian 98.8%, other 1.2%

Religion: Roman Catholic 52%, Protestant (Congregational) 40%, other (includes Seventh-Day Adventist, Muslim, Baha'i, Latter-day Saints, Church of God) 8%

Languages: I-Kiribati, English (official)

Literacy: N/A

Government Type: republic

Capital: Tarawa

Independence: 12 July 1979 (from United Kingdom)

GDP Per Capita: $3,600

Occupations: agriculture 2.7%, industry 32%, services 65.3%

Currency: Australian dollar (AUD)

Kosovo

Long Name: Republic of Kosovo

Location: Southeast Europe, between Serbia and Macedonia

Area: 4,203 sq. mi. (10,887 sq. km)

Climate: influenced by continental air masses resulting in relatively cold winters with heavy snowfall and hot, dry summers and autumns; Mediterranean and alpine influences create regional variation; maximum rainfall between October and December Terrain: flat fluvial basin with an elevation of 400-700 m above sea level surrounded by several high mountain ranges with elevations of 2,000 to 2,500 m

Population: 2,126,708

Population Growth Rate: N/A

Birth Rate (per 1,000): N/A

Death Rate (per 1,000): N/A

Life Expectancy: N/A

Ethnic Groups: Albanians 88%, Serbs 7%, other 5%

313

Religion: Muslim, Serbian Orthodox, Roman Catholic

Languages: Albanian (official), Serbian (official), Bosnian, Turkish, Roma

Literacy: N/A

Government Type: republic

Capital: Pristina

Independence: 17 February 2008

GDP Per Capita: $1,800

Occupations: agriculture 21.4%, industry and services NA

Currency: euro (EUR); Serbian Dinar (RSD) is also in circulation

Kuwait

Long Name: State of Kuwait

Location: Middle East, bordering the Persian Gulf, between Iraq and Saudi Arabia

Area: 6,880 sq. mi. (17,820 sq. km)

Climate: dry desert; intensely hot summers; short, cool winters

Terrain: flat to slightly undulating desert plain

Population: 2,596,799

Population Growth Rate: 3.591% (reflects a return to pre-Gulf crisis immigration of expatriates)

Birth Rate (per 1,000): 21.9

Death Rate (per 1,000): 2.37

Life Expectancy: 77.53 years

Ethnic Groups: Kuwaiti 45%, other Arab 35%, South Asian 9%, Iranian 4%, other 7%

Religion: Muslim 85% (Sunni 70%, Shi'a 30%), other (includes Christian, Hindu, Parsi) 15%

Languages: Arabic (official), English widely spoken

Literacy: 93.3%

Government Type: constitutional emirate

Capital: Kuwait

Independence: 19 June 1961 (from United Kingdom)

GDP Per Capita: $55,900

Occupations: N/A

Currency: Kuwaiti dinar (KD)

Kyrgyzstan

Long Name: Kyrgyz Republic

Location: Central Asia, west of China

Area: 76,641 sq. mi. (198,500 sq. km)

Climate: dry continental to polar in high Tien Shan; subtropical in southwest (Fergana Valley); temperate in northern foothill zone

Terrain: peaks of Tien Shan and associated valleys and basins encompass entire nation

Population: 5,356,869

Population Growth Rate: 1.38%

Birth Rate (per 1,000): 23.31

Death Rate (per 1,000): 6.97

Life Expectancy: 69.12 years

Ethnic Groups: Kyrgyz 64.9%, Uzbek 13.8%, Russian 12.5%, Dungan 1.1%, Ukrainian 1%, Uygur 1%, other 5.7%

Religion: Muslim 75%, Russian Orthodox 20%, other 5%

Languages: Kyrgyz 64.7% (official), Uzbek 13.6%, Russian 12.5% (official), Dungun 1%, other 8.2%

Literacy: 98.7%

Government Type: republic

Capital: Bishkek

Independence: 31 August 1991 (from Soviet Union)

GDP Per Capita: $2,000

Occupations: agriculture 55%, industry 15%, services 30%

Currency: som (KGS)

Laos

Long Name: Lao People's Democratic Republic

Location: Southeastern Asia, northeast of Thailand, west of Vietnam

Area: 91,429 sq. mi. (236,800 sq. km)

Climate: tropical monsoon; rainy season (May to November); dry season (December to April)

Terrain: mostly rugged mountains; some plains and plateaus

Population: 6,677,534

Population Growth Rate: 2.344%

Birth Rate (per 1,000): 34.46

Death Rate (per 1,000): 11.02

Life Expectancy: 56.29 years

Ethnic Groups: Lao 55%, Khmou 11%, Hmong 8%, other (over 100 minor ethnic groups) 26%

Religion: Buddhist 67%, Christian 1.5%, other and unspecified 31.5%

Languages: Lao (official), French, English, and various ethnic languages

Literacy: 68.7%

Government Type: Communist state

Capital: Vientiane

Independence: 19 July 1949 (from France)

GDP Per Capita: $2,000

Occupations: agriculture 80%, industry and services 20%

Currency: kip (LAK)

Latvia

Long Name: Republic of Latvia

Location: Eastern Europe, bordering the Baltic Sea, between Estonia and Lithuania

Area: 24,938 sq. mi. (64,589 sq. km)

Climate: maritime; wet, moderate winters

Terrain: low plain

Population: 2,245,423

Population Growth Rate: -0.629%

Birth Rate (per 1,000): 9.62

Death Rate (per 1,000): 13.63

Life Expectancy: 71.88 years

Ethnic Groups: Latvian 57.7%, Russian 29.6%, Belarusian 4.1%, Ukrainian 2.7%, Polish 2.5%, Lithuanian 1.4%, other 2%

Religion: Lutheran 19.6%, Orthodox 15.3%, other Christian 1%, other 0.4%, unspecified 63.7%

Languages: Latvian (official) 58.2%, Russian 37.5%, Lithuanian and other 4.3%

Literacy: 99.7%

Government Type: parliamentary democracy

Capital: Riga

Independence: 18 November 1918 (from Soviet Russia); 4 May 1990 is when it declared the renewal of independence; 21 August 1991 was the date of *de facto* independence from the Soviet Union

GDP Per Capita: $17,700

Occupations: agriculture 13%, industry 19%, services 68%

Currency: lat (LVL)

Lebanon

Long Name: Lebanese Republic

Location: Middle East, bordering the Mediterranean Sea, between Israel and Syria

Area: 4,015 sq. mi. (10,400 sq. km)

Climate: Mediterranean; mild to cool, wet winters with hot, dry summers; Lebanon mountains experience heavy winter snows

Terrain: narrow coastal plain; El Beqaa (Bekaa Valley) separates Lebanon and Anti-Lebanon Mountains

Population: 3,971,941

Population Growth Rate: 1.154%

Birth Rate (per 1,000): 17.61

Death Rate (per 1,000): 6.06

Life Expectancy: 73.41 years

Ethnic Groups: Arab 95%, Armenian 4%, other 1%

Religion: Muslim 59.7% (Shi'a, Sunni, Druze, Isma'ilite, Alawite or Nusayri), Christian 39% (Maronite Catholic, Greek Orthodox, Melkite Catholic, Armenian Orthodox, Syrian Catholic, Armenian Catholic, Syrian Orthodox, Roman Catholic, Chaldean, Assyrian, Copt, Protestant), other 1.3%, 17 religious sects recognized

Languages: Arabic (official), French, English, Armenian

Literacy: 87.4%

Government Type: republic

Capital: Beirut

Independence: 22 November 1943 (from League of Nations mandate under French administration)

GDP Per Capita: $10,300

Occupations: N/A

Currency: Lebanese pound (LBP)

Lesotho

Long Name: Kingdom of Lesotho

Location: Southern Africa, an enclave of South Africa

Area: 11,720 sq. mi. (30,355 sq. km)

Climate: temperate; cool to cold, dry winters; hot, wet summers

Terrain: mostly highland with plateaus, hills, and mountains

Population: 2,128,180

Population Growth Rate: 0.129%

Birth Rate (per 1,000): 24.41

Death Rate (per 1,000): 22.33

Life Expectancy: 40.17 years

Ethnic Groups: Sotho 99.7%, Europeans, Asians, and other 0.3%,

Religion: Christian 80%, indigenous beliefs 20%

Languages: Sesotho (southern Sotho), English (official), Zulu, Xhosa

Literacy: 84.8%

Government Type: parliamentary constitutional monarchy

Capital: Maseru

Independence: 4 October 1966 (from United Kingdom)

GDP Per Capita: $1,400

Occupations: agriculture 86% of resident population engaged in subsistence agriculture, roughly 35% of the active male wage earners work in South Africa, industry and services 14%

Currency: loti (LSL); South African rand (ZAR)

Liberia

Long Name: Republic of Liberia

Location: Western Africa, bordering the North Atlantic Ocean, between Cote d'Ivoire and Sierra Leone

Area: 43,000 sq. mi. (111,370 sq. km)

Climate: tropical; hot, humid; dry winters with hot days and cool to cold nights; wet, cloudy summers with frequent heavy showers

Terrain: mostly flat to rolling coastal plains rising to rolling plateau and low mountains in northeast

Population: 3,334,587

Population Growth Rate: 3.661%

Birth Rate (per 1,000): 42.92

Death Rate (per 1,000): 21.45

Life Expectancy: 41.13 years

Ethnic Groups: indigenous African 95% (including Kpelle, Bassa, Gio, Kru, Grebo, Mano, Krahn, Gola, Gbandi, Loma, Kissi, Vai, Dei, Bella, Mandingo, and Mende), Americo-Liberians 2.5% (descendants of immigrants from the US who had been slaves), Congo People 2.5% (descendants of immigrants from the Caribbean who had been slaves)

Religion: Christian 40%, Muslim 20%, indigenous beliefs 40%

Languages: English 20% (official), some 20 ethnic group languages, of which a few can be written and are used in correspondence

Literacy: 57.5%

Government Type: republic

Capital: Monrovia

Independence: 26 July 1847

GDP Per Capita: $500

Occupations: agriculture 70%, industry 8%, services 22%

Currency: Liberian dollar (LRD)

Libya

Long Name: Great Socialist People's Libyan Arab Jamahiriya

Location: Northern Africa, bordering the Mediterranean Sea, between Egypt and Tunisia

Area: 679,359 sq. mi. (1,759,540 sq. km)

Climate: Mediterranean along coast; dry, extreme desert interior

Terrain: mostly barren, flat to undulating plains, plateaus, depressions

Population: 6,173,579

Population Growth Rate: 2.216%

Birth Rate (per 1,000): 25.62

Death Rate (per 1,000): 3.46

Life Expectancy: 77.07 years

Ethnic Groups: Berber and Arab 97%, other 3% (includes Greeks, Maltese, Italians, Egyptians, Pakistanis, Turks, Indians, and Tunisians)

Religion: Sunni Muslim 97%, other 3%

Languages: Arabic, Italian, English, all are widely understood in the major cities

Literacy: 82.6%

Government Type: Jamahiriya (a state of the masses) in theory, governed by the populace through local councils; in practice, an authoritarian state

Capital: Tripoli

Independence: 24 December 1951 (from UN trusteeship)

GDP Per Capita: $12,400

Occupations: agriculture 17%, industry 23%, services 59%

Currency: Libyan dinar (LYD)

Liechtenstein

Long Name: Principality of Liechtenstein

Location: Central Europe, between Austria and Switzerland

Area: 62 sq. mi. (160 sq. km)

Climate: continental; cold, cloudy winters with frequent snow or rain; cool to moderately warm, cloudy, humid summers

Terrain: mostly mountainous (Alps) with Rhine Valley in western third

Population: 34,498

Population Growth Rate: 0.713%

Birth Rate (per 1,000): 9.86

Death Rate (per 1,000): 7.42

Life Expectancy: 79.95 years

Ethnic Groups: Liechtensteiner 65.6%, other 34.4%

Religion: Roman Catholic 76.2%, Protestant 7%, unknown 10.6%, other 6.2%

Languages: German (official), Alemannic dialect

Literacy: 100%

Government Type: constitutional monarchy

Capital: Vaduz

Independence: 23 January 1719 (Principality of Liechtenstein established); 12 July 1806 (independence from the Holy Roman Empire)

GDP Per Capita: $25,000

Occupations: agriculture 2%, industry 47%, services 51%

Currency: Swiss franc (CHF)

Lithuania

Long Name: Republic of Lithuania

Location: Eastern Europe, bordering the Baltic Sea, between Latvia and Russia

Area: 25,212 sq. mi. (65,300 sq. km)

Climate: transitional, between maritime and continental; wet, moderate winters and summers

Terrain: lowland, many scattered small lakes, fertile soil

Population: 3,565,205

Population Growth Rate: -0.284%

Birth Rate (per 1,000): 9

Death Rate (per 1,000): 11.12

Life Expectancy: 74.67 years

Ethnic Groups: Lithuanian 83.4%, Polish 6.7%, Russian 6.3%, other or unspecified 3.6%

Religion: Roman Catholic 79%, Russian Orthodox 4.1%, Protestant (including Lutheran and Evangelical Christian Baptist) 1.9%, other or unspecified 5.5%, none 9.5%

Languages: Lithuanian (official) 82%, Russian 8%, Polish 5.6%, other and unspecified 4.4%

Literacy: 99.6%

Government Type: parliamentary democracy

Capital: Vilnius

Independence: 11 March 1990 (declared); 6 September 1991 (recognized by Soviet Union)

GDP Per Capita: $16,800

Occupations: agriculture 15.8%, industry 28.2%, services 56%

Currency: litas (LTL)

Luxembourg

Long Name: Grand Duchy of Luxembourg

Location: Western Europe, between France and Germany

Area: 998 sq. mi. (2,586 sq. km)

Climate: modified continental with mild winters, cool summers

Terrain: mostly gently rolling uplands with broad, shallow valleys; uplands to slightly mountainous in the north; steep slope down to Moselle flood plain in the southeast

Population: 486,006

Population Growth Rate: 1.188%

Birth Rate (per 1,000): 11.77

Death Rate (per 1,000): 8.43

Life Expectancy: 79.18 years

Ethnic Groups: Luxembourger 63.1%, Portuguese 13.3%, French 4.5%, Italian 4.3%, German 2.3%, other EU 7.3%, other 5.2%

Religion: Roman Catholic 87%, other (includes Protestant, Jewish, and Muslim) 13%

Languages: Luxembourgish (national language), German (administrative language), French (administrative language)

Literacy: 100%

Government Type: constitutional monarchy

Capital: Luxembourg

Independence: 1839 (from the Netherlands)

GDP Per Capita: $79,400

Occupations: agriculture 1%, industry 13%, services 86%

Currency: euro (EUR)

Macedonia

Long Name: Republic of Macedonia

Location: Southeastern Europe, north of Greece

Area: 9,781 sq. mi. (25,333 sq. km)

Climate: warm, dry summers and autumns; relatively cold winters with heavy snowfall

Terrain: mountainous territory covered with deep basins and valleys; three large lakes, each divided by a frontier line; country bisected by the Vardar River

Population: 2,061,315

Population Growth Rate: 0.262%

Birth Rate (per 1,000): 12

Death Rate (per 1,000): 8.81

Life Expectancy: 74.45 years

Ethnic Groups: Macedonian 64.2%, Albanian 25.2%, Turkish 3.9%, Roma (Gypsy) 2.7%, Serb 1.8%, other 2.2%

Religion: Macedonian Orthodox 64.7%, Muslim 33.3%, other Christian 0.37%, other and unspecified 1.63%

Languages: Macedonian 66.5%, Albanian 25.1%, Turkish 3.5%, Roma 1.9%, Serbian 1.2%, other 1.8%

Literacy: 96.1%

Government Type: parliamentary democracy

Capital: Skopje

Independence: 8 September 1991 (referendum by registered voters endorsed independence from Yugoslavia)

GDP Per Capita: $8,400

Occupations: agriculture 19.6%, industry 30.4%, services 50%

Currency: Macedonian denar (MKD)

Madagascar

Long Name: Republic of Madagascar

Location: Southern Africa, island in the Indian Ocean, east of Mozambique

Area: 226,656 sq. mi. (587,040 sq. km)

Climate: tropical along coast, temperate inland, arid in south

Terrain: narrow coastal plain, high plateau and mountains in center

Population: 20,042,552

Population Growth Rate: 3.005%

Birth Rate (per 1,000): 38.38

Death Rate (per 1,000): 8.32

Life Expectancy: 62.52 years

Ethnic Groups: Malayo-Indonesian (Merina and related Betsileo), Cotiers (mixed African, Malayo-Indonesian, and Arab ancestry - Betsimisaraka, Tsimihety, Antaisaka, Sakalava), French, Indian, Creole, Comoran

Religion: indigenous beliefs 52%, Christian 41%, Muslim 7%

Languages: English (official), French (official), Malagasy (official)

Literacy: 68.9%

Government Type: republic

Capital: Antananarivo

Independence: 26 June 1960 (from France)

GDP Per Capita: $900

Occupations: N/A

Currency: ariary (MGA)

Malawi

Long Name: Republic of Malawi

Location: Southern Africa, east of Zambia

Area: 45,745 sq. mi. (118,480 sq. km)

Climate: sub-tropical; rainy season (November to May); dry season (May to November)

Terrain: narrow elongated plateau with rolling plains, rounded hills, some mountains

Population: 13,931,831

Population Growth Rate: 2.39%

Birth Rate (per 1,000): 41.79

Death Rate (per 1,000): 17.89

Life Expectancy: 43.45 years

Ethnic Groups: Chewa, Nyanja, Tumbuka, Yao, Lomwe, Sena, Tonga, Ngoni, Ngonde, Asian, European

Religion: Christian 79.9%, Muslim 12.8%, other 3%, none 4.3%

Languages: Chichewa 57.2% (official), Chinyanja 12.8%, Chiyao 10.1%, Chitumbuka 9.5%, Chisena 2.7%, Chilomwe 2.4%, Chitonga 1.7%, other 3.6%

Literacy: 62.7%

Government Type: multiparty democracy

Capital: Lilongwe

Independence: 6 July 1964 (from United Kingdom)

GDP Per Capita: $800

Occupations: agriculture 90%, industry and services 10%

Currency: Malawian kwacha (MWK)

Malaysia

Long Name: Malaysia

Location: Southeastern Asia, peninsula bordering Thailand and northern one-third of the island of Borneo, bordering Indonesia, Brunei, and the South China Sea, south of Vietnam

Area: 127,317 sq. mi. (329,750 sq. km)

Climate: tropical; annual southwest (April to October) and northeast (October to February) monsoons

Terrain: coastal plains rising to hills and mountains

Population: 25,274,132

Population Growth Rate: 1.742%

Birth Rate (per 1,000): 22.44

Death Rate (per 1,000): 5.02

Life Expectancy: 73.03 years

Ethnic Groups: Malay 50.4%, Chinese 23.7%, indigenous 11%, Indian 7.1%, others 7.8%

Religion: Muslim 60.4%, Buddhist 19.2%, Christian 9.1%, Hindu 6.3%, Confucianism, Taoism, other traditional Chinese religions 2.6%, other or unknown 1.5%, none 0.8%

Languages: Bahasa Malaysia (official), English, Chinese (Cantonese, Mandarin, Hokkien, Hakka, Hainan, Foochow), Tamil, Telugu, Malayalam, Panjabi, Thai, in East Malaysia there are several indigenous languages; most widely spoken are Iban and Kadazan

Literacy: 88.7%

Government Type: constitutional monarchy

Capital: Kuala Lumpur

Independence: 31 August 1957 (from United Kingdom)

GDP Per Capita: $14,500

Occupations: agriculture 13%, industry 36%, services 51%

Currency: ringgit (MYR)

Maldives

Long Name: Republic of Maldives

Location: Southern Asia, group of atolls in the Indian Ocean, south-southwest of India

Area: 116 sq. mi. (300 sq. km)

Climate: tropical; hot, humid; dry, northeast monsoon (November to March); rainy, southwest monsoon (June to August)

Terrain: flat, with white sandy beaches

Population: 385,925

Population Growth Rate: 5.566%

Birth Rate (per 1,000): 14.84

Death Rate (per 1,000): 3.66

Life Expectancy: 73.72 years

Ethnic Groups: South Indians, Sinhalese, Arabs

Religion: Sunni Muslim

Languages: Maldivian Dhivehi (dialect of Sinhala, script derived from Arabic), English spoken by most government officials

Literacy: 96.3%

Government Type: republic

Capital: Male

Independence: 26 July 1965 (from United Kingdom)

GDP Per Capita: $4,600

Occupations: agriculture 22%, industry 18%, services 60%

Currency: rufiyaa (MVR)

Mali

Long Name: Republic of Mali

Location: Western Africa, southwest of Algeria

Area: 478,764 sq. mi. (1,240,000 sq. km)

Climate: subtropical to arid; hot and dry (February to June); rainy, humid, and mild (June to November); cool and dry (November to February)

Terrain: mostly flat to rolling northern plains covered by sand; savanna in south, rugged hills in northeast

Population: 12,324,029

Population Growth Rate: 2.725%

Birth Rate (per 1,000): 49.38

Death Rate (per 1,000): 16.16

Life Expectancy: 49.94 years

Ethnic Groups: Mande 50% (Bambara, Malinke, Soninke), Peul 17%, Voltaic 12%, Songhai 6%, Tuareg and Moor 10%, other 5%

Religion: Muslim 90%, Christian 1%, indigenous beliefs 9%

Languages: French (official), Bambara 80%, numerous African languages

Literacy: 46.4%

Government Type: republic

Capital: Bamako

Independence: 22 September 1960 (from France)

GDP Per Capita: $1,100

Occupations: agriculture 80%, industry and services 20%

Currency: Communaute Financiere Africaine franc (XOF)

Malta

Long Name: Republic of Malta

Location: Southern Europe, islands in the Mediterranean Sea, south of Sicily (Italy)

Area: 122 sq. mi. (316 sq. km)

Climate: Mediterranean; mild, rainy winters; hot, dry summers

Terrain: mostly low, rocky, flat to dissected plains; many coastal cliffs

Population: 403,532

Population Growth Rate: 0.407%

Birth Rate (per 1,000): 10.33

Death Rate (per 1,000): 8.29

Life Expectancy: 79.3 years

Ethnic Groups: Maltese (descendants of ancient Carthaginians and Phoenicians with strong elements of Italian and other Mediterranean stock)

Religion: Roman Catholic 98%, 2% other

Languages: Maltese (official) 90.2%, English (official) 6%, multilingual 3%, other 0.8%

Literacy: 92.8%

Government Type: republic

Capital: Valletta

Independence: 21 September 1964 (from United Kingdom)

GDP Per Capita: $23,400

Occupations: agriculture 3%, industry 22%, services 75%

Currency: euro (EUR)

Marshall Islands

Long Name: Republic of the Marshall Islands

Location: Oceania, two archipelagic island chains of 29 atolls, each made up of many small islets, and five single islands in the North Pacific Ocean, about half way between Hawaii and Australia

Area: 70 sq. mi. (181 sq. km)

Climate: tropical; hot and humid; wet season May to November; islands border typhoon belt

Terrain: low coral limestone and sand islands

Population: 63,174

Population Growth Rate: 2.142%

Birth Rate (per 1,000): 31.52

Death Rate (per 1,000): 4.57

Life Expectancy: 70.9 years

Ethnic Groups: Marshallese 92.1%, mixed Marshallese 5.9%, other 2%

Religion: Protestant 54.8%, Assembly of God 25.8%, Roman Catholic 8.4%, Bukot nan Jesus 2.8%, Mormon 2.1%, other Christian 3.6%, other 1%, none 1.5%

Languages: Marshallese (official) 98.2%, other languages 1.8%, English widely spoken as a second language

Literacy: 93.7%

Government Type: constitutional government in free association with the US; the Compact of Free Association entered into force 21 October 1986 and the Amended Compact entered into force in May 2004

Capital: Majuro

Independence: 21 October 1986 (from the US-administered UN trusteeship)

GDP Per Capita: $2,900

Occupations: agriculture 21.4%, industry 20.9%, services 57.7%

Currency: US dollar (USD)

Mauritania

Long Name: Islamic Republic of Mauritania

Location: Northern Africa, bordering the North Atlantic Ocean, between Senegal and Western Sahara

Area: 397,954 sq. mi. (1,030,700 sq. km)

Climate: desert; constantly hot, dry, dusty

Terrain: mostly barren, flat plains of the Sahara; some central hills

Population: 3,364,940

Population Growth Rate: 2.852%

Birth Rate (per 1,000): 40.14

Death Rate (per 1,000): 11.61

Life Expectancy: 53.91 years

Ethnic Groups: mixed Moor/black 40%, Moor 30%, black 30%

Religion: Muslim 100%

Languages: Arabic (official and national), Pulaar, Soninke, Wolof (all national languages), French, Hassaniya

Literacy: 51.2%

Government Type: Democratic Republic

Capital: Nouakchott

Independence: 28 November 1960 (from France)

GDP Per Capita: $1,800

Occupations: agriculture 50%, industry 10%, services 40%

Currency: ouguiya (MRO)

Mauritius

Long Name: Republic of Mauritius

Location: Southern Africa, island in the Indian Ocean, east of Madagascar

Area: 788 sq. mi. (2,040 sq. km)

Climate: tropical, modified by southeast trade winds; warm, dry winter (May to November); hot, wet, humid summer (November to May)

Terrain: small coastal plain rising to discontinuous mountains encircling central plateau

Population: 1,274,189

Population Growth Rate: 0.8%

Birth Rate (per 1,000): 14.64

Death Rate (per 1,000): 6.55

Life Expectancy: 73.75 years

Ethnic Groups: Indo-Mauritian 68%, Creole 27%, Sino-Mauritian 3%, Franco-Mauritian 2%

Religion: Hindu 48%, Roman Catholic 23.6%, Muslim 16.6%, other Christian 8.6%, other 2.5%, unspecified 0.3%, none 0.4%

Languages: Creole 80.5%, Bhojpuri 12.1%, French 3.4%, English (official; spoken by less than 1% of the population), other 3.7%, unspecified 0.3%

Literacy: 84.4%

Government Type: parliamentary democracy

Capital: Port Louis

Independence: 12 March 1968 (from United Kingdom)

GDP Per Capita: $11,300

Occupations: agriculture and fishing 9%, construction and industry 30%, transportation and communication 7%, trade, restaurants, hotels 22%, finance 6%, other services 25%

Currency: Mauritian rupee (MUR)

Mexico

Long Name: United Mexican States

Location: Middle America, bordering the Caribbean Sea and the Gulf of Mexico, between Belize and the US and bordering the North Pacific Ocean, between Guatemala and the US

Area: 761,602 sq. mi. (1,972,550 sq. km)

Climate: varies from tropical to desert

Terrain: high, rugged mountains; low coastal plains; high plateaus; desert

Population: 109,955,400

Population Growth Rate: 1.142%

Birth Rate (per 1,000): 20.04

Death Rate (per 1,000): 4.78

Life Expectancy: 75.84 years

Ethnic Groups: mestizo (Amerindian-Spanish) 60%, Amerindian or predominantly Amerindian 30%, white 9%, other 1%

Religion: Roman Catholic 76.5%, Protestant 6.3% (Pentecostal 1.4%, Jehovah's Witnesses 1.1%, other 3.8%), other 0.3%, unspecified 13.8%, none 3.1%

Languages: Spanish only 92.7%, Spanish and indigenous languages 5.7%, indigenous only 0.8%, unspecified 0.8%; note - indigenous languages include various Mayan, Nahuatl, and other regional languages

Literacy: 91%

Government Type: federal republic

Capital: Mexico (Distrito Federal)

Independence: 16 September 1810 (declared); 27 September 1821 (recognized by Spain)

GDP Per Capita: $12,400

Occupations: agriculture 18%, industry 24%, services 58%

Currency: Mexican peso (MXN)

Micronesia

Long Name: Federated States of Micronesia

Location: Oceania, island group in the North Pacific Ocean, about three-quarters of the way from Hawaii to Indonesia

Area: 271 sq. mi. (702 sq. km)

Climate: tropical; heavy year-round rainfall, especially in the eastern islands; located on southern edge of the typhoon belt with occasionally severe damage

Terrain: islands vary geologically from high mountainous islands to low, coral atolls; volcanic outcroppings on Pohnpei, Kosrae, and Chuuk

Population: 107,665

Population Growth Rate: -0.191%

Birth Rate (per 1,000): 23.66

Death Rate (per 1,000): 4.53

Life Expectancy: 70.65 years

Ethnic Groups: Chuukese 48.8%, Pohnpeian 24.2%, Kosraean 6.2%, Yapese 5.2%, Yap outer islands 4.5%, Asian 1.8%, Polynesian 1.5%, other 6.4%, unknown 1.4%

Religion: Roman Catholic 50%, Protestant 47%, other 3%

Languages: English (official and common language), Chuukese, Kosrean, Pohnpeian, Yapese, Ulithian, Woleaian, Nukuoro, Kapingamarangi

Literacy: 89%

Government Type: constitutional government in free association with the US; the Compact of Free Association entered into force 3 November 1986 and the Amended Compact entered into force May 2004

Capital: Palikir

Independence: 3 November 1986 (from the US-administered UN trusteeship)

GDP Per Capita: $2,300

Occupations: agriculture 0.9%, industry 34.4%, services 64.7%

Currency: US dollar (USD)

Moldova

Long Name: Republic of Moldova

Location: Eastern Europe, northeast of Romania

Area: 13,067 sq. mi. (33,843 sq. km)

Climate: moderate winters, warm summers

Terrain: rolling steppe, gradual slope south to Black Sea

Population: 4,324,450

Population Growth Rate: -0.092%

Birth Rate (per 1,000): 11.01

Death Rate (per 1,000): 10.8

Life Expectancy: 70.5 years

Ethnic Groups: Moldovan/Romanian 78.2%, Ukrainian 8.4%, Russian 5.8%, Gagauz 4.4%, Bulgarian 1.9%, other 1.3%

Religion: Eastern Orthodox 98%, Jewish 1.5%, Baptist and other 0.5%

Languages: Moldovan (official, virtually the same as the Romanian language), Russian, Gagauz (a Turkish dialect)

Literacy: 99.1%

Government Type: republic

Capital: Chisinau

Independence: 27 August 1991 (from Soviet Union)

GDP Per Capita: $2,300

Occupations: agriculture 40.7%, industry 12.1%, services 47.2%

Currency: Moldovan leu (MDL)

Monaco

Long Name: Principality of Monaco

Location: Western Europe, bordering the Mediterranean Sea on the southern coast of France, near the border with Italy

Area: 0.75 sq. mi. (1.95 sq. km)

Climate: Mediterranean with mild, wet winters and hot, dry summers

Terrain: hilly, rugged, rocky

Population: 32,796

Population Growth Rate: 0.375%

Birth Rate (per 1,000): 9.09

Death Rate (per 1,000): 12.96

Life Expectancy: 79.96 years

Ethnic Groups: French 47%, Monegasque 16%, Italian 16%, other 21%

Religion: Roman Catholic 90%, other 10%

Languages: French (official), English, Italian, Monegasque

Literacy: 99%

Government Type: constitutional monarchy

Capital: Monaco

Independence: 1419 (beginning of rule by the House of Grimaldi)

325

GDP Per Capita: $30,000

Occupations: primarily services

Currency: euro (EUR)

Mongolia

Long Name: Mongolia

Location: Northern Asia, between China and Russia

Area: 603,906 sq. mi. (1,564,116 sq. km)

Climate: desert; continental (large daily and seasonal temperature ranges)

Terrain: vast semidesert and desert plains, grassy steppe, mountains in west and southwest; Gobi Desert in south-central

Population: 2,996,081

Population Growth Rate: 1.493%

Birth Rate (per 1,000): 21.09

Death Rate (per 1,000): 6.16

Life Expectancy: 67.32 years

Ethnic Groups: Mongol (mostly Khalkha) 94.9%, Turkic (mostly Kazakh) 5%, other (including Chinese and Russian) 0.1%

Religion: Buddhist Lamaist 50%, Shamanist and Christian 6%, Muslim 4%, none 40%

Languages: Khalkha Mongol 90%, Turkic, Russian

Literacy: 97.8%

Government Type: mixed parliamentary/presidential

Capital: Ulaanbaatar

Independence: 11 July 1921 (from China)

GDP Per Capita: $2,900

Occupations: agriculture 39.9%, industry 11.7%, services 49.4%

Currency: togrog/tugrik (MNT)

Montenegro

Long Name: Montenegro

Location: Southeastern Europe, between the Adriatic Sea and Serbia

Area: 5,415 sq. mi. (14,026 sq. km)

Climate: Mediterranean climate, hot dry summers and autumns and relatively cold winters with heavy snowfalls inland

Terrain: highly indented coastline with narrow coastal plain backed by rugged high limestone mountains and plateaus

Population: 678,177

Population Growth Rate: -0.925%

Birth Rate (per 1,000): 11.17

Death Rate (per 1,000): 8.51

Life Expectancy: N/A

Ethnic Groups: Montenegrin 43%, Serbian 32%, Bosniak 8%, Albanian 5%, other (Muslims, Croats, Roma (Gypsy)) 12%

Religion: Orthodox 74.2%, Muslim 17.7%, Catholic 3.5%, other 0.6%, unspecified 3%, atheist 1%

Languages: Serbian 63.6%, Montenegrin (official) 22%, Bosnian 5.5%, Albanian 5.3%, unspecified 3.7%

Literacy: N/A

Government Type: republic

Capital: Podgorica

Independence: 3 June 2006 (from Serbia and Montenegro)

GDP Per Capita: $3,800

Occupations: agriculture 2%, industry 30%, services 68%

Currency: euro (EUR)

Morocco

Long Name: Kingdom of Morocco

Location: Northern Africa, bordering the North Atlantic Ocean and the Mediterranean Sea, between Algeria and Western Sahara

Area: 172,413 sq. mi. (446,550 sq. km)

Climate: Mediterranean, becoming more extreme in the interior

Terrain: northern coast and interior are mountainous with large areas of bordering plateaus, intermontane valleys, and rich coastal plains

Population: 34,343,220

Population Growth Rate: 1.505%

Birth Rate (per 1,000): 21.31

Death Rate (per 1,000): 5.49

Life Expectancy: 71.52 years

Ethnic Groups: Arab-Berber 99.1%, other 0.7%, Jewish 0.2%

Religion: Muslim 98.7%, Christian 1.1%, Jewish 0.2%

Languages: Arabic (official), Berber dialects, French often the language of business, government, and diplomacy

Literacy: 52.3%

Government Type: constitutional monarchy

Capital: Rabat

Independence: 2 March 1956 (from France)

GDP Per Capita: $3,700

Occupations: agriculture 40%, industry 15%, services 45%

Currency: Moroccan dirham (MAD)

Mozambique

Long Name: Republic of Mozambique

Location: Southeastern Africa, bordering the Mozambique Channel, between South Africa and Tanzania

Area: 309,494 sq. mi. (801,590 sq. km)

Climate: tropical to subtropical

Terrain: mostly coastal lowlands, uplands in center, high plateaus in northwest, mountains in west

Population: 21,284,700

Population Growth Rate: 1.792%

Birth Rate (per 1,000): 38.21

Death Rate (per 1,000): 20.29

Life Expectancy: 41.04 years

Ethnic Groups: African 99.66% (Makhuwa, Tsonga, Lomwe, Sena, and others), Europeans 0.06%, Euro-Africans 0.2%, Indians 0.08%

Religion: Catholic 23.8%, Muslim 17.8%, Zionist Christian 17.5%, other 17.8%, none 23.1%

Languages: Emakhuwa 26.1%, Xichangana 11.3%, Portuguese 8.8% (official; spoken by 27% of population as a second language), Elomwe 7.6%, Cisena 6.8%, Echuwabo 5.8%, other Mozambican languages 32%, other foreign languages 0.3%, unspecified 1.3%

Literacy: 47.8%

Government Type: republic

Capital: Maputo

Independence: 25 June 1975 (from Portugal)

GDP Per Capita: $800

Occupations: agriculture 81% , industry 6%, services 13%

Currency: metical (MZM)

Myanmar

Long Name: Union of Myanmar

Location: Southeastern Asia, bordering the Andaman Sea and the Bay of Bengal, between Bangladesh and Thailand

Area: 261,969 sq. mi. (678,500 sq. km)

Climate: tropical monsoon; cloudy, rainy, hot, humid summers (southwest monsoon, June to September); less cloudy, scant rainfall, mild temperatures, lower humidity during winter (northeast monsoon, December to April)

Terrain: central lowlands ringed by steep, rugged highlands

Population: 47,758,180

Population Growth Rate: 0.8%

Birth Rate (per 1,000): 17.23

Death Rate (per 1,000): 9.23

Life Expectancy: 62.94 years

Ethnic Groups: Burman 68%, Shan 9%, Karen 7%, Rakhine 4%, Chinese 3%, Indian 2%, Mon 2%, other 5%

Religion: Buddhist 89%, Christian 4% (Baptist 3%, Roman Catholic 1%), Muslim 4%, animist 1%, other 2%

Languages: Burmese, minority ethnic groups have their own languages

Literacy: 89.9%

Government Type: military junta

Capital: Rangoon

Independence: 4 January 1948 (from United Kingdom)

GDP Per Capita: $1,900

Occupations: agriculture 70%, industry 7%, services 23%

Currency: kyat (MMK)

Namibia

Long Name: Republic of Namibia

Location: Southern Africa, bordering the South Atlantic Ocean, between Angola and South Africa

Area: 318,694 sq. mi. (825,418 sq. km)

Climate: desert; hot, dry; rainfall sparse and erratic

Terrain: mostly high plateau; Namib Desert along coast; Kalahari Desert in east

Population: 2,088,669

Population Growth Rate: 0.947%

Birth Rate (per 1,000): 23.19

Death Rate (per 1,000): 14.07

Life Expectancy: 49.89 years

Ethnic Groups: black 87.5%, white 6%, mixed 6.5% (about 50% of the population belong to the Ovambo tribe and 9% to the Kavangos tribe; other ethnic groups include Herero 7%, Damara 7%, Nama 5%, Caprivian 4%, Bushmen 3%, Baster 2%, Tswana 0.5%)

Religion: Christian 80% to 90% (Lutheran 50% at least), indigenous beliefs 10% to 20%

Languages: English 7% (official), Afrikaans common language of most of the population and about 60% of the white population, German 32%, indigenous languages 1% (includes Oshivambo, Herero, Nama)

Literacy: 85%

Government Type: republic

Capital: Windhoek

Independence: 21 March 1990 (from South African mandate)

GDP Per Capita: $5,200

Occupations: agriculture 47%, industry 20%, services 33%

Currency: Namibian dollar (NAD); South African rand (ZAR)

Nauru

Long Name: Republic of Nauru

Location: Oceania, island in the South Pacific Ocean, south of the Marshall Islands

Area: 8 sq. mi. (21 sq. km)

Climate: tropical with a monsoonal pattern; rainy season (November to February)

Terrain: sandy beach rises to fertile ring around raised coral reefs with phosphate plateau in center

Population: 13,770

Population Growth Rate: 1.772%

Birth Rate (per 1,000): 24.26

Death Rate (per 1,000): 6.54

Life Expectancy: 63.81 years

Ethnic Groups: Nauruan 58%, other Pacific Islander 26%, Chinese 8%, European 8%

Religion: Nauru Congregational 35.4%, Roman Catholic 33.2%, Nauru Independent Church 10.4%, other 14.1%, none 4.5%, unspecified 2.4%

Languages: Nauruan (official; a distinct Pacific Island language), English widely understood, spoken, and used for most government and commercial purposes

Literacy: N/A

Government Type: republic

Capital: no official capital; government offices in Yaren District

Independence: 31 January 1968 (from the Australia-, NZ-, and United Kingdom-administered UN trusteeship)

GDP Per Capita: $5,000

Occupations: employed in mining phosphates, public administration, education, and transportation

Currency: Australian dollar (AUD)

Nepál

Long Name: Federal Democratic Republic of Nepál

Location: Southern Asia, between China and India

Area: 56,827 sq. mi. (147,181 sq. km)

Climate: varies from cool summers and severe winters in north to subtropical summers and mild winters in south

Terrain: Tarai or flat river plain of the Ganges in south, central hill region, rugged Himalayas in north

Population: 29,519,114

Population Growth Rate: 2.095%

Birth Rate (per 1,000): 29.92

Death Rate (per 1,000): 8.97

Life Expectancy: 60.94 years

Ethnic Groups: Chhettri 15.5%, Brahman-Hill 12.5%, Magar 7%, Tharu 6.6%, Tamang 5.5%, Newar 5.4%, Muslim 4.2%, Kami 3.9%, Yadav 3.9%, other 32.7%, unspecified 2.8%

Religion: Hindu 80.6%, Buddhist 10.7%, Muslim 4.2%, Kirant 3.6%, other 0.9%

Languages: Nepali 47.8%, Maithali 12.1%, Bhojpuri 7.4%, Tharu (Dagaura/Rana) 5.8%, Tamang 5.1%, Newar 3.6%, Magar 3.3%, Awadhi 2.4%, other 10%, unspecified 2.5%

Literacy: 48.6%

Government Type: democratic republic

Capital: Kathmandu

Independence: 1768 (unified by Prithvi Narayan Shah)

GDP Per Capita: $1,000

Occupations: agriculture 76%, industry 6%, services 18%

Currency: Nepalese rupee (NPR)

The Netherlands

Long Name: Kingdom of the Netherlands

Location: Western Europe, bordering the North Sea, between Belgium and Germany

329

Area: 16,033 sq. mi. (41,526 sq. km)

Climate: temperate; marine; cool summers and mild winters

Terrain: mostly coastal lowland and reclaimed land (polders); some hills in southeast

Population: 16,645,313

Population Growth Rate: 0.436%

Birth Rate (per 1,000): 10.53

Death Rate (per 1,000): 8.71

Life Expectancy: 79.25 years

Ethnic Groups: Dutch 80.7%, EU 5%, Indonesian 2.4%, Turkish 2.2%, Surinamese 2%, Moroccan 2%, Netherlands Antilles & Aruba 0.8%, other 4.8%

Religion: Roman Catholic 30%, Dutch Reformed 11%, Calvinist 6%, other Protestant 3%, Muslim 5.8%, other 2.2%, none 42%

Languages: Dutch (official), Frisian (official)

Literacy: 99%

Government Type: constitutional monarchy

Capital: Amsterdam

Independence: 23 January 1579 (the northern provinces of the Low Countries conclude the Union of Utrecht breaking with Spain; on 26 July 1581 they formally declared their independence with an Act of Abjuration; however, it was not until 30 January 1648 and the Peace of Westphalia that Spain recognized this independence)

GDP Per Capita: $39,000

Occupations: agriculture 3%, industry 21%, services 76%

Currency: euro (EUR)

New Zealand

Long Name: New Zealand

Location: Oceania, islands in the South Pacific Ocean, southeast of Australia

Area: 103,737 sq. mi. (268,680 sq. km)

Climate: temperate with sharp regional contrasts

Terrain: predominately mountainous with some large coastal plains

Population: 4,173,460

Population Growth Rate: 0.971%

Birth Rate (per 1,000): 14.09

Death Rate (per 1,000): 7

Life Expectancy: 80.24 years

Ethnic Groups: European 69.8%, Maori 7.9%, Asian 5.7%, Pacific islander 4.4%, other 0.5%, mixed 7.8%, unspecified 3.8%

Religion: Anglican 14.9%, Roman Catholic 12.4%, Presbyterian 10.9%, Methodist 2.9%, Pentecostal 1.7%, Baptist 1.3%, other Christian 9.4%, other 3.3%, unspecified 17.2%, none 26%

Languages: English (official), Maori (official), Sign Language (official)

Literacy: 99%

Government Type: parliamentary democracy

Capital: Wellington

Independence: 26 September 1907 (from United Kingdom)

GDP Per Capita: $27,200

Occupations: agriculture 7%, industry 19%, services 74%

Currency: New Zealand dollar (NZD)

Nicaragua

Long Name: Republic of Nicaragua

Location: Central America, bordering both the Caribbean Sea and the North

Pacific Ocean, between Costa Rica and Honduras

Area: 49,998 sq. mi. (129,494 sq. km)

Climate: tropical in lowlands, cooler in highlands

Terrain: extensive Atlantic coastal plains rising to central interior mountains; narrow Pacific coastal plain interrupted by volcanoes

Population: 5,785,846

Population Growth Rate: 1.825%

Birth Rate (per 1,000): 23.7

Death Rate (per 1,000): 4.33

Life Expectancy: 71.21 years

Ethnic Groups: mestizo (mixed Amerindian and white) 69%, white 17%, black 9%, Amerindian 5%

Religion: Roman Catholic 58.5%, Evangelical 21.6%, Moravian 1.6%, Jehovah's Witness 0.9%, other 1.7%, none 15.7%

Languages: Spanish 97.5% (official), Miskito 1.7%, other 0.8%

Literacy: 67.5%

Government Type: republic

Capital: Managua

Independence: 15 September 1821 (from Spain)

GDP Per Capita: $2,800

Occupations: agriculture 29%, industry 19%, services 52%

Currency: gold cordoba (NIO)

Niger

Long Name: Republic of Niger

Location: Western Africa, southeast of Algeria

Area: 489,189 sq. mi. (1,267,000 sq. km)

Climate: desert; mostly hot, dry, dusty; tropical in extreme south

Terrain: predominately desert plains and sand dunes; flat to rolling plains in south; hills in north

Population: 13,272,679

Population Growth Rate: 2.878%

Birth Rate (per 1,000): 49.62

Death Rate (per 1,000): 20.26

Life Expectancy: 44.28 years

Ethnic Groups: Haoussa 55.4%, Djerma Sonrai 21%, Tuareg 9.3%, Peuhl 8.5%, Kanouri Manga 4.7%, other 1.2%

Religion: Muslim 80%, other (includes indigenous beliefs and Christian) 20%

Languages: French (official), Hausa, Djerma

Literacy: 28.7%

Government Type: republic

Capital: Niamey

Independence: 3 August 1960 (from France)

GDP Per Capita: $700

Occupations: agriculture 90%, industry 6%, services 4%

Currency: Communaute Financiere Africaine franc (XOF)

Nigeria

Long Name: Federal Republic of Nigeria

Location: Western Africa, bordering the Gulf of Guinea, between Benin and Cameroon

Area: 356,667 sq. mi. (923,768 sq. km)

Climate: varies; equatorial in south, tropical in center, arid in north

Terrain: southern lowlands merge into central hills and plateaus; mountains in southeast, plains in north

Population: 146,255,312

Population Growth Rate: 2.025%

Birth Rate (per 1,000): 37.23

Death Rate (per 1,000): 16.88

Life Expectancy: 46.53 years

Ethnic Groups: Nigeria, Africa's most populous country, is composed of more than 250 ethnic groups; the following are the most populous and politically influential: Hausa and Fulani 29%, Yoruba 21%, Igbo (Ibo) 18%, Ijaw 10%, Kanuri 4%, Ibibio 3.5%, Tiv 2.5%

Religion: Muslim 50%, Christian 40%, indigenous beliefs 10%

Languages: English (official), Hausa, Yoruba, Igbo (Ibo), Fulani

Literacy: 68%

Government Type: federal republic

Capital: Abuja

Independence: 1 October 1960 (from United Kingdom)

GDP Per Capita: $2,100

Occupations: agriculture 70%, industry 10%, services: 20%

Currency: naira (NGN)

North Korea

Long Name: Democratic People's Republic of Korea

Location: Eastern Asia, northern half of the Korean Peninsula bordering the Korea Bay and the Sea of Japan, between China and South Korea

Area: 46,541 sq. mi. (120,540 sq. km)

Climate: temperate with rainfall concentrated in summer

Terrain: mostly hills and mountains separated by deep, narrow valleys; coastal plains wide in west, discontinuous in east

Population: 23,479,088

Population Growth Rate: 0.732%

Birth Rate (per 1,000): 14.61

Death Rate (per 1,000): 7.29

Life Expectancy: 72.2 years

Ethnic Groups: racially homogeneous; there is a small Chinese community and a few ethnic Japanese

Religion: traditionally Buddhist and Confucianist, some Christian and syncretic Chondogyo (Religion of the Heavenly Way)

Languages: Korean

Literacy: 99%

Government Type: Communist state one-man dictatorship

Capital: Pyongyang

Independence: 15 August 1945 (from Japan)

GDP Per Capita: $1,700

Occupations: agriculture 37%, industry and services 63%

Currency: North Korean won (KPW)

Norway

Long Name: Kingdom of Norway

Location: Northern Europe, bordering the North Sea and the North Atlantic Ocean, west of Sweden

Area: 125,020 sq. mi. (323,802 sq. km)

Climate: temperate along coast, modified by North Atlantic Current; colder interior with increased precipitation and colder summers; rainy year-round on west coast

Terrain: glaciated; mostly high plateaus and rugged mountains broken by fertile valleys; small, scattered plains; coastline deeply indented by fjords; arctic tundra in north

Population: 4,644,457

Population Growth Rate: 0.35%

Birth Rate (per 1,000): 11.12

Death Rate (per 1,000): 9.33

Life Expectancy: 79.81 years

Ethnic Groups: Norwegian 94.4% (includes Sami, about 60,000), other European 3.6%, other 2%

Religion: Church of Norway 85.7%, Pentecostal 1%, Roman Catholic 1%, other Christian 2.4%, Muslim 1.8%, other 8.1%

Languages: Bokmal Norwegian (official), Nynorsk Norwegian (official), small Sami- and Finnish-speaking minorities; note - Sami is official in six municipalities

Literacy: 100%

Government Type: constitutional monarchy

Capital: Oslo

Independence: 7 June 1905 (Norway declared the union with Sweden dissolved); 26 October 1905 (Sweden agreed to the repeal of the union)

GDP Per Capita: $53,300

Occupations: agriculture 4%, industry 22%, services 74%

Currency: Norwegian krone (NOK)

Oman

Long Name: Sultanate of Oman

Location: Middle East, bordering the Arabian Sea, Gulf of Oman, and Persian Gulf, between Yemen and UAE

Area: 82,031 sq. mi. (212,460 sq. km)

Climate: dry desert; hot, humid along coast; hot, dry interior; strong southwest summer monsoon (May to September) in far south

Terrain: central desert plain, rugged mountains in north and south

Population: 3,311,640

Population Growth Rate: 3.19%

Birth Rate (per 1,000): 35.26

Death Rate (per 1,000): 3.68

Life Expectancy: 73.91 years

Ethnic Groups: Arab, Baluchi, South Asian (Indian, Pakistani, Sri Lankan, Bangladeshi), African

Religion: Ibadhi Muslim 75%, other (includes Sunni Muslim, Shi'a Muslim, Hindu) 25%

Languages: Arabic (official), English, Baluchi, Urdu, Indian dialects

Literacy: 81.4%

Government Type: monarchy

Capital: Muscat

Independence: 1650 (expulsion of the Portuguese)

GDP Per Capita: $19,000

Occupations: N/A

Currency: Omani rial (OMR)

Pakistan

Long Name: Islamic Republic of Pakistan

Location: Southern Asia, bordering the Arabian Sea, between India on the east and Iran and Afghanistan on the west and China in the north

Area: 310,402 sq. mi. (803,940 sq. km)

Climate: mostly hot, dry desert; temperate in northwest; arctic in north

Terrain: flat Indus plain in east; mountains in north and northwest; Balochistan plateau in west

Population: 172,800,048

Population Growth Rate: 1.999%

Birth Rate (per 1,000): 28.35

Death Rate (per 1,000): 7.85

Life Expectancy: 64.13 years

Ethnic Groups: Punjabi 44.68%, Pashtun (Pathan) 15.42%, Sindhi 14.1%, Sariaki 8.38%, Muhagirs 7.57%, Balochi 3.57%, other 6.28%

Religion: Muslim 95% (Sunni 75%, Shi'a 20%), other (includes Christian and Hindu) 5%

Languages: Punjabi 48%, Sindhi 12%, Siraiki (a Punjabi variant) 10%, Pashtu 8%, Urdu (official) 8%, Balochi 3%, Hindko 2%, Brahui 1%, English (official; lingua franca of Pakistani elite and most government ministries), Burushaski and other 8%

Literacy: 49.9%

Government Type: federal republic

Capital: Islamabad

Independence: 14 August 1947 (from British India)

GDP Per Capita: $2,400

Occupations: agriculture 42%, industry 20%, services 38%

Currency: Pakistani rupee (PKR)

Palau

Long Name: Republic of Palau

Location: Oceania, group of islands in the North Pacific Ocean, southeast of the Philippines

Area: 177 sq. mi. (458 sq. km)

Climate: tropical; hot and humid; wet season May to November

Terrain: varying geologically from the high, mountainous main island of Babelthuap to low, coral islands usually fringed by large barrier reefs

Population: 21,093

Population Growth Rate: 1.157%

Birth Rate (per 1,000): 17.4

Death Rate (per 1,000): 6.73

Life Expectancy: 71 years

Ethnic Groups: Palauan (Micronesian with Malayan and Melanesian admixtures) 69.9%, Filipino 15.3%, Chinese 4.9%, other Asian 2.4%, white 1.9%, Carolinian 1.4%, other Micronesian 1.1%, other or unspecified 3.2%

Religion: Roman Catholic 41.6%, Protestant 23.3%, Modekngei 8.8% (indigenous to Palau), Seventh-Day Adventist 5.3%, Jehovah's Witness 0.9%, Latter-Day Saints 0.6%, other 3.1%, unspecified or none 16.4%

Languages: Palauan 64.7% official in all islands except Sonsoral (Sonsoralese and English are official), Tobi (Tobi and English are official), and Angaur (Angaur, Japanese, and English are official), Filipino 13.5%, English 9.4%, Chinese 5.7%, Carolinian 1.5%, Japanese 1.5%, other Asian 2.3%, other languages 1.5%

Literacy: 92%

Government Type: constitutional government in free association with the US; the Compact of Free Association entered into force 1 October 1994

Capital: Melekeok

Independence: 1 October 1994 (from the US-administered UN trusteeship)

GDP Per Capita: $7,600

Occupations: agriculture 20%, other 80%

Currency: US dollar (USD)

Panama

Long Name: Republic of Panama

Location: Central America, bordering both the Caribbean Sea and the North Pacific Ocean, between Colombia and Costa Rica

Area: 30,193 sq. mi. (78,200 sq. km)

Climate: tropical maritime; hot, humid, cloudy; prolonged rainy season (May to January), short dry season (January to May)

Terrain: interior mostly steep, rugged mountains and dissected, upland plains; coastal areas largely plains and rolling hills

Population: 3,309,679

Population Growth Rate: 1.544%

Birth Rate (per 1,000): 20.68

Death Rate (per 1,000): 4.71

Life Expectancy: 76.88 years

Ethnic Groups: mestizo (mixed Amerindian and white) 70%, Amerindian and mixed (West Indian) 14%, white 10%, Amerindian 6%

Religion: Roman Catholic 85%, Protestant 15%

Languages: Spanish (official), English 14%; note - many Panamanians bilingual

Literacy: 91.9%

Government Type: constitutional democracy

Capital: Panama

Independence: 3 November 1903 (from Colombia; became independent from Spain 28 November 1821)

GDP Per Capita: $10,700

Occupations: agriculture 15%, industry 18%, services 67%

Currency: balboa (PAB); US dollar (USD)

Papua New Guinea

Long Name: Independent State of Papua New Guinea

Location: Oceania, group of islands including the eastern half of the island of New Guinea between the Coral Sea and the South Pacific Ocean, east of Indonesia

Area: 178,703 sq. mi. (462,840 sq. km)

Climate: tropical; northwest monsoon (December to March), southeast monsoon (May to October); slight seasonal temperature variation

Terrain: mostly mountains with coastal lowlands and rolling foothills

Population: 5,931,769

Population Growth Rate: 2.118%

Birth Rate (per 1,000): 28.14

Death Rate (per 1,000): 6.96

Life Expectancy: 66 years

Ethnic Groups: Melanesian, Papuan, Negrito, Micronesian, Polynesian

Religion: Roman Catholic 27%, Evangelical Lutheran 19.5%, United Church 11.5%, Seventh-Day Adventist 10%, Pentecostal 8.6%, Evangelical Alliance 5.2%, Anglican 3.2%, Baptist 2.5%, other Protestant 8.9%, Bahai 0.3%, indigenous beliefs and other 3.3%

Languages: Melanesian Pidgin serves as the lingua franca, English spoken by 1%-2%, Motu spoken in Papua region, 820 indigenous languages spoken

Literacy: 57.3%

Government Type: constitutional parliamentary democracy

Capital: Port Moresby

Independence: 16 September 1975 (from the Australian-administered UN trusteeship)

GDP Per Capita: $2,100

Occupations: agriculture 85%, other 15%

Currency: kina (PGK)

Paraguay

Long Name: Republic of Paraguay

Location: Central South America, northeast of Argentina

Area: 157,046 sq. mi. (406,750 sq. km)

Climate: subtropical to temperate; substantial rainfall in the eastern portions, becoming semiarid in the far west

Terrain: grassy plains and wooded hills east of Rio Paraguay; Gran Chaco region west of Rio Paraguay mostly low, marshy plain near the river, and dry forest and thorny scrub elsewhere

Population: 6,831,306

Population Growth Rate: 2.39%

Birth Rate (per 1,000): 28.47

Death Rate (per 1,000): 4.49

Life Expectancy: 75.56 years

Ethnic Groups: mestizo (mixed Spanish and Amerindian) 95%, other 5%

Religion: Roman Catholic 89.6%, Protestant 6.2%, other Christian 1.1%, other or unspecified 1.9%, none 1.1%

Languages: Spanish (official), Guarani (official)

Literacy: 94%

Government Type: constitutional republic

Capital: Asuncion

Independence: 14 May 1811 (from Spain)

GDP Per Capita: $4,000

Occupations: agriculture 31%, industry 17%, services 52%

Currency: guarani (PYG)

Peru

Long Name: Republic of Peru

Location: Western South America, bordering the South Pacific Ocean, between Chile and Ecuador

Area: 496,224 sq. mi. (1,285,220 sq. km)

Climate: varies from tropical in east to dry desert in west; temperate to frigid in Andes

Terrain: western coastal plain (costa), high and rugged Andes in center (sierra), eastern lowland jungle of Amazon Basin (selva)

Population: 29,180,900

Population Growth Rate: 1.264%

Birth Rate (per 1,000): 19.77

Death Rate (per 1,000): 6.16

Life Expectancy: 70.44 years

Ethnic Groups: Amerindian 45%, mestizo (mixed Amerindian and white) 37%, white 15%, black, Japanese, Chinese, and other 3%

Religion: Roman Catholic 81%, Seventh Day Adventist 1.4%, other Christian 0.7%, other 0.6%, unspecified or none 16.3%

Languages: Spanish (official), Quechua (official), Aymara, and a large number of minor Amazonian languages

Literacy: 87.7%

Government Type: constitutional republic

Capital: Lima

Independence: 28 July 1821 (from Spain)

GDP Per Capita: $7,600

Occupations: agriculture 9%, industry 18%, services 73%

Currency: nuevo sol (PEN)

The Philippines

Long Name: Republic of the Philippines

Location: Southeastern Asia, archipelago between the Philippine Sea and the South China Sea, east of Vietnam

Area: 115,830 sq. mi. (300,000 sq. km)

Climate: tropical marine; northeast monsoon (November to April); southwest monsoon (May to October)

Terrain: mostly mountains with narrow to extensive coastal lowlands

Population: 96,061,680

Population Growth Rate: 1.991%

Birth Rate (per 1,000): 26.42

Death Rate (per 1,000): 5.15

Life Expectancy: 70.8 years

Ethnic Groups: Tagalog 28.1%, Cebuano 13.1%, Ilocano 9%, Bisaya/Binisaya 7.6%, Hiligaynon Ilonggo 7.5%, Bikol 6%, Waray 3.4%, other 25.3%

Religion: Roman Catholic 80.9%, Muslim 5%, Evangelical 2.8%, Iglesia ni Kristo 2.3%, Aglipayan 2%, other Christian 4.5%, other 1.8%, unspecified 0.6%, none 0.1%

Languages: Filipino (official; based on Tagalog) and English (official); eight major dialects - Tagalog, Cebuano, Ilocano, Hiligaynon or Ilonggo, Bicol, Waray, Pampango, and Pangasinan

Literacy: 92.6%

Government Type: republic

Capital: Manila

Independence: 12 June 1898 (independence proclaimed from Spain); 4 July 1946 (from the US)

GDP Per Capita: $3,200

Occupations: agriculture 35%, industry 15%, services 50%

Currency: Philippine peso (PHP)

Poland

Long Name: Republic of Poland

Location: Central Europe, east of Germany

Area: 120,725 sq. mi. (312,679 sq. km)

Climate: temperate with cold, cloudy, moderately severe winters with frequent precipitation; mild summers with frequent showers and thundershowers

Terrain: mostly flat plain; mountains along southern border

Population: 38,500,696

Population Growth Rate: -0.045%

Birth Rate (per 1,000): 10.01

Death Rate (per 1,000): 9.99

Life Expectancy: 75.41 years

Ethnic Groups: Polish 96.7%, German 0.4%, Belarusian 0.1%, Ukrainian 0.1%, other and unspecified 2.7%

Religion: Roman Catholic 89.8% (about 75% practicing), Eastern Orthodox 1.3%, Protestant 0.3%, other 0.3%, unspecified 8.3%

Languages: Polish 97.8%, other and unspecified 2.2%

Literacy: 99.8%

Government Type: republic

Capital: Warsaw

Independence: 11 November 1918 (republic proclaimed)

GDP Per Capita: $16,200

Occupations: agriculture 16.1%, industry 29%, services 54.9%

Currency: zloty (PLN)

Portugal

Long Name: Portuguese Republic

Location: Southwestern Europe, bordering the North Atlantic Ocean, west of Spain

Area: 35,672 sq. mi. (92,391 sq. km)

Climate: maritime temperate; cool and rainy in north, warmer and drier in south

Terrain: mountainous north of the Tagus River, rolling plains in south

Population: 10,676,910

Population Growth Rate: 0.305%

Birth Rate (per 1,000): 10.45

Death Rate (per 1,000): 10.62

Life Expectancy: 78.04 years

Ethnic Groups: homogeneous Mediterranean stock; citizens of black African descent who immigrated to mainland during decolonization number less than 100,000; since 1990 East Europeans have entered Portugal

Religion: Roman Catholic 84.5%, other Christian 2.2%, other 0.3%, unknown 9%, none 3.9%

Languages: Portuguese (official), Mirandese (official - but locally used)

Literacy: 93.3%

Government Type: republic; parliamentary democracy

Capital: Lisbon

Independence: 1143 (Kingdom of Portugal recognized); 5 October 1910 (republic proclaimed)

GDP Per Capita: $21,800

Occupations: agriculture 10%, industry 30%, services 60%

Currency: euro (EUR)

Qatar

Long Name: State of Qatar

Location: Middle East, peninsula bordering the Persian Gulf and Saudi Arabia

Area: 4,416 sq. mi. (11,437 sq. km)

Climate: arid; mild, pleasant winters; very hot, humid summers

Terrain: mostly flat and barren desert covered with loose sand and gravel

Population: 824,789

Population Growth Rate: 1.093%

Birth Rate (per 1,000): 15.69

Death Rate (per 1,000): 2.47

Life Expectancy: 75.19 years

Ethnic Groups: Arab 40%, Indian 18%, Pakistani 18%, Iranian 10%, other 14%

Religion: Muslim 77.5%, Christian 8.5%, other 14%

Languages: Arabic (official), English commonly used as a second language

Literacy: 89%

Government Type: emirate

Capital: Doha

Independence: 3 September 1971 (from United Kingdom)

GDP Per Capita: $87,600

Occupations: primarily industry

Currency: Qatari rial (QAR)

Romania

Long Name: Romania

Location: Southeastern Europe, bordering the Black Sea, between Bulgaria and Ukraine

Area: 91,699 sq. mi. (237,500 sq. km)

Climate: temperate; cold, cloudy winters with frequent snow and fog; sunny summers with frequent showers and thunderstorms

Terrain: central Transylvanian Basin is separated from the Plain of Moldavia on the east by the Carpathian Mountains and separated from the Walachian Plain on the south by the Transylvanian Alps

Population: 22,246,862

Population Growth Rate: -0.136%

Birth Rate (per 1,000): 10.61

Death Rate (per 1,000): 11.84

Life Expectancy: 72.18 years

Ethnic Groups: Romanian 89.5%, Hungarian 6.6%, Roma 2.5%, Ukrainian 0.3%, German 0.3%, Russian 0.2%, Turkish 0.2%, other 0.4%

Religion: Eastern Orthodox (including all sub-denominations) 86.8%, Protestant (various denominations including Reformate and Pentecostal) 7.5%, Roman Catholic 4.7%, other (mostly Muslim) and unspecified 0.9%, none 0.1%

Languages: Romanian 91% (official), Hungarian 6.7%, Romany (Gypsy) 1.1%, other 1.2%

Literacy: 97.3%

Government Type: republic

Capital: Bucharest

Independence: 9 May 1877 (independence proclaimed from the Ottoman Empire; independence recognized 13 July 1878 by the Treaty of Berlin); 26 March 1881 (kingdom proclaimed); 30 December 1947 (republic proclaimed)

GDP Per Capita: $11,100

Occupations: agriculture 29.7%, industry 23.2%, services 47.1%

Currency: "new" leu (RON) was introduced in 2005; "old" leu (ROL) was phased out in 2006

Russia

Long Name: Russian Federation

Location: Northern Asia (the area west of the Urals is considered part of Europe), bordering the Arctic Ocean, between Europe and the North Pacific Ocean

Area: 6,592,741 sq. mi. (17,075,200 sq. km)

Climate: ranges from steppes in the south through humid continental in much of European Russia; subarctic in Siberia to tundra climate in the polar north; winters vary from cool along Black Sea coast to frigid in Siberia; summers vary from warm in the steppes to cool along Arctic coast

Terrain: broad plain with low hills west of Urals; vast coniferous forest and tundra in Siberia; uplands and mountains along southern border regions

Population: 140,702,096

Population Growth Rate: -0.474%

Birth Rate (per 1,000): 11.03

Death Rate (per 1,000): 16.06

Life Expectancy: 65.94 years

Ethnic Groups: Russian 79.8%, Tatar 3.8%, Ukrainian 2%, Bashkir 1.2%, Chuvash 1.1%, other or unspecified 12.1%

Religion: Russian Orthodox 15-20%, Muslim 10-15%, other Christian 2% (2006 est.), large populations of non-practicing believers and non-believers, a legacy of over seven decades of Soviet rule

Languages: Russian, many minority languages

Literacy: 99.4%

Government Type: federation

Capital: Moscow

339

Independence: 24 August 1991 (from Soviet Union)

GDP Per Capita: $14,800

Occupations: agriculture 10.8%, industry 28.8%, services 60.5%

Currency: Russian ruble (RUB)

Rwanda

Long Name: Republic of Rwanda

Location: Central Africa, east of Democratic Republic of the Congo

Area: 10,169 sq. mi. (26,338 sq. km)

Climate: temperate; two rainy seasons (February to April, November to January); mild in mountains with frost and snow possible

Terrain: mostly grassy uplands and hills; relief is mountainous with altitude declining from west to east

Population: 10,186,063

Population Growth Rate: 2.779%

Birth Rate (per 1,000): 39.97

Death Rate (per 1,000): 14.46

Life Expectancy: 49.76 years

Ethnic Groups: Hutu (Bantu) 84%, Tutsi (Hamitic) 15%, Twa (Pygmy) 1%

Religion: Roman Catholic 56.5%, Protestant 26%, Adventist 11.1%, Muslim 4.6%, indigenous beliefs 0.1%, none 1.7%

Languages: Kinyarwanda (official) universal Bantu vernacular, French (official), English (official), Kiswahili (Swahili) used in commercial centers

Literacy: 70.4%

Government Type: republic; presidential, multiparty system

Capital: Kigali

Independence: 1 July 1962 (from Belgium-administered UN trusteeship)

GDP Per Capita: $800

Occupations: agriculture 90%, industry and services 10%

Currency: Rwandan franc (RWF)

Saint Kitts and Nevis

Long Name: Federation of Saint Kitts and Nevis

Location: Caribbean, islands in the Caribbean Sea, about one-third of the way from Puerto Rico to Trinidad and Tobago

Area: 101 sq. mi. (261 sq. km)

Climate: tropical, tempered by constant sea breezes; little seasonal temperature variation; rainy season (May to November)

Terrain: volcanic with mountainous interiors

Population: 39,817

Population Growth Rate: 0.723%

Birth Rate (per 1,000): 17.73

Death Rate (per 1,000): 8.19

Life Expectancy: 72.94 years

Ethnic Groups: predominantly black; some British, Portuguese, and Lebanese

Religion: Anglican, other Protestant, Roman Catholic

Languages: English

Literacy: 97.8%

Government Type: parliamentary democracy

Capital: Basseterre

Independence: 19 September 1983 (from United Kingdom)

GDP Per Capita: $13,900

Occupations: primarily services and industry

Currency: East Caribbean dollar (XCD)

Saint Lucia

Long Name: Saint Lucia

Location: Caribbean, island between the Caribbean Sea and North Atlantic Ocean, north of Trinidad and Tobago

Area: 238 sq. mi. (616 sq. km)

Climate: tropical, moderated by northeast trade winds; dry season January to April, rainy season May to August

Terrain: volcanic and mountainous with some broad, fertile valleys

Population: 159,585

Population Growth Rate: 0.436%

Birth Rate (per 1,000): 15.4

Death Rate (per 1,000): 6.71

Life Expectancy: 76.25 years

Ethnic Groups: black 82.5%, mixed 11.9%, East Indian 2.4%, other or unspecified 3.1%

Religion: Roman Catholic 67.5%, Seventh Day Adventist 8.5%, Pentecostal 5.7%, Rastafarian 2.1%, Anglican 2%, Evangelical 2%, other Christian 5.1%, other 1.1%, unspecified 1.5%, none 4.5%

Languages: English (official), French patois

Literacy: 90.1%

Government Type: parliamentary democracy

Capital: Castries

Independence: 22 February 1979 (from United Kingdom)

GDP Per Capita: $10,700

Occupations: agriculture 21.7%, industry 24.7%, services 53.6%

Currency: East Caribbean dollar (XCD)

Saint Vincent and the Grenadines

Long Name: Saint Vincent and the Grenadines

Location: Caribbean, islands between the Caribbean Sea and North Atlantic Ocean, north of Trinidad and Tobago

Area: 150 sq. mi. (389 sq. km)

Climate: tropical; little seasonal temperature variation; rainy season (May to November)

Terrain: volcanic, mountainous

Population: 118,432

Population Growth Rate: 0.231%

Birth Rate (per 1,000): 15.82

Death Rate (per 1,000): 5.96

Life Expectancy: 74.34 years

Ethnic Groups: black 66%, mixed 19%, East Indian 6%, European 4%, Carib Amerindian 2%, other 3%

Religion: Anglican 47%, Methodist 28%, Roman Catholic 13%, other (includes Hindu, Seventh-Day Adventist, other Protestant) 12%

Languages: English, French patois

Literacy: 96%

Government Type: parliamentary democracy

Capital: Kingstown

Independence: 27 October 1979 (from United Kingdom)

GDP Per Capita: $9,800

Occupations: agriculture 26%, industry 17%, services 57%

Currency: East Caribbean dollar (XCD)

Samoa

Long Name: Independent State of Samoa

Location: Oceania, group of islands in the South Pacific Ocean, about half way between Hawaii and New Zealand

Area: 1,137 sq. mi. (2,944 sq. km)

Climate: tropical; rainy season (November to April), dry season (May to October)

Terrain: two main islands (Savaii, Upolu) and several smaller islands and uninhabited islets; narrow coastal plain with volcanic, rocky, rugged mountains in interior

Population: 217,083

Population Growth Rate: 1.322%

Birth Rate (per 1,000): 28.2

Death Rate (per 1,000): 5.84

Life Expectancy: 71.58 years

Ethnic Groups: Samoan 92.6%, Euronesians (persons of European and Polynesian blood) 7%, Europeans 0.4%

Religion: Congregationalist 34.8%, Roman Catholic 19.6%, Methodist 15%, Latter-Day Saints 12.7%, Assembly of God 6.6%, Seventh-Day Adventist 3.5%, Worship Centre 1.3%, other Christian 4.5%, other 1.9%, unspecified 0.1%

Languages: Samoan (Polynesian), English

Literacy: 99.7%

Government Type: parliamentary democracy

Capital: Apia

Independence: 1 January 1962 (from New Zealand-administered UN trusteeship)

GDP Per Capita: $5,400

Occupations: N/A

Currency: tala (SAT)

San Marino

Long Name: Most Serene Republic of San Marino

Location: Southern Europe, an enclave in central Italy

Area: 24 sq. mi. (61 sq. km)

Climate: Mediterranean; mild to cool winters; warm, sunny summers

Terrain: rugged mountains

Population: 29,973

Population Growth Rate: 1.181%

Birth Rate (per 1,000): 9.74

Death Rate (per 1,000): 8.37

Life Expectancy: 81.88 years

Ethnic Groups: Sammarinese, Italian

Religion: Roman Catholic

Languages: Italian

Literacy: 96%

Government Type: republic

Capital: San Marino

Independence: 3 September AD 301

GDP Per Capita: $34,100

Occupations: agriculture 0.2%, industry 40.1%, services 59.7%

Currency: euro (EUR)

São Tomé and Príncipe

Long Name: Democratic Republic of São Tomé and Príncipe

Location: Western Africa, islands in the Gulf of Guinea, straddling the Equator, west of Gabon

Area: 386 sq. mi. (1,001 sq. km)

Climate: tropical; hot, humid; one rainy season (October to May)

Terrain: volcanic, mountainous

Population: 206,178

Population Growth Rate: 3.116%

Birth Rate (per 1,000): 39.12

Death Rate (per 1,000): 5.98

Life Expectancy: 68 years

Ethnic Groups: mestico, angolares (descendants of Angolan slaves), forros (descendants of freed slaves), servicais (contract laborers from Angola, Mozambique, and Cape Verde), tongas (children of servicais born on the islands), Europeans (primarily Portuguese)

Religion: Catholic 70.3%, Evangelical 3.4%, New Apostolic 2%, Adventist 1.8%, other 3.1%, none 19.4%

Languages: Portuguese (official)

Literacy: 84.9%

Government Type: republic

Capital: São Tomé

Independence: 12 July 1975 (from Portugal)

GDP Per Capita: $1,600

Occupations: population mainly engaged in subsistence agriculture and fishing; shortages of skilled workers

Currency: dobra (STD)

Saudi Arabia

Long Name: Kingdom of Saudi Arabia

Location: Middle East, bordering the Persian Gulf and the Red Sea, north of Yemen

Area: 829,996 sq. mi. (2,149,690 sq. km)

Climate: harsh, dry desert with great temperature extremes

Terrain: mostly uninhabited, sandy desert

Population: 28,146,656

Population Growth Rate: 1.954%

Birth Rate (per 1,000): 28.85

Death Rate (per 1,000): 2.49

Life Expectancy: 76.09 years

Ethnic Groups: Arab 90%, Afro-Asian 10%

Religion: Muslim 100%

Languages: Arabic

Literacy: 78.8%

Government Type: monarchy

Capital: Riyadh

Independence: 23 September 1932 (unification of the kingdom)

GDP Per Capita: $19,800

Occupations: agriculture 12%, industry 25%, services 63%

Currency: Saudi riyal (SAR)

Senegal

Long Name: Republic of Senegal

Location: Western Africa, bordering the North Atlantic Ocean, between Guinea-Bissau and Mauritania

Area: 75,749 sq. mi. (196,190 sq. km)

Climate: tropical; hot, humid; rainy season (May to November) has strong southeast winds; dry season (December to April) dominated by hot, dry, harmattan wind

Terrain: generally low, rolling, plains rising to foothills in southeast

Population: 12,853,259

Population Growth Rate: 2.58%

Birth Rate (per 1,000): 36.52

Death Rate (per 1,000): 10.72

Life Expectancy: 57.08 years

Ethnic Groups: Wolof 43.3%, Pular 23.8%, Serer 14.7%, Jola 3.7%, Mandinka 3%, Soninke 1.1%, European and Lebanese 1%, other 9.4%

Religion: Muslim 94%, Christian 5% (mostly Roman Catholic), indigenous beliefs 1%

Languages: French (official), Wolof, Pulaar, Jola, Mandinka

Literacy: 39.3%

343

Government Type: republic

Capital: Dakar

Independence: 4 April 1960 (from France); complete independence achieved upon dissolution of federation with Mali on 20 August 1960

GDP Per Capita: $1,700

Occupations: agriculture 77.5%, industry and services 22.5%

Currency: Communaute Financiere Africaine franc (XOF)

Serbia

Long Name: Republic of Serbia

Location: Southeastern Europe, between Macedonia and Hungary

Area: 29,913 sq. mi. (77,474 sq. km)

Climate: in the north, continental climate (cold winters and hot, humid summers with well distributed rainfall); in other parts, continental and Mediterranean climate (relatively cold winters with heavy snowfall and hot, dry summers and autumns)

Terrain: extremely varied; to the north, rich fertile plains; to the east, limestone ranges and basins; to the southeast, ancient mountains and hills

Population: 10,159,046

Population Growth Rate: N/A

Birth Rate (per 1,000): N/A

Death Rate (per 1,000): N/A

Life Expectancy: 75.29 years

Ethnic Groups: Serb 82.9%, Hungarian 3.9%, Romany (Gypsy) 1.4%, Yugoslavs 1.1%, Bosniaks 1.8%, Montenegrin 0.9%, other 8%

Religion: Serbian Orthodox 85%, Catholic 5.5%, Protestant 1.1%, Muslim 3.2%, unspecified 2.6%, other, unknown, or atheist 2.6%

Languages: Serbian 88.3% (official), Hungarian 3.8%, Bosniak 1.8%, Romany (Gypsy) 1.1%, other 4.1%, unknown 0.9%, Romanian, Hungarian, Slovak, Ukrainian, and Croatian all official in Vojvodina

Literacy: 96.4%

Government Type: republic

Capital: Belgrade

Independence: 5 June 2006 (from Serbia and Montenegro)

GDP Per Capita: $10,400

Occupations: agriculture 30%, industry 46%, services 24%

Currency: Serbian dinar (RSD)

Seychelles

Long Name: Republic of Seychelles

Location: archipelago in the Indian Ocean, northeast of Madagascar

Area: 176 sq. mi. (455 sq. km)

Climate: tropical marine; humid; cooler season during southeast monsoon (late May to September); warmer season during northwest monsoon (March to May)

Terrain: Mahe Group is granitic, narrow coastal strip, rocky, hilly; others are coral, flat, elevated reefs

Population: 82,247

Population Growth Rate: 0.428%

Birth Rate (per 1,000): 15.6

Death Rate (per 1,000): 6.21

Life Expectancy: 72.6 years

Ethnic Groups: mixed French, African, Indian, Chinese, and Arab

Religion: Roman Catholic 82.3%, Anglican 6.4%, Seventh Day Adventist 1.1%, other Christian 3.4%, Hindu 2.1%, Muslim 1.1%, other non-Chris-

tian 1.5%, unspecified 1.5%, none
0.6%

Languages: Creole 91.8%, English 4.9%
(official), other 3.1%, unspecified
0.2%

Literacy: 91.8%

Government Type: republic

Capital: Victoria

Independence: 29 June 1976 (from United Kingdom)

GDP Per Capita: $16,600

Occupations: agriculture 3%, industry
23%, services 74%

Currency: Seychelles rupee (SCR)

Sierra Leone

Long Name: Republic of Sierra Leone

Location: Western Africa, bordering the
North Atlantic Ocean, between
Guinea and Liberia

Area: 27,699 sq. mi. (71,740 sq. km)

Climate: tropical; hot, humid; summer
rainy season (May to December); winter dry season (December to April)

Terrain: coastal belt of mangrove
swamps, wooded hill country, upland
plateau, mountains in east

Population: 6,294,774

Population Growth Rate: 2.282%

Birth Rate (per 1,000): 45.08

Death Rate (per 1,000): 22.26

Life Expectancy: 40.93 years

Ethnic Groups: 20 African ethnic groups
90% (Temne 30%, Mende 30%, other
30%), Creole (Krio) 10% (descendants of freed Jamaican slaves who
were settled in the Freetown area in
the late-18th century), refugees from
Liberia's recent civil war, small num-

bers of Europeans, Lebanese, Pakistanis, and Indians

Religion: Muslim 60%, Christian 10%,
indigenous beliefs 30%

Languages: English (official, regular use
limited to literate minority), Mende
(principal vernacular in the south),
Temne (principal vernacular in the
north), Krio (English-based Creole,
spoken by the descendants of freed
Jamaican slaves who were settled in
the Freetown area, a lingua franca
and a first language for 10% of the
population but understood by 95%)

Literacy: 35.1%

Government Type: constitutional
democracy

Capital: Freetown

Independence: 27 April 1961 (from United Kingdom)

GDP Per Capita: $600

Occupations: N/A

Currency: leone (SLL)

Singapore

Long Name: Republic of Singapore

Location: Southeastern Asia, islands
between Malaysia and Indonesia

Area: 268 sq. mi. (693 sq. km)

Climate: tropical; hot, humid, rainy; two
distinct monsoon seasons - Northeastern monsoon (December to
March) and Southwestern monsoon
(June to September); inter-monsoon -
frequent afternoon and early evening
thunderstorms

Terrain: lowland; gently undulating central plateau contains water catchment
area and nature preserve

Population: 4,608,167

Population Growth Rate: 1.135%

Birth Rate (per 1,000): 8.99

Death Rate (per 1,000): 4.53

Life Expectancy: 81.89 years

Ethnic Groups: Chinese 76.8%, Malay 13.9%, Indian 7.9%, other 1.4%

Religion: Buddhist 42.5%, Muslim 14.9%, Taoist 8.5%, Hindu 4%, Catholic 4.8%, other Christian 9.8%, other 0.7%, none 14.8%

Languages: Mandarin 35%, English 23%, Malay 14.1%, Hokkien 11.4%, Cantonese 5.7%, Teochew 4.9%, Tamil 3.2%, other Chinese dialects 1.8%, other 0.9%

Literacy: 92.5%

Government Type: parliamentary republic

Capital: Singapore

Independence: 9 August 1965 (from Malaysian Federation)

GDP Per Capita: $49,900

Occupations: manufacturing 21%, construction 5%, transportation and communication 7%, financial, business, and other services 42%, other 25%

Currency: Singapore dollar (SGD)

Slovakia

Long Name: Slovak Republic

Location: Central Europe, south of Poland

Area: 18,859 sq. mi. (48,845 sq. km)

Climate: temperate; cool summers; cold, cloudy, humid winters

Terrain: rugged mountains in the central and northern part and lowlands in the south

Population: 5,455,407

Population Growth Rate: 0.143%

Birth Rate (per 1,000): 10.64

Death Rate (per 1,000): 9.5

Life Expectancy: 75.17 years

Ethnic Groups: Slovak 85.8%, Hungarian 9.7%, Roma 1.7%, Ruthenian/Ukrainian 1%, other and unspecified 1.8%

Religion: Roman Catholic 68.9%, Protestant 10.8%, Greek Catholic 4.1%, other or unspecified 3.2%, none 13%

Languages: Slovak (official) 83.9%, Hungarian 10.7%, Roma 1.8%, Ukrainian 1%, other or unspecified 2.6%

Literacy: 99.6%

Government Type: parliamentary democracy

Capital: Bratislava

Independence: 1 January 1993 (Czechoslovakia split into the Czech Republic and Slovakia)

GDP Per Capita: $20,200

Occupations: agriculture 5.8%, industry 29.3%, construction 9%, services 55.9%

Currency: Slovak koruna (SKK)

Slovenia

Long Name: Republic of Slovenia

Location: Central Europe, eastern Alps bordering the Adriatic Sea, between Austria and Croatia

Area: 7,827 sq. mi. (20,273 sq. km)

Climate: Mediterranean climate on the coast, continental climate with mild to hot summers and cold winters in the plateaus and valleys to the east

Terrain: a short coastal strip on the Adriatic, an alpine mountain region adjacent to Italy and Austria, mixed mountains and valleys with numerous rivers to the east

Population: 2,007,711

Population Growth Rate: -0.088%

Birth Rate (per 1,000): 8.99

Death Rate (per 1,000): 10.51

Life Expectancy: 76.73 years

Ethnic Groups: Slovene 83.1%, Serb 2%, Croat 1.8%, Bosniak 1.1%, other or unspecified 12%

Religion: Catholic 57.8%, Muslim 2.4%, Orthodox 2.3%, other Christian 0.9%, unaffiliated 3.5%, other or unspecified 23%, none 10.1%

Languages: Slovenian 91.1%, Serbo-Croatian 4.5%, other or unspecified 4.4%

Literacy: 99.7%

Government Type: parliamentary republic

Capital: Ljubljana

Independence: 25 June 1991 (from Yugoslavia)

GDP Per Capita: $28,000

Occupations: agriculture 2.5%, industry 36%, services 61.5%

Currency: euro (EUR)

Solomon Islands

Long Name: Solomon Islands

Location: Oceania, group of islands in the South Pacific Ocean, east of Papua New Guinea

Area: 10,985 sq. mi. (28,450 sq. km)

Climate: tropical monsoon; few extremes of temperature and weather

Terrain: mostly rugged mountains with some low coral atolls

Population: 581,318

Population Growth Rate: 2.467%

Birth Rate (per 1,000): 28.48

Death Rate (per 1,000): 3.81

Life Expectancy: 73.44 years

Ethnic Groups: Melanesian 94.5%, Polynesian 3%, Micronesian 1.2%, other 1.1%, unspecified 0.2%

Religion: Church of Melanesia 32.8%, Roman Catholic 19%, South Seas Evangelical 17%, Seventh-Day Adventist 11.2%, United Church 10.3%, Christian Fellowship Church 2.4%, other Christian 4.4%, other 2.4%, unspecified 0.3%, none 0.2%

Languages: Melanesian pidgin in much of the country is lingua franca; English (official; but spoken by only 1%-2% of the population); 120 indigenous languages

Literacy: N/A

Government Type: parliamentary democracy

Capital: Honiara

Independence: 7 July 1978 (from United Kingdom)

GDP Per Capita: $1,900

Occupations: agriculture 75%, industry 5%, services 20%

Currency: Solomon Islands dollar (SBD)

Somalia

Long Name: Somalia

Location: Eastern Africa, bordering the Gulf of Aden and the Indian Ocean, east of Ethiopia

Area: 246,200 sq. mi. (637,657 sq. km)

Climate: principally desert; northeast monsoon (December to February), moderate temperatures in north and hot in south; southwest monsoon (May to October), torrid in the north and hot in the south, irregular rainfall, hot and humid periods (tangambili) between monsoons

Terrain: mostly flat to undulating plateau rising to hills in north

Population: 9,558,666

Population Growth Rate: 2.824%

Birth Rate (per 1,000): 44.12

Death Rate (per 1,000): 15.89

Life Expectancy: 49.25 years

Ethnic Groups: Somali 85%, Bantu and other non-Somali 15% (including Arabs 30,000)

Religion: Sunni Muslim

Languages: Somali (official), Arabic, Italian, English

Literacy: 37.8%

Government Type: no permanent national government; transitional, parliamentary federal government

Capital: Mogadishu

Independence: 1 July 1960 (from a merger of British Somaliland, which became independent from the United Kingdom on 26 June 1960, and Italian Somaliland, which became independent from the Italian-administered UN trusteeship on 1 July 1960, to form the Somali Republic)

GDP Per Capita: $600

Occupations: agriculture 71%, industry and services 29%

Currency: Somali shilling (SOS)

South Africa

Long Name: Republic of South Africa

Location: Southern Africa, at the southern tip of the continent of Africa

Area: 471,008 sq. mi. (1,219,912 sq. km)

Climate: mostly semiarid; subtropical along east coast; sunny days, cool nights

Terrain: vast interior plateau rimmed by rugged hills and narrow coastal plain

Population: 48,782,756

Population Growth Rate: 0.828%

Birth Rate (per 1,000): 20.23

Death Rate (per 1,000): 16.94

Life Expectancy: 48.89 years

Ethnic Groups: black African 79%, white 9.6%, colored 8.9%, Indian/Asian 2.5%

Religion: Zion Christian 11.1%, Pentecostal/Charismatic 8.2%, Catholic 7.1%, Methodist 6.8%, Dutch Reformed 6.7%, Anglican 3.8%, Muslim 1.5%, other Christian 36%, other 2.3%, unspecified 1.4%, none 15.1%

Languages: IsiZulu 23.8%, IsiXhosa 17.6%, Afrikaans 13.3%, Sepedi 9.4%, English 8.2%, Setswana 8.2%, Sesotho 7.9%, Xitsonga 4.4%, other 7.2%

Literacy: 86.4%

Government Type: republic

Capital: Pretoria

Independence: 31 May 1910 (Union of South Africa formed from four British colonies: Cape Colony, Natal, Transvaal, and Orange Free State); 31 May 1961 (republic declared) 27 April 1994 (majority rule)

GDP Per Capita: $9,700

Occupations: agriculture 9%, industry 26%, services 65%

Currency: rand (ZAR)

South Korea

Long Name: Republic of Korea

Location: Eastern Asia, southern half of the Korean Peninsula bordering the Sea of Japan and the Yellow Sea

Area: 38,023 sq. mi. (98,480 sq. km)

Climate: temperate, with rainfall heavier in summer than winter

Terrain: mostly hills and mountains; wide coastal plains in west and south

Population: 48,379,392

Population Growth Rate: 0.269%

Birth Rate (per 1,000): 9.09

Death Rate (per 1,000): 5.73

Life Expectancy: 78.64 years

Ethnic Groups: homogeneous (except for about 20,000 Chinese)

Religion: Christian 26.3% (Protestant 19.7%, Roman Catholic 6.6%), Buddhist 23.2%, other or unknown 1.3%, none 49.3%

Languages: Korean, English widely taught in junior high and high school

Literacy: 97.9%

Government Type: republic

Capital: Seoul

Independence: 15 August 1945 (from Japan)

GDP Per Capita: $25,000

Occupations: agriculture 7.5%, industry 17.3%, services 75.2%

Currency: South Korean won (KRW)

Spain

Long Name: Kingdom of Spain

Location: Southwestern Europe, bordering the Bay of Biscay, Mediterranean Sea, North Atlantic Ocean, and Pyrenees Mountains, southwest of France

Area: 194,897 sq. mi. (504,782 sq. km)

Climate: temperate; clear, hot summers in interior, more moderate and cloudy along coast; cloudy, cold winters in interior, partly cloudy and cool along coast

Terrain: large, flat to dissected plateau surrounded by rugged hills; Pyrenees in north

Population: 40,491,052

Population Growth Rate: 0.096%

Birth Rate (per 1,000): 9.87

Death Rate (per 1,000): 9.9

Life Expectancy: 79.92 years

Ethnic Groups: composite of Mediterranean and Nordic types

Religion: Roman Catholic 94%, other 6%

Languages: Castilian Spanish (official) 74%, Catalan 17%, Galician 7%, Basque 2%, are official regionally

Literacy: 97.9%

Government Type: parliamentary monarchy

Capital: Madrid

Independence: the Iberian peninsula was characterized by a variety of independent kingdoms prior to the Muslim occupation that began in the early 8th century A.D. and lasted nearly seven centuries; the small Christian redoubts of the north began the reconquest almost immediately, culminating in the seizure of Granada in 1492; this event completed the unification of several kingdoms and is traditionally considered the forging of present-day Spain

GDP Per Capita: $33,600

Occupations: agriculture 5.3%, industry 30.1%, services 64.6%

Currency: euro (EUR)

Sri Lanka

Long Name: Democratic Socialist Republic of Sri Lanka

Location: Southern Asia, island in the Indian Ocean, south of India

Area: 25,332 sq. mi. (65,610 sq. km)

Climate: tropical monsoon; northeast monsoon (December to March); southwest monsoon (June to October)

Terrain: mostly low, flat to rolling plain; mountains in south-central interior

Population: 21,128,772

Population Growth Rate: 0.943%

Birth Rate (per 1,000): 16.63

Death Rate (per 1,000): 6.07

Life Expectancy: 74.97 years

Ethnic Groups: Sinhalese 73.8%, Sri Lankan Moors 7.2%, Indian Tamil 4.6%, Sri Lankan Tamil 3.9%, other 0.5%, unspecified 10%

Religion: Buddhist 69.1%, Muslim 7.6%, Hindu 7.1%, Christian 6.2%, unspecified 10%

Languages: Sinhala (official and national language) 74%, Tamil (national language) 18%, other 8%, English is commonly used in government and is spoken competently by about 10% of the population

Literacy: 90.7%

Government Type: republic

Capital: Colombo

Independence: 4 February 1948 (from United Kingdom)

GDP Per Capita: $4,000

Occupations: agriculture 34.3%, industry 25.3%, services 40.4%

Currency: Sri Lankan rupee (LKR)

Sudan

Long Name: Republic of the Sudan

Location: Northern Africa, bordering the Red Sea, between Egypt and Eritrea

Area: 967,494 sq. mi. (2,505,810 sq. km)

Climate: tropical in south; arid desert in north; rainy season varies by region (April to November)

Terrain: generally flat, featureless plain; mountains in far south, northeast and west; desert dominates the north

Population: 40,218,456

Population Growth Rate: 2.134%

Birth Rate (per 1,000): 34.31

Death Rate (per 1,000): 13.64

Life Expectancy: 50.28 years

Ethnic Groups: black 52%, Arab 39%, Beja 6%, foreigners 2%, other 1%

Religion: Sunni Muslim 70% (in north), Christian 5% (mostly in south and Khartoum), indigenous beliefs 25%

Languages: Arabic (official), Nubian, Ta Bedawie, diverse dialects of Nilotic, Nilo-Hamitic, Sudanic languages, English

Literacy: 61.1%

Government Type: Government of National Unity (GNU) - the National Congress Party (NCP) and Sudan People's Liberation Movement (SPLM) formed a power-sharing government under the 2005 Comprehensive Peace Agreement (CPA); the NCP, which came to power by military coup in 1989, is the majority partner; the agreement stipulates national elections in 2009

Capital: Khartoum

Independence: 1 January 1956 (from Egypt and United Kingdom)

GDP Per Capita: $1,900

Occupations: agriculture 80%, industry 7%, services 13%

Currency: Sudanese pounds (SDG)

Suriname

Long Name: Republic of Suriname

Location: Northern South America, bordering the North Atlantic Ocean, between French Guiana and Guyana

Area: 63,039 sq. mi. (163,270 sq. km)

Climate: tropical; moderated by trade winds

Terrain: mostly rolling hills; narrow coastal plain with swamps

Population: 475,996

Population Growth Rate: 1.099%

Birth Rate (per 1,000): 17.02

Death Rate (per 1,000): 5.51

Life Expectancy: 73.48 years

Ethnic Groups: Hindustani (also known locally as "East Indians"; their ancestors emigrated from northern India in the latter part of the 19th century) 37%, Creole (mixed white and black) 31%, Javanese 15%, "Maroons" (their African ancestors were brought to the country in the 17th and 18th centuries as slaves and escaped to the interior) 10%, Amerindian 2%, Chinese 2%, white 1%, other 2%

Religion: Hindu 27.4%, Protestant 25.2% (predominantly Moravian), Roman Catholic 22.8%, Muslim 19.6%, indigenous beliefs 5%

Languages: Dutch (official), English (widely spoken), Sranang Tongo (Surinamese, sometimes called Taki-Taki, is native language of Creoles and much of the younger population and is lingua franca among others), Caribbean Hindustani (a dialect of Hindi), Javanese

Literacy: 89.6%

Government Type: constitutional democracy

Capital: Paramaribo

Independence: 25 November 1975 (from the Netherlands)

GDP Per Capita: $8,700

Occupations: agriculture 8%, industry 14%, services 78%

Currency: Surinam dollar (SRD)

Swaziland

Long Name: Kingdom of Swaziland

Location: Southern Africa, between Mozambique and South Africa

Area: 6,704 sq. mi. (17,363 sq. km)

Climate: varies from tropical to near temperate

Terrain: mostly mountains and hills; some moderately sloping plains

Population: 1,128,814

Population Growth Rate: -0.41%

Birth Rate (per 1,000): 26.6

Death Rate (per 1,000): 30.7

Life Expectancy: 31.99 years

Ethnic Groups: African 97%, European 3%

Religion: Zionist 40% (a blend of Christianity and indigenous ancestral worship), Roman Catholic 20%, Muslim 10%, other (includes Anglican, Bahai, Methodist, Mormon, Jewish) 30%

Languages: English (official, government business conducted in English), siSwati (official)

Literacy: 81.6%

Government Type: monarchy

Capital: Mbabane

Independence: 6 September 1968 (from United Kingdom)

GDP Per Capita: $4,700

Occupations: N/A

Currency: lilangeni (SZL)

Sweden

Long Name: Kingdom of Sweden

Location: Northern Europe, bordering the Baltic Sea, Gulf of Bothnia, Kattegat, and Skagerrak, between Finland and Norway

Area: 173,731 sq. mi. (449,964 sq. km)

Climate: temperate in south with cold, cloudy winters and cool, partly cloudy summers; subarctic in north

Terrain: mostly flat or gently rolling lowlands; mountains in west

Population: 9,045,389

Population Growth Rate: 0.157%

Birth Rate (per 1,000): 10.15

Death Rate (per 1,000): 10.24

Life Expectancy: 80.74 years

Ethnic Groups: indigenous population: Swedes with Finnish and Sami minorities; foreign-born or first-generation immigrants: Finns, Yugoslavs, Danes, Norwegians, Greeks, Turks

Religion: Lutheran 87%, other (includes Roman Catholic, Orthodox, Baptist, Muslim, Jewish, and Buddhist) 13%

Languages: Swedish, small Sami- and Finnish-speaking minorities

Literacy: 99%

Government Type: constitutional monarchy

Capital: Stockholm

Independence: 6 June 1523 (Gustav Vasa elected king)

GDP Per Capita: $37,500

Occupations: agriculture 2%, industry 24%, services 74%

Currency: Swedish krona (SEK)

Switzerland

Long Name: Swiss Confederation

Location: Central Europe, east of France, north of Italy

Area: 15,942 sq. mi. (41,290 sq. km)

Climate: temperate, but varies with altitude; cold, cloudy, rainy/snowy winters; cool to warm, cloudy, humid summers with occasional showers

Terrain: mostly mountains (Alps in south, Jura in northwest) with a central plateau of rolling hills, plains, and large lakes

Population: 7,581,520

Population Growth Rate: 0.329%

Birth Rate (per 1,000): 9.62

Death Rate (per 1,000): 8.54

Life Expectancy: 80.74 years

Ethnic Groups: German 65%, French 18%, Italian 10%, Romansch 1%, other 6%

Religion: Roman Catholic 41.8%, Protestant 35.3%, Muslim 4.3%, Orthodox 1.8%, other Christian 0.4%, other 1%, unspecified 4.3%, none 11.1%

Languages: German (official) 63.7%, French (official) 20.4%, Italian (official) 6.5%, Serbo-Croatian 1.5%, Albanian 1.3%, Portuguese 1.2%, Spanish 1.1%, English 1%, Romansch (official) 0.5%, other 2.8%

Literacy: 99%

Government Type: formally a confederation but similar in structure to a federal republic

Capital: Bern

Independence: 1 August 1291 (founding of the Swiss Confederation)

GDP Per Capita: $40,100

Occupations: agriculture 4.6%, industry 26.3%, services 69.1%

Currency: Swiss franc (CHF)

Syria

Long Name: Syrian Arab Republic

Location: Middle East, bordering the Mediterranean Sea, between Lebanon and Turkey

Area: 71,498 sq. mi. (185,180 sq. km)

Climate: mostly desert; hot, dry, sunny summers (June to August) and mild, rainy winters (December to February) along coast; cold weather with snow or sleet periodically in Damascus

Terrain: primarily semiarid and desert plateau; narrow coastal plain; mountains in west

Population: 19,747,586

Population Growth Rate: 2.189%

Birth Rate (per 1,000): 26.57

Death Rate (per 1,000): 4.68

Life Expectancy: 70.9 years

Ethnic Groups: Arab 90.3%, Kurds, Armenians, and other 9.7%

Religion: Sunni Muslim 74%, other Muslim (includes Alawite, Druze) 16%, Christian (various denominations) 10%, Jewish (tiny communities in Damascus, Al Qamishli, and Aleppo)

Languages: Arabic (official); Kurdish, Armenian, Aramaic, Circassian widely understood; French, English somewhat understood

Literacy: 79.6%

Government Type: republic under an authoritarian military-dominated regime

Capital: Damascus

Independence: 17 April 1946 (from League of Nations mandate under French administration)

GDP Per Capita: $4,700

Occupations: agriculture 19.2%, industry 14.5%, services 66.3%

Currency: Syrian pound (SYP)

Taiwan

Long Name: Taiwan

Location: Eastern Asia, islands bordering the East China Sea, Philippine Sea, South China Sea, and Taiwan Strait, north of the Philippines, off the southeastern coast of China

Area: 13,892 sq. mi. (35,980 sq. km)

Climate: tropical; marine; rainy season during southwest monsoon (June to August); cloudiness is persistent and extensive all year

Terrain: eastern two-thirds mostly rugged mountains; flat to gently rolling plains in west

Population: 22,920,946

Population Growth Rate: 0.238%

Birth Rate (per 1,000): 8.99

Death Rate (per 1,000): 6.65

Life Expectancy: 77.76 years

Ethnic Groups: Taiwanese (including Hakka) 84%, mainland Chinese 14%, indigenous 2%

Religion: mixture of Buddhist and Taoist 93%, Christian 4.5%, other 2.5%

Languages: Mandarin Chinese (official), Taiwanese (Min), Hakka dialects

Literacy: 96.1%

Government Type: multiparty democracy

Capital: Taipei

Independence: Following the Communist victory on the mainland in 1949, two million Nationalists fled to Taiwan and established a government using the 1946 constitution drawn up for all of China. Over the next five decades, the ruling authorities gradually democratized and incorporated the local population within the governing structure. Taiwan is functionally independent of China and has diplomatic relations with many countries, but China still claims it as officially part of the People's Republic of China.

GDP Per Capita: $30,100

Occupations: agriculture 5.3%, industry 36.8%, services 57.9%

Currency: New Taiwan dollar (TWD)

Tajikistan

Long Name: Republic of Tajikistan

Location: Central Asia, west of China

Area: 55,251 sq. mi. (143,100 sq. km)

Climate: mid-latitude continental, hot summers, mild winters; semiarid to polar in Pamir Mountains

Terrain: Pamir and Alay Mountains dominate landscape; western Fergana Valley in north, Kofarnihon and Vakhsh Valleys in southwest

Population: 7,211,884

Population Growth Rate: 1.893%

Birth Rate (per 1,000): 27.18

Death Rate (per 1,000): 6.94

Life Expectancy: 64.97 years

Ethnic Groups: Tajik 79.9%, Uzbek 15.3%, Russian 1.1%, Kyrgyz 1.1%, other 2.6%

Religion: Sunni Muslim 85%, Shi'a Muslim 5%, other 10%

Languages: Tajik (official), Russian widely used in government and business

Literacy: 99.5%

Government Type: republic

Capital: Dushanbe

Independence: 9 September 1991 (from Soviet Union)

GDP Per Capita: $1,600

Occupations: agriculture 67.2%, industry 7.5%, services 25.3%

Currency: somoni (TJS)

Tanzania

Long Name: United Republic of Tanzania

Location: Eastern Africa, bordering the Indian Ocean, between Kenya and Mozambique

Area: 364,898 sq. mi. (945,087 sq. km)

Climate: varies from tropical along coast to temperate in highlands

Terrain: plains along coast; central plateau; highlands in north, south

Population: 40,213,160

Population Growth Rate: 2.072%

Birth Rate (per 1,000): 35.12

Death Rate (per 1,000): 12.92

Life Expectancy: 51.45 years

Ethnic Groups: mainland - African 99% (of which 95% are Bantu consisting of more than 130 tribes), other 1% (consisting of Asian, European, and Arab); Zanzibar - Arab, African, mixed Arab and African

Religion: mainland - Christian 30%, Muslim 35%, indigenous beliefs 35%; Zanzibar - more than 99% Muslim

Languages: Kiswahili or Swahili (official), Kiunguja (name for Swahili in Zanzibar), English (official, primary language of commerce, administration, and higher education), Arabic (widely spoken in Zanzibar), many local languages

Literacy: 69.4%

Government Type: republic

Capital: Dar es Salaam

Independence: 26 April 1964; Tanganyika became independent 9 December 1961 (from United Kingdom-administered UN trusteeship); Zanzibar became independent 19 December 1963 (from United Kingdom); Tanganyika united with Zanzibar 26 April 1964 to form the United Republic of Tanganyika and Zanzibar; renamed United Republic of Tanzania 29 October 1964

GDP Per Capita: $1,300

Occupations: agriculture 80%, industry and services 20%

Currency: Tanzanian shilling (TZS)

Thailand

Long Name: Kingdom of Thailand

Location: Southeastern Asia, bordering the Andaman Sea and the Gulf of Thailand, southeast of Burma

Area: 198,456 sq. mi. (514,000 sq. km)

Climate: tropical; rainy, warm, cloudy southwest monsoon (mid-May to September); dry, cool northeast monsoon (November to mid-March); southern isthmus always hot and humid

Terrain: central plain; Khorat Plateau in the east; mountains elsewhere

Population: 65,493,296

Population Growth Rate: 0.64%

Birth Rate (per 1,000): 13.57

Death Rate (per 1,000): 7.17

Life Expectancy: 72.83 years

Ethnic Groups: Thai 75%, Chinese 14%, other 11%

Religion: Buddhist 94.6%, Muslim 4.6%, Christian 0.7%, other 0.1%

Languages: Thai, English (secondary language of the elite), ethnic and regional dialects

Literacy: 92.6%

Government Type: constitutional monarchy

Capital: Bangkok

Independence: 1238 (traditional founding date; never colonized)

GDP Per Capita: $8,000

Occupations: agriculture 49%, industry 14%, services 37%

Currency: baht (THB)

Timor-Leste

Long Name: Democratic Republic of Timor-Leste

Location: Southeastern Asia, northwest of Australia in the Lesser Sunda Islands at the eastern end of the Indonesian archipelago; note - Timor-Leste includes the eastern half of the island of Timor, the Oecussi (Ambeno) region on the northwest portion of the island of Timor, and the islands of Pulau Atauro and Pulau Jaco

Area: 5,794 sq. mi. (15,007 sq. km)

Climate: tropical; hot, humid; distinct rainy and dry seasons

Terrain: mountainous

Population: 1,108,777

Population Growth Rate: 2.05%

Birth Rate (per 1,000): 26.52

Death Rate (per 1,000): 6.02

Life Expectancy: 66.94 years

Ethnic Groups: Austronesian (Malayo-Polynesian), Papuan, small Chinese minority

Religion: Roman Catholic 98%, Muslim 1%, Protestant 1%

Languages: Tetum (official), Portuguese (official), Indonesian, English, there are about 16 indigenous languages (Tetum, Galole, Mambae, and Kemak are spoken by significant numbers of people)

Literacy: 58.6%

Government Type: republic

Capital: Dili

Independence: 28 November 1975 (independence proclaimed from Portugal); 20 May 2002 is the official date of international recognition of Timor-Leste's independence from Indonesia

GDP Per Capita: $2,500

Occupations: N/A

Currency: US dollar (USD)

Togo

Long Name: Togolese Republic

Location: Western Africa, bordering the Bight of Benin, between Benin and Ghana

Area: 21,925 sq. mi. (56,785 sq. km)

Climate: tropical; hot, humid in south; semiarid in north

Terrain: gently rolling savanna in north; central hills; southern plateau; low coastal plain with extensive lagoons and marshes

Population: 5,858,673

Population Growth Rate: 2.717%

Birth Rate (per 1,000): 36.66

Death Rate (per 1,000): 9.48

Life Expectancy: 58.28 years

Ethnic Groups: African (37 tribes; largest and most important are Ewe,

Mina, and Kabre) 99%, European and Syrian-Lebanese less than 1%

Religion: Christian 29%, Muslim 20%, indigenous beliefs 51%

Languages: French (official and the language of commerce), Ewe and Mina (the two major African languages in the south), Kabye (sometimes spelled Kabiye) and Dagomba (the two major African languages in the north)

Literacy: 60.9%

Government Type: republic under transition to multiparty democratic rule

Capital: Lome

Independence: 27 April 1960 (from French-administered UN trusteeship)

GDP Per Capita: $900

Occupations: agriculture 65%, industry 5%, services 30%

Currency: Communaute Financiere Africaine franc (XOF)

Tonga

Long Name: Kingdom of Tonga

Location: Oceania, archipelago in the South Pacific Ocean, about two-thirds of the way from Hawaii to New Zealand

Area: 289 sq. mi. (748 sq. km)

Climate: tropical; modified by trade winds; warm season (December to May), cool season (May to December)

Terrain: most islands have limestone base formed from uplifted coral formation; others have limestone overlying volcanic base

Population: 119,009

Population Growth Rate: 1.669%

Birth Rate (per 1,000): 21.81

Death Rate (per 1,000): 5.12

Life Expectancy: 70.44 years

Ethnic Groups: Polynesian, Europeans

Religion: Christian (Free Wesleyan Church claims over 30,000 adherents)

Languages: Tongan, English

Literacy: 98.9%

Government Type: constitutional monarchy

Capital: Nukmu'alofa

Independence: 4 June 1970 (from United Kingdom protectorate)

GDP Per Capita: $5,100

Occupations: agriculture 65%, industry and services 35%

Currency: pa'anga (TOP)

Trinidad and Tobago

Long Name: Republic of Trinidad and Tobago

Location: Caribbean, islands between the Caribbean Sea and the North Atlantic Ocean, northeast of Venezuela

Area: 1,980 sq. mi. (5,128 sq. km)

Climate: tropical; rainy season (June to December)

Terrain: mostly plains with some hills and low mountains

Population: 1,047,366

Population Growth Rate: -0.891%

Birth Rate (per 1,000): 13.22

Death Rate (per 1,000): 10.93

Life Expectancy: 67 years

Ethnic Groups: Indian (South Asian) 40%, African 37.5%, mixed 20.5%, other 1.2%, unspecified 0.8%

Religion: Roman Catholic 26%, Hindu 22.5%, Anglican 7.8%, Baptist 7.2%, Pentecostal 6.8%, Muslim 5.8%, Seventh Day Adventist 4%, other Christ-

ian 5.8%, other 10.8%, unspecified 1.4%, none 1.9%

Languages: English (official), Caribbean Hindustani (a dialect of Hindi), French, Spanish, Chinese

Literacy: 98.6%

Government Type: parliamentary democracy

Capital: Port-of-Spain

Independence: 31 August 1962 (from United Kingdom)

GDP Per Capita: $25,400

Occupations: agriculture 4%, manufacturing, mining, and quarrying 12.9%, construction and utilities 17.5%, services 65.6%

Currency: Trinidad and Tobago dollar (TTD)

Tunisia

Long Name: Tunisian Republic

Location: Northern Africa, bordering the Mediterranean Sea, between Algeria and Libya

Area: 63,170 sq. mi. (163,610 sq. km)

Climate: temperate in north with mild, rainy winters and hot, dry summers; desert in south

Terrain: mountains in north; hot, dry central plain; semiarid south merges into the Sahara

Population: 10,383,577

Population Growth Rate: 0.989%

Birth Rate (per 1,000): 15.5

Death Rate (per 1,000): 5.17

Life Expectancy: 75.56 years

Ethnic Groups: Arab 98%, European 1%, Jewish and other 1%

Religion: Muslim 98%, Christian 1%, Jewish and other 1%

Languages: Arabic (official and one of the languages of commerce), French (commerce)

Literacy: 74.3%

Government Type: republic

Capital: Tunis

Independence: 20 March 1956 (from France)

GDP Per Capita: $7,400

Occupations: agriculture 55%, industry 23%, services 22%

Currency: Tunisian dinar (TND)

Turkey

Long Name: Republic of Turkey

Location: Southeastern Europe and Southwestern Asia (that portion of Turkey west of the Bosporus is geographically part of Europe), bordering the Black Sea, between Bulgaria and Georgia, and bordering the Aegean Sea and the Mediterranean Sea, between Greece and Syria

Area: 301,382 sq. mi. (780,580 sq. km)

Climate: temperate; hot, dry summers with mild, wet winters; harsher in interior

Terrain: high central plateau (Anatolia); narrow coastal plain; several mountain ranges

Population: 71,892,808

Population Growth Rate: 1.013%

Birth Rate (per 1,000): 16.15

Death Rate (per 1,000): 6.02

Life Expectancy: 73.14 years

Ethnic Groups: Turkish 80%, Kurdish 20%

Religion: Muslim 99.8% (mostly Sunni), other 0.2%

Languages: Turkish (official), Kurdish, Dimli (or Zaza), Azeri, Kabardian, there is also a substantial Gagauz population in the European part of Turkey

Literacy: 87.4%

Government Type: republican parliamentary democracy

Capital: Ankara

Independence: 29 October 1923

GDP Per Capita: $12,000

Occupations: agriculture 35.9%, industry 22.8%, services 41.2%

Currency: Turkish lira (TRY)

Turkmenistan

Long Name: Turkmenistan

Location: Central Asia, bordering the Caspian Sea, between Iran and Kazakhstan

Area: 188,456 sq. mi. (488,100 sq. km)

Climate: subtropical desert

Terrain: flat-to-rolling sandy desert with dunes rising to mountains in the south; low mountains along border with Iran; borders Caspian Sea in west

Population: 5,179,571

Population Growth Rate: 1.596%

Birth Rate (per 1,000): 25.07

Death Rate (per 1,000): 6.11

Life Expectancy: 68.6 years

Ethnic Groups: Turkmen 85%, Uzbek 5%, Russian 4%, other 6%

Religion: Muslim 89%, Eastern Orthodox 9%, unknown 2%

Languages: Turkmen 72%, Russian 12%, Uzbek 9%, other 7%

Literacy: 98.8%

Government Type: republic; authoritarian presidential rule, with little power outside the executive branch

Capital: Ashgabat

Independence: 27 October 1991 (from Soviet Union)

GDP Per Capita: $5,300

Occupations: agriculture 48.2%, industry 14%, services 37.8%

Currency: Turkmen manat (TMM)

Tuvalu

Long Name: Tuvalu

Location: Oceania, island group consisting of nine coral atolls in the South Pacific Ocean, about one-half of the way from Hawaii to Australia

Area: 10 sq. mi. (26 sq. km)

Climate: tropical; moderated by easterly trade winds (March to November); westerly gales and heavy rain (November to March)

Terrain: very low-lying and narrow coral atolls

Population: 12,177

Population Growth Rate: 1.577%

Birth Rate (per 1,000): 22.75

Death Rate (per 1,000): 6.98

Life Expectancy: 68.97 years

Ethnic Groups: Polynesian 96%, Micronesian 4%

Religion: Church of Tuvalu (Congregationalist) 97%, Seventh-Day Adventist 1.4%, Baha'i 1%, other 0.6%

Languages: Tuvaluan, English, Samoan, Kiribati (on the island of Nui)

Literacy: N/A

Government Type: constitutional monarchy with a parliamentary democracy

Capital: Funafuti

Independence: 1 October 1978 (from United Kingdom)

GDP Per Capita: $1,600

Occupations: people make a living mainly through exploitation of the sea, reefs, and atolls and from wages sent home by those abroad (mostly workers in the phosphate industry and sailors)

Currency: Australian dollar (AUD)

Uganda

Long Name: Republic of Uganda

Location: Eastern Africa, west of Kenya

Area: 91,135 sq. mi. (236,040 sq. km)

Climate: tropical; generally rainy with two dry seasons (December to February, June to August); semiarid in northeast

Terrain: mostly plateau with rim of mountains

Population: 31,367,972

Population Growth Rate: 3.603%

Birth Rate (per 1,000): 48.15

Death Rate (per 1,000): 12.32

Life Expectancy: 52.34 years

Ethnic Groups: Baganda 16.9%, Banyakole 9.5%, Basoga 8.4%, Bakiga 6.9%, Iteso 6.4%, Langi 6.1%, Acholi 4.7%, Bagisu 4.6%, Lugbara 4.2%, Bunyoro 2.7%, other 29.6%

Religion: Roman Catholic 41.9%, Protestant 42% (Anglican 35.9%, Pentecostal 4.6%, Seventh Day Adventist 1.5%), Muslim 12.1%, other 3.1%, none 0.9%

Languages: English (official national language, taught in grade schools, used in courts of law and by most newspapers and some radio broad-

359

casts), Ganda or Luganda (most widely used of the Niger-Congo languages, preferred for native language publications in the capital and may be taught in school), other Niger-Congo languages, Nilo-Saharan languages, Swahili, Arabic

Literacy: 66.8%

Government Type: republic

Capital: Kampala

Independence: 9 October 1962 (from United Kingdom)

GDP Per Capita: $1,000

Occupations: agriculture 82%, industry 5%, services 13%

Currency: Ugandan shilling (UGX)

Ukraine

Long Name: Ukraine

Location: Eastern Europe, bordering the Black Sea, between Poland, Romania, and Moldova in the west and Russia in the east

Area: 233,089 sq. mi. (603,700 sq. km)

Climate: temperate continental; Mediterranean only on the southern Crimean coast; precipitation disproportionately distributed, highest in west and north, lesser in east and southeast; winters vary from cool along the Black Sea to cold farther inland; summers are warm across the greater part of the country, hot in the south

Terrain: most of Ukraine consists of fertile plains (steppes) and plateaus, mountains being found only in the west (the Carpathians), and in the Crimean Peninsula in the extreme south

Population: 45,994,288

Population Growth Rate: -0.651%

Birth Rate (per 1,000): 9.55

Death Rate (per 1,000): 15.93

Life Expectancy: 68.06 years

Ethnic Groups: Ukrainian 77.8%, Russian 17.3%, Belarusian 0.6%, Moldovan 0.5%, Crimean Tatar 0.5%, Bulgarian 0.4%, Hungarian 0.3%, Romanian 0.3%, Polish 0.3%, Jewish 0.2%, other 1.8%

Religion: Ukrainian Orthodox-Kyiv Patriarchate 50.4%, Ukrainian Orthodox-Moscow Patriarchate 26.1%, Ukrainian Greek Catholic 8%, Ukrainian Autocephalous Orthodox 7.2%, Roman Catholic 2.2%, Protestant 2.2%, Jewish 0.6%, other 3.2%

Languages: Ukrainian (official) 67%, Russian 24%, other 9% (includes small Romanian-, Polish-, and Hungarian-speaking minorities)

Literacy: 99.4%

Government Type: republic

Capital: Kyiv

Independence: 24 August 1991 (from Soviet Union)

GDP Per Capita: $7,000

Occupations: agriculture 25%, industry 20%, services 55%

Currency: hryvnia (UAH)

United Arab Emirates

Long Name: United Arab Emirates

Location: Middle East, bordering the Gulf of Oman and the Persian Gulf, between Oman and Saudi Arabia

Area: 32,278 sq. mi. (83,600 sq. km)

Climate: desert; cooler in eastern mountains

Terrain: flat, barren coastal plain merging into rolling sand dunes of vast desert wasteland; mountains in east

Population: 4,621,399

Population Growth Rate: 3.833%

Birth Rate (per 1,000): 16.06

Death Rate (per 1,000): 2.13

Life Expectancy: 75.89 years

Ethnic Groups: Emirati 19%, other Arab and Iranian 23%, South Asian 50%, other expatriates (includes Westerners and East Asians) 8%

Religion: Muslim 96% (Shi'a 16%), other (includes Christian, Hindu) 4%

Languages: Arabic (official), Persian, English, Hindi, Urdu

Literacy: 77.9%

Government Type: federation with specified powers delegated to the UAE federal government and other powers reserved to member emirates

Capital: Abu Dhabi

Independence: 2 December 1971 (from United Kingdom)

GDP Per Capita: $37,000

Occupations: agriculture 7%, industry 15%, services 78%

Currency: Emirati dirham (AED)

United Kingdom

Long Name: United Kingdom of Great Britain and Northern Ireland

Location: Western Europe, islands including the northern one-sixth of the island of Ireland between the North Atlantic Ocean and the North Sea, northwest of France

Area: 94,525 sq. mi. (244,820 sq. km)

Climate: temperate; moderated by prevailing southwest winds over the North Atlantic Current; more than one-half of the days are overcast

Terrain: mostly rugged hills and low mountains; level to rolling plains in east and southeast

Population: 60,943,912

Population Growth Rate: 0.276%

Birth Rate (per 1,000): 10.65

Death Rate (per 1,000): 10.05

Life Expectancy: 78.85 years

Ethnic Groups: white (of which English 83.6%, Scottish 8.6%, Welsh 4.9%, Northern Irish 2.9%) 92.1%, black 2%, Indian 1.8%, Pakistani 1.3%, mixed 1.2%, other 1.6%

Religion: Christian (Anglican, Roman Catholic, Presbyterian, Methodist) 71.6%, Muslim 2.7%, Hindu 1%, other 1.6%, unspecified or none 23.1%

Languages: English, Welsh (about 26% of the population of Wales), Scottish form of Gaelic (about 60,000 in Scotland)

Literacy: 99%

Government Type: constitutional monarchy

Capital: London

Independence: England has existed as a unified entity since the 10th century; the union between England and Wales, begun in 1284 with the Statute of Rhuddlan, was not formalized until 1536 with an Act of Union; in another Act of Union in 1707, England and Scotland agreed to permanently join as Great Britain; the legislative union of Great Britain and Ireland was implemented in 1801, with the adoption of the name the United Kingdom of Great Britain and Ireland; the Anglo-Irish treaty of 1921 formalized a partition of Ireland; six northern Irish counties remained part of the United Kingdom as Northern Ireland

and the current name of the country, the United Kingdom of Great Britain and Northern Ireland, was adopted in 1927

GDP Per Capita: $35,000

Occupations: agriculture 1.4%, industry 18.2%, services 80.4%

Currency: British pound (GBP)

United States

Long Name: United States of America

Location: North America, bordering both the North Atlantic Ocean and the North Pacific Ocean, between Canada and Mexico

Area: 3,794,066 sq. mi. (9,826,630 sq. km)

Climate: mostly temperate, but tropical in Hawaii and Florida, arctic in Alaska, semiarid in the great plains west of the Mississippi River, and arid in the Great Basin of the southwest; low winter temperatures in the northwest are ameliorated occasionally in January and February by warm chinook winds from the eastern slopes of the Rocky Mountains

Terrain: vast central plain, mountains in west, hills and low mountains in east; rugged mountains and broad river valleys in Alaska; rugged, volcanic topography in Hawaii

Population: 303,824,640

Population Growth Rate: 0.883%

Birth Rate (per 1,000): 14.18

Death Rate (per 1,000): 8.27

Life Expectancy: 78.14 years

Ethnic Groups: white 79.96%, black 12.85%, Asian 4.43%, Amerindian and Alaska native 0.97%, native Hawaiian and other Pacific islander 0.18%, two or more races 1.61%

Religion: Protestant 51.3%, Roman Catholic 23.9%, Mormon 1.7%, other Christian 1.6%, Jewish 1.7%, Buddhist 0.7%, Muslim 0.6%, other or unspecified 2.5%, unaffiliated 12.1%, none 4%

Languages: English 82.1%, Spanish 10.7%, other Indo-European 3.8%, Asian and Pacific island 2.7%, other 0.7%

Literacy: 99%

Government Type: Constitution-based federal republic; strong democratic tradition

Capital: Washington, DC

Independence: 4 July 1776 (from Great Britain)

GDP Per Capita: $45,800

Occupations: farming, forestry, and fishing 0.6%, manufacturing, extraction, transportation, and crafts 22.6%, managerial, professional, and technical 35.5%, sales and office 24.8%, other services 16.5%

Currency: US dollar (USD)

Uruguay

Long Name: Oriental Republic of Uruguay

Location: Southern South America, bordering the South Atlantic Ocean, between Argentina and Brazil

Area: 68,039 sq. mi. (176,220 sq. km)

Climate: warm temperate; freezing temperatures almost unknown

Terrain: mostly rolling plains and low hills; fertile coastal lowland

Population: 3,477,778

Population Growth Rate: 0.486%

Birth Rate (per 1,000): 14.17

Death Rate (per 1,000): 9.12

Life Expectancy: 76.14 years

Ethnic Groups: white 88%, mestizo 8%, black 4%, Amerindian (practically nonexistent)

Religion: Roman Catholic 47.1%, non-Catholic Christians 11.1%, nondenominational 23.2%, Jewish 0.3%, atheist or agnostic 17.2%, other 1.1%

Languages: Spanish, Portunol, or Brazilero (Portuguese-Spanish mix on the Brazilian frontier)

Literacy: 98%

Government Type: constitutional republic

Capital: Montevideo

Independence: 25 August 1825

GDP Per Capita: $10,800

Occupations: agriculture 9%, industry 15%, services 76%

Currency: Uruguayan peso (UYU)

Uzbekistan

Long Name: Republic of Uzbekistan

Location: Central Asia, north of Afghanistan

Area: 172,741 sq. mi. (447,400 sq. km)

Climate: mostly mid-latitude desert, long, hot summers, mild winters; semiarid grassland in east

Terrain: mostly flat-to-rolling sandy desert with dunes; broad, flat intensely irrigated river valleys along course of Amu Darya, Syr Darya (Sirdaryo), and Zarafshon; Fergana Valley in east surrounded by mountainous Tajikistan and Kyrgyzstan; shrinking Aral Sea in west

Population: 27,345,026

Population Growth Rate: 0.965%

Birth Rate (per 1,000): 17.99

Death Rate (per 1,000): 5.3

Life Expectancy: 71.69 years

Ethnic Groups: Uzbek 80%, Russian 5.5%, Tajik 5%, Kazakh 3%, Karakalpak 2.5%, Tatar 1.5%, other 2.5%

Religion: Muslim 88% (mostly Sunnis), Eastern Orthodox 9%, other 3%

Languages: Uzbek 74.3%, Russian 14.2%, Tajik 4.4%, other 7.1%

Literacy: 99.3%

Government Type: republic; authoritarian presidential rule, with little power outside the executive branch

Capital: Tashkent

Independence: 1 September 1991 (from Soviet Union)

GDP Per Capita: $2,400

Occupations: agriculture 44%, industry 20%, services 36%

Currency: soum (UZS)

Vanuatu

Long Name: Republic of Vanuatu

Location: Oceania, group of islands in the South Pacific Ocean, about three-quarters of the way from Hawaii to Australia

Area: 4,710 sq. mi. (12,200 sq. km)

Climate: tropical; moderated by southeast trade winds from May to October; moderate rainfall from November to April; may be affected by cyclones from December to April

Terrain: mostly mountainous islands of volcanic origin; narrow coastal plains

Population: 215,446

Population Growth Rate: 1.434%

Birth Rate (per 1,000): 21.95

Death Rate (per 1,000): 7.61

Life Expectancy: 63.61 years

Ethnic Groups: Ni-Vanuatu 98.5%, other 1.5%

Religion: Presbyterian 31.4%, Anglican 13.4%, Roman Catholic 13.1%, Seventh-Day Adventist 10.8%, other Christian 13.8%, indigenous beliefs 5.6% (including Jon Frum cargo cult), other 9.6%, none 1%, unspecified 1.3%

Languages: local languages (more than 100) 72.6%, pidgin (known as Bislama or Bichelama) 23.1%, English 1.9%, French 1.4%, other 0.3%, unspecified 0.7%

Literacy: 74%

Government Type: parliamentary republic

Capital: Port-Vila

Independence: 30 July 1980 (from France and United Kingdom)

GDP Per Capita: $3,900

Occupations: agriculture 65%, industry 5%, services 30%

Currency: vatu (VUV)

Vatican City

Long Name: State of the Vatican City, The Holy See

Location: Southern Europe, an enclave of Rome (Italy)

Area: 0.17 sq. mi. (0.44 sq. km)

Climate: temperate; mild, rainy winters (September to May) with hot, dry summers (May to September)

Terrain: urban; low hill

Population: 824

Population Growth Rate: 0.003%

Birth Rate (per 1,000): N/A

Death Rate (per 1,000): N/A

Life Expectancy: N/A

Ethnic Groups: Italians, Swiss, other

Religion: Roman Catholic

Languages: Italian, Latin, French, various other languages

Literacy: 100%

Government Type: ecclesiastical

Capital: Vatican City

Independence: 11 February 1929 (from Italy)

GDP Per Capita: N/A

Occupations: essentially services with a small amount of industry; nearly all dignitaries, priests, nuns, guards, and the approximately 3,000 lay workers live outside the Vatican

Currency: euro (EUR)

Venezuela

Long Name: Bolivarian Republic of Venezuela

Location: Northern South America, bordering the Caribbean Sea and the North Atlantic Ocean, between Colombia and Guyana

Area: 352,143 sq. mi. (912,050 sq. km)

Climate: tropical; hot, humid; more moderate in highlands

Terrain: Andes Mountains and Maracaibo Lowlands in northwest; central plains (llanos); Guiana Highlands in southeast

Population: 26,414,816

Population Growth Rate: 1.498%

Birth Rate (per 1,000): 20.92

Death Rate (per 1,000): 5.1

Life Expectancy: 73.45 years

Ethnic Groups: Spanish, Italian, Portuguese, Arab, German, African, indigenous people

Religion: nominally Roman Catholic 96%, Protestant 2%, other 2%

Languages: Spanish (official), numerous indigenous dialects

Literacy: 93%

Government Type: federal republic

Capital: Caracas

Independence: 5 July 1811 (from Spain)

GDP Per Capita: $12,800

Occupations: agriculture 13%, industry 23%, services 64%

Currency: bolivar (VEB)

Vietnam

Long Name: Socialist Republic of Vietnam

Location: Southeastern Asia, bordering the Gulf of Thailand, Gulf of Tonkin, and South China Sea, alongside China, Laos, and Cambodia

Area: 127,243 sq. mi. (329,560 sq. km)

Climate: tropical in south; monsoonal in north with hot, rainy season (May to September) and warm, dry season (October to March)

Terrain: low, flat delta in south and north; central highlands; hilly, mountainous in far north and northwest

Population: 86,116,560

Population Growth Rate: 0.99%

Birth Rate (per 1,000): 16.47

Death Rate (per 1,000): 6.18

Life Expectancy: 71.33 years

Ethnic Groups: Kinh (Viet) 86.2%, Tay 1.9%, Thai 1.7%, Muong 1.5%, Khome 1.4%, Hoa 1.1%, Nun 1.1%, Hmong 1%, others 4.1%

Religion: Buddhist 9.3%, Catholic 6.7%, Hoa Hao 1.5%, Cao Dai 1.1%, Protestant 0.5%, Muslim 0.1%, none 80.8%

Languages: Vietnamese (official), English (increasingly favored as a second language), some French, Chinese, and Khmer; mountain area languages (Mon-Khmer and Malayo-Polynesian)

Literacy: 90.3%

Government Type: Communist state

Capital: Hanoi

Independence: 2 September 1945 (from France)

GDP Per Capita: $2,600

Occupations: agriculture 55.6%, industry 18.9%, services: 25.5%

Currency: dong (VND)

Yemen

Long Name: Republic of Yemen

Location: Middle East, bordering the Arabian Sea, Gulf of Aden, and Red Sea, between Oman and Saudi Arabia

Area: 203,849 sq. mi. (527,970 sq. km)

Climate: mostly desert; hot and humid along west coast; temperate in western mountains affected by seasonal monsoon; extraordinarily hot, dry, harsh desert in east

Terrain: narrow coastal plain backed by flat-topped hills and rugged mountains; dissected upland desert plains in center slope into the desert interior of the Arabian Peninsula

Population: 23,013,376

Population Growth Rate: 3.46%

Birth Rate (per 1,000): 42.42

Death Rate (per 1,000): 7.83

Life Expectancy: 62.9 years

Ethnic Groups: predominantly Arab; but also Afro-Arab, South Asians, Europeans

Religion: Muslim including Shaf'i (Sunni) and Zaydi (Shi'a), small numbers of Jewish, Christian, and Hindu

Languages: Arabic

Literacy: 50.2%

Government Type: republic

Capital: Sanaa

Independence: 22 May 1990 (Republic of Yemen was established with the merger of the Yemen Arab Republic [Yemen (Sanaa) or North Yemen] and the Marxist-dominated People's Democratic Republic of Yemen [Yemen (Aden) or South Yemen])

GDP Per Capita: $2,500

Occupations: most people are employed in agriculture and herding; services, construction, industry, and commerce account for less than one-fourth of the labor force

Currency: Yemeni rial (YER)

Zambia

Long Name: Republic of Zambia

Location: Southern Africa, east of Angola

Area: 290,585 sq. mi. (752,614 sq. km)

Climate: tropical; modified by altitude; rainy season (October to April)

Terrain: mostly high plateau with some hills and mountains

Population: 11,669,534

Population Growth Rate: 1.654%

Birth Rate (per 1,000): 40.52

Death Rate (per 1,000): 21.35

Life Expectancy: 38.59 years

Ethnic Groups: African 98.7%, European 1.1%, other 0.2%

Religion: Christian 50%-75%, Muslim and Hindu 24%-49%, indigenous beliefs 1%

Languages: English (official), major vernaculars - Bemba, Kaonda, Lozi, Lunda, Luvale, Nyanja, Tonga, and about 70 other indigenous languages

Literacy: 80.6%

Government Type: republic

Capital: Lusaka

Independence: 24 October 1964 (from United Kingdom)

GDP Per Capita: $1,400

Occupations: agriculture 85%, industry 6%, services 9%

Currency: Zambian kwacha (ZMK)

Zimbabwe

Long Name: Republic of Zimbabwe

Location: Southern Africa, between South Africa and Zambia

Area: 150,803 sq. mi. (390,580 sq. km)

Climate: tropical; moderated by altitude; rainy season (November to March)

Terrain: mostly high plateau with higher central plateau (high veld); mountains in east

Population: 11,350,111

Population Growth Rate: -0.787%

Birth Rate (per 1,000): 31.62

Death Rate (per 1,000): 17.29

Life Expectancy: 44.28 years

Ethnic Groups: African 98% (Shona 82%, Ndebele 14%, other 2%), mixed and Asian 1%, white less than 1%

Religion: syncretic (part Christian, part indigenous beliefs) 50%, Christian 25%, indigenous beliefs 24%, Muslim and other 1%

Languages: English (official), Shona, Sindebele (the language of the Ndebele, sometimes called Ndebele), numerous but minor tribal dialects

Literacy: 90.7%

Government Type: parliamentary democracy

Capital: Harare

Independence: 18 April 1980 (from United Kingdom)

GDP Per Capita: $200

Occupations: agriculture 66%, industry 10%, services 24%

Currency: Zimbabwean dollar (ZWD)

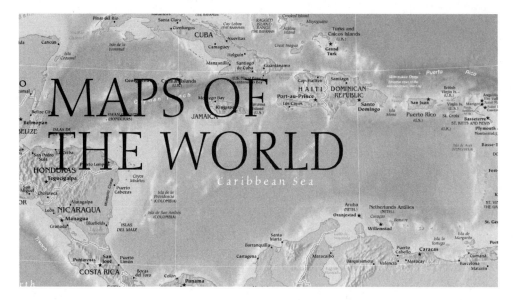

MAPS OF THE WORLD

Caribbean Sea

Maps reprinted from *The 2008 World Fact Book,* courtesy of the U.S. Central Intelligence Agency.

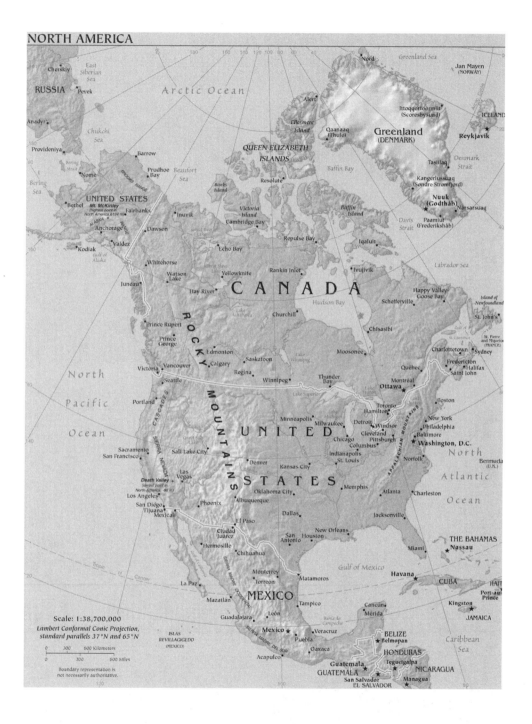

NORTH AMERICA

RUSSIA
Cherskiy
East Siberian Sea
Pevek
Anadyr
Providenya
Chukchi Sea
Arctic Ocean
Barrow
Prudhoe Bay
Beaufort Sea
Nome
Bering Strait
UNITED STATES
Bethel
Mt. McKinley
Highest point in North America 6194 m
Fairbanks
Anchorage
Inuvik
Dawson
Valdez
Kodiak
Gulf of Alaska
Whitehorse
Watson Lake
Juneau
Prince Rupert
Prince George
Hay River
Echo Bay
Yellowknife
Repulse Bay
Rankin Inlet
Ivujivik
Banks Island
Resolute
Victoria Island
Cambridge Bay
Baffin Island
Baffin Bay
Ellesmere Island
Alert
QUEEN ELIZABETH ISLANDS
Qaanaaq (Thule)
Nord
Greenland Sea
Jan Mayen (NORWAY)
Ittoqqortoormiit (Scoresbysund)
ICELAND
Greenland (DENMARK)
Reykjavik
Tasilaq
Denmark Strait
Kangerlussuaq (Søndre Strømfjord)
Nuuk (Godthåb)
Narsarsuaq
Paamiut (Frederikshåb)
Davis Strait
Iqaluit
Labrador Sea
North Pacific Ocean
ROCKY MOUNTAINS
CASCADES
SIERRA NEVADA
Edmonton
Calgary
Saskatoon
Regina
Winnipeg
CANADA
Churchill
Chisasibi
Moosonee
Schefferville
Happy Valley Goose Bay
Island of Newfoundland
St. John's
St. Pierre and Miquelon (FRANCE)
Sydney
Charlottetown
Fredericton
Halifax
Saint John
Hudson Bay
Lake Winnipeg
Thunder Bay
Lake Superior
Québec
Montréal
Ottawa
Toronto
Hamilton
Boston
Victoria
Vancouver
Seattle
Portland
Sacramento
San Francisco
Salt Lake City
Denver
UNITED STATES
Minneapolis
Milwaukee
Chicago
Detroit
Windsor
Cleveland
Pittsburgh
Columbus
Indianapolis
St. Louis
APPALACHIAN MOUNTAINS
New York
Philadelphia
Baltimore
Washington, D.C.
Norfolk
Bermuda (U.K.)
North Atlantic Ocean
Las Vegas
Death Valley
lowest point in North America 86 m
Los Angeles
San Diego
Tijuana
Mexicali
Phoenix
Albuquerque
Kansas City
Oklahoma City
Memphis
Atlanta
Charleston
Jacksonville
Dallas
El Paso
Ciudad Juárez
San Antonio
Houston
New Orleans
Miami
THE BAHAMAS
Nassau
Hermosillo
Chihuahua
Gulf of Mexico
Havana
CUBA
HAITI
Port-au-Prince
Tropic of Cancer
La Paz
Monterrey
Torreón
Matamoros
Kingston
JAMAICA
Mazatlán
MEXICO
Tampico
Cancún
Mérida
Guadalajara
León
Bahía de Campeche
Veracruz
México
Puebla
Oaxaca
SIERRA MADRE DEL SUR
Acapulco
ISLAS REVILLAGIGEDO (MEXICO)
BELIZE
Belmopan
HONDURAS
Tegucigalpa
NICARAGUA
Managua
Guatemala
GUATEMALA
San Salvador
EL SALVADOR
Caribbean Sea

Scale: 1:38,700,000
Lambert Conformal Conic Projection, standard parallels 37°N and 65°N
0   300   600 Kilometers
0   300   600 Miles
Boundary representation is not necessarily authoritative.

369

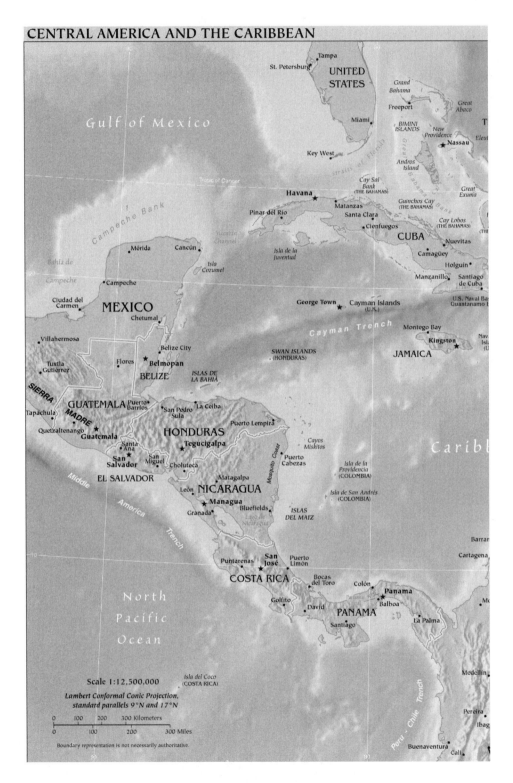

Tampa
St. Petersburg
UNITED STATES
Grand Bahama
Freeport
Great Abaco

Gulf of Mexico
Miami
BIMINI ISLANDS
New Providence
Eleut

Key West
Nassau

Andros Island

Tropic of Cancer

Campeche Bank
Cay Sal Bank (THE BAHAMAS)
Great Exuma

Havana
Matanzas
Guinchos Cay (THE BAHAMAS)

Pinar del Río
Santa Clara
Cay Lobos (THE BAHAMAS)
THE

Cienfuegos
CUBA
Nuevitas

Bahía de Campeche

Mérida
Cancún
Isla de la Juventud
Camagüey

Isla Cozumel
Holguín
Manzanillo
Santiago de Cuba

Campeche

Ciudad del Carmen

MEXICO
Chetumal
George Town
Cayman Islands (U.K.)
U.S. Naval Ba Guantanamo E

Villahermosa
Belize City
Cayman Trench
Montego Bay

Tuxtla Gutiérrez
Flores
Belmopan
SWAN ISLANDS (HONDURAS)
Kingston
Nava Isla (U

BELIZE
ISLAS DE LA BAHÍA
JAMAICA

SIERRA
GUATEMALA
Puerto Barrios
San Pedro Sula
La Ceiba

Tapachula

MADRE
Puerto Lempira

Quetzaltenango
Guatemala
HONDURAS
Cayos Miskitos
Caribb

Santa Ana
Tegucigalpa
Isla de la Providencia (COLOMBIA)

San Salvador
San Miguel
Choluteca
Puerto Cabezas

EL SALVADOR
Matagalpa
Isla de San Andrés (COLOMBIA)

Middle
León
NICARAGUA

America
Managua
Bluefields
ISLAS DEL MAIZ

Granada
Lago de Nicaragua

Trench
Barrar
Cartagena

Puntarenas
San José
Puerto Limón

COSTA RICA
Bocas del Toro
Colón
Panama

North
Golfito
David
Balboa

Pacific
Santiago
PANAMA
La Palma

Ocean

Medellín

Scale 1:12,500,000
Isla del Coco (COSTA RICA)

Lambert Conformal Conic Projection,
standard parallels 9°N and 17°N
Pereira
Ibag

0   100   200   300 Kilometers
0   100   200   300 Miles
Buenaventura
Cali

Boundary representation is not necessarily authoritative.

370

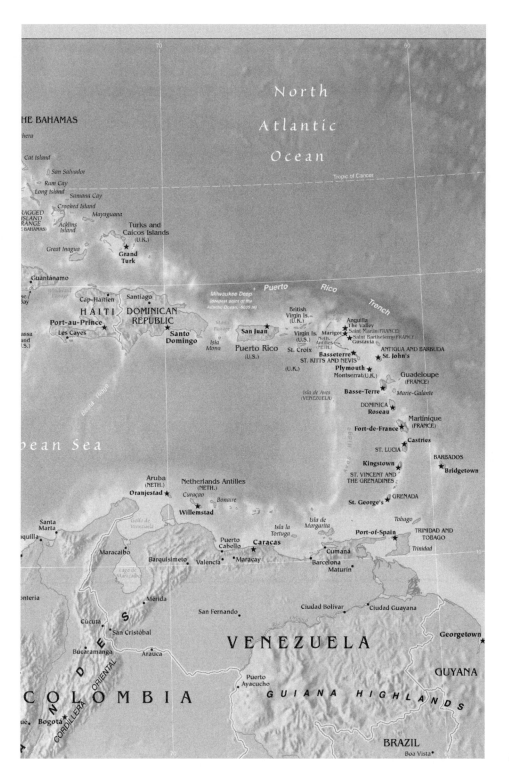

North

Atlantic

Ocean

HE BAHAMAS

hera

Cat Island

San Salvador
Rum Cay
Long Island
Samaná Cay
Crooked Island
RAGGED
ISLAND
RANGE
BAHAMAS)
Acklins
Island
Mayaguana

Turks and
Caicos Islands
(U.K.)

Great Inagua

★ Grand
Turk

Tropic of Cancer

Guantánamo

se
ay

Cap-Haitien   Santiago

Milwaukee Deep
(deepest point of the
Atlantic Ocean, -8605 m)

Puerto   Rico

Trench

HAITI   DOMINICAN
REPUBLIC

Port-au-Prince
Les Cayes

Santo
Domingo ★

Isla
Mona

San Juan ★

Puerto Rico
(U.S.)

British
Virgin Is.
(U.K.)

Virgin Is.
(U.S.)

St. Croix

Anguilla
The Valley
★ Saint Martin (FRANCE)
Marigot ★ Saint Barthelemy (FRANCE)
Neth.
Antilles
(NETH.)   Gustavia

Basseterre ★
ST. KITTS AND NEVIS
(U.K.)

ANTIGUA AND BARBUDA
★ St. John's

assa
and
S.)

Plymouth ★
Montserrat (U.K.)

Guadeloupe
(FRANCE)

Beata Ridge

Isla de Aves
(VENEZUELA)

Basse-Terre ★   Marie-Galante

DOMINICA
Roseau

Martinique
(FRANCE)

ean Sea

Fort-de-France ★

Castries ★
ST. LUCIA

BARBADOS
★ Bridgetown

Kingstown ★
ST. VINCENT AND
THE GRENADINES

Aruba
(NETH.)
Oranjestad ★

Netherlands Antilles
(NETH.)
Curaçao

Bonaire

St. George's ★   GRENADA

Willemstad ★

Golfo de
Venezuela

Isla la
Tortuga

Isla de
Margarita

Port-of-Spain ★

Tobago

TRINIDAD AND
TOBAGO

Santa
Marta

Puerto
Cabello   Caracas

Cumaná

Trinidad

quilla

Maracaibo

Barquisimeto   Valencia   Maracay

Lago de
Maracaibo

Barcelona
Maturín

onteria

Mérida

San Fernando

Ciudad Bolívar   Ciudad Guayana

Cúcuta
San Cristóbal

Georgetown ★

Bucaramanga   Arauca

VENEZUELA

GUYANA

S

COLOMBIA

CORDILLERA   ORIENTAL

ué.   Bogotá ★

Puerto
Ayacucho

GUIANA   HIGHLANDS

BRAZIL

Boa Vista •

371

SOUTH AMERICA

Caribbean Sea

HONDURAS
Tegucigalpa
NICARAGUA
Managua
San José
COSTA RICA
Panama
PANAMA

Isla de San Andrés (COLOMBIA)

Martinique (FRANCE)
ST. LUCIA
ST. VINCENT AND
THE GRENADINES
GRENADA
BARBADOS

Aruba
(NETH.)
Netherlands
Antilles
(NETH.)

Barranquilla
Cartagena
Maracaibo
Caracas
Port-of-Spain
TRINIDAD AND
TOBAGO

Valencia
Barquisimeto
Cúcuta
San Cristóbal
VENEZUELA
Ciudad
Guayana
Georgetown
Paramaribo
Cayenne

Medellín
Bogotá
COLOMBIA
GUIANA
HIGHLANDS
GUYANA
SURINAME
French
Guiana
(FRANCE)

Isla de Malpelo
(COLOMBIA)

Cali
Boa Vista
Macapá

Quito
ECUADOR
Guayaquil

Equator

A M A Z O N
Manaus
Santarém
Belém
São Luís
Fortaleza

Iquitos

Piura

B A S I N
Teresina
Natal

Trujillo
Rio
Branco
Pôrto
Velho
Recife

Huánuco

PERU
BRAZIL
BRAZILIAN
Maceió

Lima
MATO GROSSO
PLATEAU
Salvador

Cusco
Trinidad
Cuiabá
Brasília
Goiânia

South
Pacific
Ocean

Arequipa
La Paz
BOLIVIA
HIGHLANDS

Arica
Cochabamba
Santa Cruz
Uberlândia
Belo
Horizonte

Sucre
Potosí

Iquique
Campo
Grande
Vitória

Antofagasta
PARAGUAY
Rio de Janeiro

Tropic of Capricorn
Salta
Asunción
São Paulo
Santos
Curitiba

Isla San Ambrosio
(CHILE)
San Miguel
de Tucumán
Resistencia
Florianópolis

Isla San Félix
(CHILE)

CHILE
Pôrto Alegre

ARCHIPIÉLAGO
JUAN FERNÁNDEZ
(CHILE)

Cerro Aconcagua
(highest point in
South America, 6962 m)
Córdoba
Santa Fe
Salto
URUGUAY

Valparaíso
Mendoza
Rosario

Santiago
Buenos Aires
Montevideo
La Plata

Concepción
ARGENTINA
Mar del Plata

Bahía Blanca

San Carlos de
Bariloche

Puerto Montt

South
Atlantic
Ocean

Comodoro Rivadavia

Laguna del Carbón
(lowest point in South America and
the Western Hemisphere, -105 m)

Río
Gallegos
Strait of
Magellan
Stanley
Falkland Islands
(Islas Malvinas)
(administered by U.K.,
claimed by ARGENTINA)

Punta Arenas

Ushuaia
Cape
Horn

South Georgia and the
South Sandwich Islands
(administered by U.K.,
claimed by ARGENTINA)

Scale 1:35,000,000
Azimuthal Equal-Area Projection

500 Kilometers
500 Miles

Boundary representation is
not necessarily authoritative.

372

AFRICA

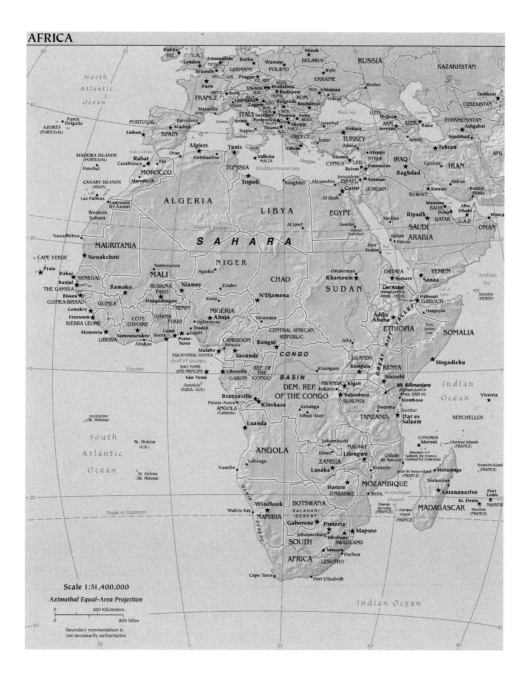

Scale 1:51,400,000

Azimuthal Equal-Area Projection

Boundary representation is
not necessarily authoritative.

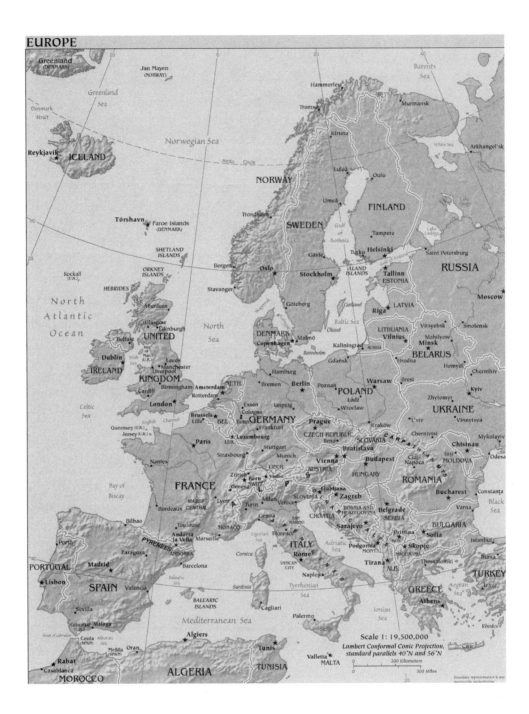

EUROPE

374

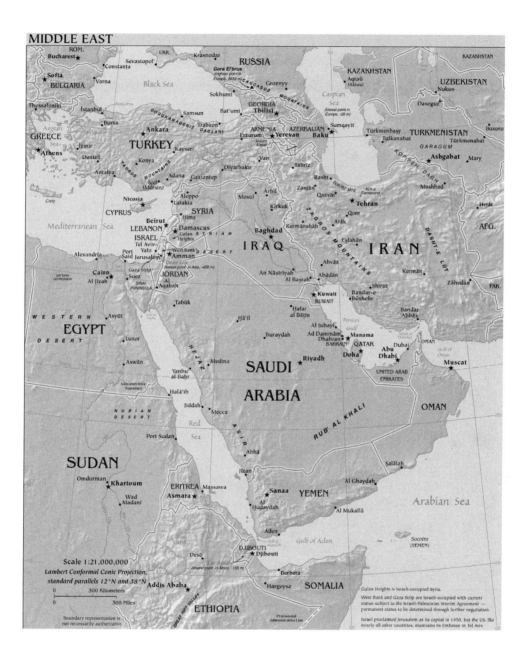

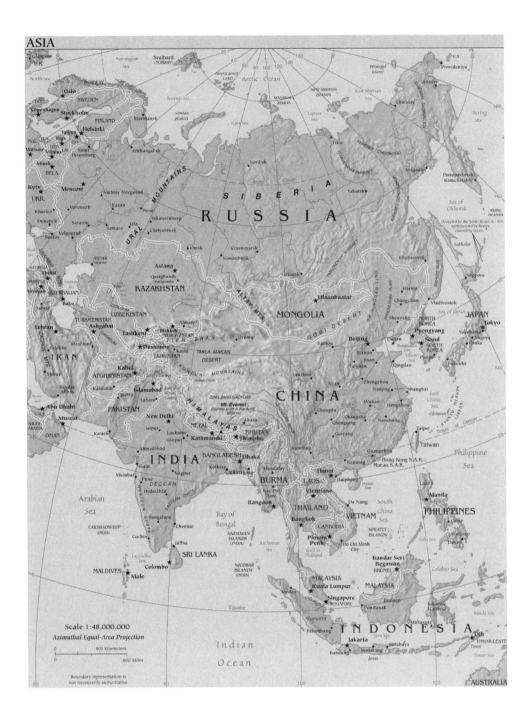

Scale 1:48,000,000
Azimuthal Equal-Area Projection

800 Kilometers

600 Miles

Boundary representation is
not necessarily authoritative.

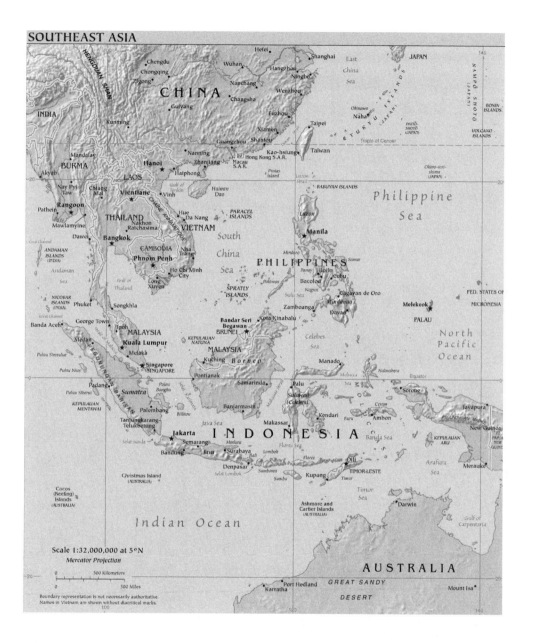

Hengduan Shan

Chengdu
Chongqing
Zigong

Hefei
Wuhan
Hangzhou
Shanghai
Ningbo

JAPAN

East
China
Sea

NAMPŌ SHOTŌ (JAPAN)

BONIN
ISLANDS

INDIA

CHINA

Nanchang
Changsha

Wenzhou

Okinawa
Naha

RYUKYU ISLANDS (JAPAN)

Tropic of Cancer

DAITŌ-
SHOTŌ
(JAPAN)

VOLCANO
ISLANDS

Kunming

Guiyang

Fuzhou

Xiamen
Guangzhou Shantou

Taipei

Taiwan

Mandalay

BURMA

Akyab

Hanoi

Nanning

Zhanjiang

Hong Kong S.A.R.
Macau
S.A.R.

Kao-hsiung
Hong Kong S.A.R.

Olimo-tori-
shima
(JAPAN)

Nay Pyi
Taw

Chiang
Mai

LAOS

Haiphong

Gulf of
Tonkin

Hainan
Dao

Pentas
Island

Luzon
Strait

BABUYAN ISLANDS

Philippine
Sea

Pathein

Rangoon

Vientiane

Vinh

Hue

PARACEL
ISLANDS

Luzon

Mawlamyine

THAILAND

Nakhon
Ratchasima

Da Nang

VIETNAM

Manila

Dawei

Bangkok

CAMBODIA

Phnom Penh

South
China
Sea

PHILIPPINES

Mindoro

Samar

ANDAMAN
ISLANDS
(INDIA)

Nha
Trang

Ho Chi Minh
City

Panay Iloilo
Cebu

Andaman
Sea

Gulf of
Thailand

Long
Xuyen

Palawan

Bacolod

Negros

Cagayan de Oro

FED. STATES OF

MICRONESIA

NICOBAR
ISLANDS
(INDIA)

Phuket

Songkhla

SPRATLY
ISLANDS

Sulu Sea

Zamboanga

Mindanao

Davao

Melekeok

PALAU

Banda Aceh

George Town

Ipoh

MALAYSIA

Bandar Seri
Begawan

Kota Kinabalu

Celebes
Sea

North
Pacific
Ocean

Medan

Kuala Lumpur

KEPULAUAN
NATUNA

BRUNEI

Pulau Simeulue

Melaka

MALAYSIA

Kuching

Borneo

Manado

Halmahera

Pulau Nias

Singapore
SINGAPORE

Pontianak

Equator

Padang

Pulau Siberut

Sumatra

Palau
Bangka

Samarinda

Palu

Sulawesi
(Celebes)

Ceram

Sorong

Jayapura

PEGUNUNGAN MAOKE

New Guinea

KEPULAUAN
MENTAWAI

Palembang

Biliton

Banjarmasin

Kendari

Buru

Ambon

KEPULAUAN
ARU

PAPUA
NEW
GUINEA

Tanjungkarang-
Telukbetung

Java Sea

Makassar

Banda Sea

Selat Sunda

Jakarta

Semarang

Madura

INDONESIA

Bandung

Jawa

Surabaya

Bali

Lombok

Flores

Flores Sea

DILI

Arafura
Sea

Merauke

Denpasar

Selat Lombok

Sumbawa

Sumba

Kupang

Timor

TIMOR-LESTE

Christmas Island
(AUSTRALIA)

Timor
Sea

Cocos
(Keeling)
Islands
(AUSTRALIA)

Ashmore and
Cartier Islands
(AUSTRALIA)

Darwin

Gulf of
Carpentaria

Indian Ocean

Scale 1:32,000,000 at 5°N
Mercator Projection

0   500 Kilometers

0   500 Miles

AUSTRALIA

Boundary representation is not necessarily authoritative.
Names in Vietnam are shown without diacritical marks.

Port Hedland
Karratha

GREAT SANDY

DESERT

Mount Isa

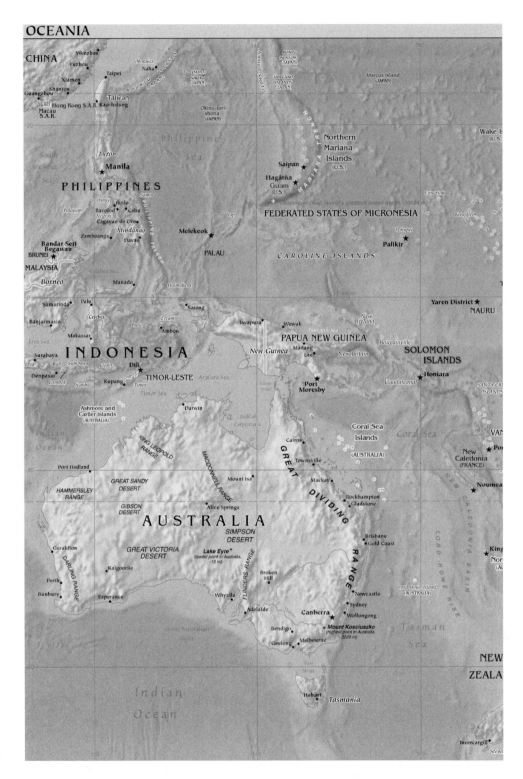

CHINA
Wenzhou
Fuzhou
Xiamen
Shantou
Guangzhou
Macau S.A.R.
Hong Kong S.A.R.

Okinawa
Naha
Taipei
Taiwan
Kao-hsiung
Taiwan Strait

South China Sea

DAITO-SHOTO (JAPAN)
OKINO-SHOTO

BONIN ISLANDS (JAPAN)
VOLCANO ISLANDS (JAPAN)

Marcus Island (JAPAN)

Okino-tori-shima (JAPAN)

Wake Is (U.S.)

Philippine Sea

Luzon
Manila
PHILIPPINES

Northern Mariana Islands (U.S.)
Saipan
Hagåtña
Guam (U.S.)

MARIANA TRENCH

Challenger Deep (world's greatest ocean depth, 10,924 m)

Emwetak

Panay
Iloilo
Bacolod
Cebu
Negros
Cagayan de Oro
Zamboanga
Mindanao
Davao

Samar

Yap

FEDERATED STATES OF MICRONESIA

Kwajalein

Melekeok
Bandar Seri Begawan
BRUNEI
MALAYSIA
Borneo

PHILIPPINE TRENCH

PALAU

CAROLINE ISLANDS

Pohnpei
Palikir

Manado

Samarinda
Palu
Celebes
Banjarmasin
Makassar
Surabaya
Bali
Denpasar
Lombok
Sumbawa
Sumba
Flores

Buru
Ceram
Ambon
Halmahera

Sorong

Jayapura

New Ireland

Yaren District
NAURU

INDONESIA
Java
Java Sea

Dili
Kupang
Timor
Timor Sea
TIMOR-LESTE
Arafura Sea

Wewak
New Guinea
Madang
Lae

PAPUA NEW GUINEA

Bougainville
New Britain

SOLOMON ISLANDS
Honiara

Port Moresby

Guadalcanal

SANTA CRUZ ISLANDS

Ashmore and Cartier Islands (AUSTRALIA)

Darwin
Gulf of Carpentaria

Coral Sea Islands
(AUSTRALIA)

Coral Sea

VAN

Indian Ocean

KING LEOPOLD RANGE

MACDONNELL RANGE

Cairns

GREAT

Townsville

New Caledonia (FRANCE)

Noumea

Port Hedland

GREAT SANDY DESERT

Mount Isa

Mackay

DIVIDING

Rockhampton
Gladstone

HAMMERSLEY RANGE

GIBSON DESERT

Alice Springs

AUSTRALIA

SIMPSON DESERT

Brisbane
Gold Coast

LORD HOWE RISE

NEW CALEDONIA BASIN

Geraldton

GREAT VICTORIA DESERT

Lake Eyre
(lowest point in Australia, -15 m)

RANGE

King
Nor
(AU

Kalgoorlie

FLINDERS RANGE

Broken Hill

Lord Howe Island (AUSTRALIA)

Perth
DARLING RANGE

Newcastle

Bunbury

Esperance

Whyalla

Sydney
Wollongong

Adelaide
Canberra

Tasman Sea

Great Australian Bight

Bendigo

Mount Kosciuszko
(highest point in Australia, 2229 m)

Geelong
Melbourne

NEW ZEALA

Bass Strait

Hobart
Tasmania

Invercargill
Stew

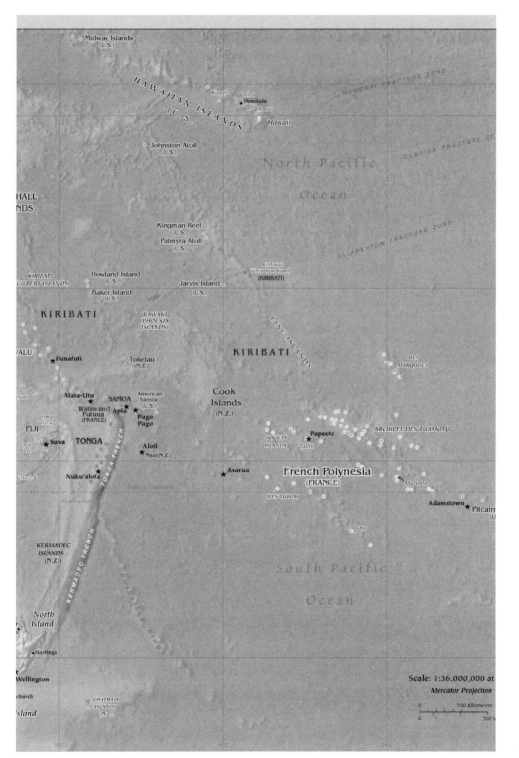

Midway Islands
(U.S.)

HAWAIIAN ISLANDS
(U.S.)

Kauai
Oahu
Honolulu
Maui

Hawaii

MOLOKAI FRACTURE ZONE

Johnston Atoll
(U.S.)

North Pacific

Ocean

CLARION FRACTURE ZO

HALL
NDS

Kingman Reef
(U.S.)
Palmyra Atoll
(U.S.)

CLIPPERTON FRACTURE ZONE

Kiritimati
(Christmas Island)
(KIRIBATI)

KIRIBATI
(GILBERT ISLANDS)

Howland Island
(U.S.)

Jarvis Island
(U.S.)

Baker Island
(U.S.)

KIRIBATI

RAWAKI
(PHOENIX
ISLANDS)

LINE ISLANDS

KIRIBATI

ÎLES
MARQUISES

VALU

★Funafuti

Tokelau
(N.Z.)

Cook
Islands
(N.Z.)

atuma

Mata-Utu
★

Wallis and
Futuna
(FRANCE)

SAMOA
★Apia

American
Samoa
(U.S.)

Pago
Pago

ARCHIPEL DES TUAMOTU

Vanua
Levu

SOCIETY
ISLANDS

Papeete
★

FIJI

Vity
Levu

★Suva

TONGA

★Alofi
Niue(N.Z.)

Tahiti

★Avarua

French Polynesia
(FRANCE)

Mururoa

Lewe-te-Ra

Nuku'alofa
★

TONGA TRENCH

ÎLES TUBUAI

Adamstown
★Pitcairn
(U

Minerva Reef

Tropic of Capricorn

KERMADEC
ISLANDS
(N.Z.)

KERMADEC TRENCH

Raya

South Pacific

Ocean

North
Island

Hastings

Wellington

Scale: 1:36,000,000 at

Mercator Projection

church

CHATHAM
ISLANDS
(N.Z.)

0   500 Kilometers

sland

0   500 M

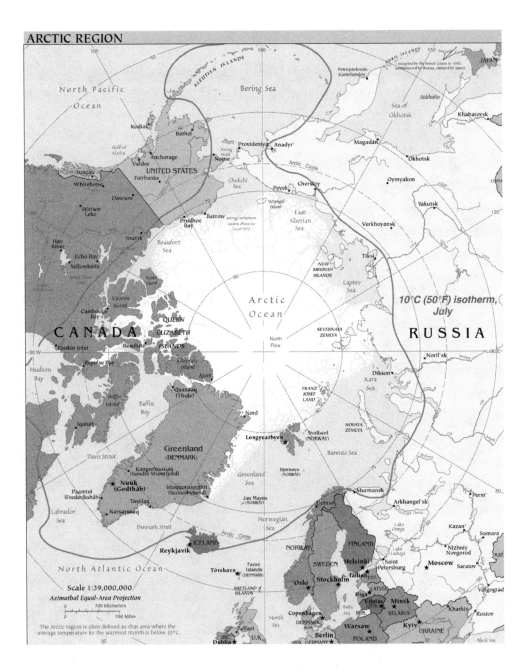

Scale 1:39,000,000
Azimuthal Equal-Area Projection
0   500 Kilometers
0   500 Miles

The Arctic region is often defined as that area where the
average temperature for the warmest month is below 10°C.

ANTARCTIC REGION

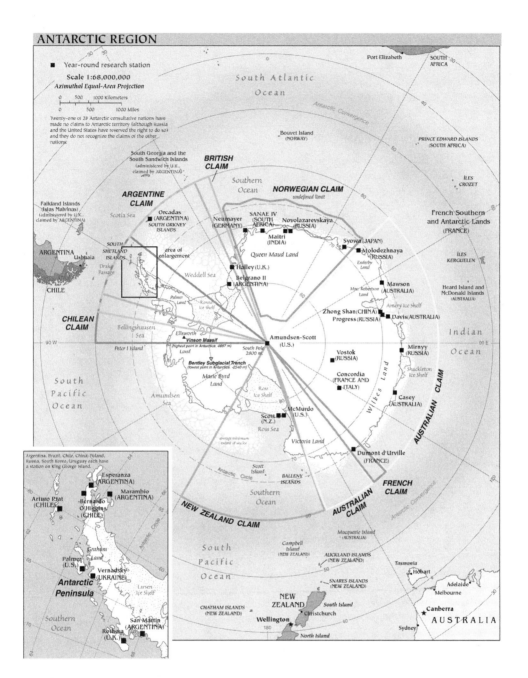

■ Year-round research station

Scale 1:68,000,000
Azimuthal Equal-Area Projection

Twenty-one of 28 Antarctic consultative nations have made no claims to Antarctic territory (although Russia and the United States have reserved the right to do so) and they do not recognize the claims of the other nations.

Argentina, Brazil, Chile, China, Poland, Russia, South Korea, Uruguay each have a station on King George Island.

Antarctic Peninsula

Political Map of the World, April 2007

AUSTRALIA — Independent state
Bermuda — Dependency or area of special sovereignty
Sicily / AZORES — Island / island group
★ — Capital

Scale 1:35,000,000
Robinson Projection
standard parallels 38°N and 38°S

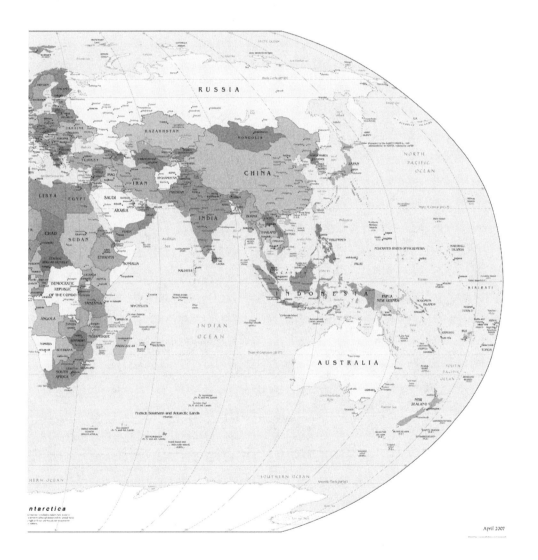

April 2007

Index

Note: (ill.) indicates photos and illustrations.

H

Hadley, John, 21
Hadrian, Emperor, 193
Hadrian's Wall, 193
hail, 55
Haile Selassie, 253
Haiti, 176, 305
Haley, Alex, 259
Hammurabi, King, 237
Hanging Gardens of Babylon, 10
Haram, 239
Hartford Courant, 162
Hartsfield-Jackson Atlanta International Airport, 96
Harvard University (Massachusetts), 158
Havel, Vaclav, 210
Hawaii
    highest mountain on Earth, 9–10
    interstate highways, 99
    as largest island in United States, 150
    leper colony, 153
    number of islands, 153
    Sandwich Islands, 153
    tsunamis, 81
hazard, 73
heat, 69
Heathrow Airport (London, England), 97
Hebrew calendar, 133
hemisphere, 3–4
Henry the Navigator, Prince, 141
Henson, Mathew, 147–48
high low temperature, 68
Highlands, 190
highway, 98
Hillary, Edmund, 220
Hilo, Hawaii, 156
Himalayas, 34, 56
Hindus, 120
Hindustani language, 119
Hiroshima, Japan, 225
Hitler, Adolf, 99, 107, 108, 192
Hokkaido, Japan, 223
Holocene Epoch, 2

Holy Land, 28
holy sites in Jerusalem, 120, 120 (ill.)
Home Insurance Company Building (Chicago, Illinois), 96
homosexuals, 119
Honduras, 113, 305–6
Hong Kong, 218
Honshu, Japan, 223
Horn of Africa, 258
horse latitudes, 67
horse meat, 124–25
Horseshoe Falls, 171
hot springs, 35
Hot Springs, New Mexico, 157
Houphouet-Boigny, Felix, 259
Humboldt, Alexander von, 140
Hungary
    Budapest, 210
    Maygar, 210
    overview, 306
Hurricane Katrina, 84, 156
hurricanes
    Hurricane Katrina, 84
    most damaging part of, 83
    ranks of, 83
    willy-willy, 83
    wind speed, 83
husbands, 125
Hussein, Saddam, 243
Hutus, 260
hydrologic cycle, 43–44

I

Ibn-Batuta, 141
Ice Age, 2, 45
ice core samples, 66–67
Iceland
    mid-Atlantic ridge, 5
    overview, 306–7
    volcanoes, 191
    world's first legislature, 199
igneous rocks, 34
Imperial Valley, 35

Incan civilization, 62, 180
incidence maps, 75
India
    Bollywood, 221
    Bombay, 221
    caste system, 222
    dispute with Pakistan, 113
    disputes over maps, 21
    Dum Dum airport, 221
    East Pakistan, 222
    Kashmir, 222
    New Delhi, 221
    overview, 307
    population, 13, 221–22
    Taj Mahal, 221
Indian Ocean, 10, 48, 81–82
Indo-Australian Plate, 33
Indochina, 227
Indonesia
    Islamic nation, 241
    islands, 231
    Java, 231
    oil, 232
    overview, 307–8
    population, 13
industrial revolution, 12
infanticide, 124
influenza pandemic, 121
Ingushetia, 205
Institute at Sagres, Portugal, 141
International Date Line, 130
Internet, 16, 159
interstate highways, 98–99
Iran, 246, 308
Iraq
    Kuwait and maps, 21
    overview, 308–9
    Persian Gulf War, 242
    Strait of Hormuz, 106
Iraq War, 243, 243 (ill.)
Ireland
    Great Starvation, 193–94
    overview, 309
    as part of United Kingdom, 196
irredentism, 108
irrigation, 58
Isabella, Queen, 144–45

395

Prester John, 141

Prime Meridian, 3, 24, 25, 26, 128

primogeniture, 125

Professional Geographer, 2

projections, 28

Puerto Rico, 154–55, 155 (ill.)

Puerto Rico Trench, 10

Purtscheller, Ludwig, 250

Putrid Sea, 209

pyramids, 237–38, 238 (ill.)

Pyramids of Egypt, 10

Q

Qatar

as country with only one border, 14

highest per capita GDP, 114

overview, 338

Qin Shi Huang, 216

Qinghai-Tibet railroad, 217

Québec, Canada, 170

R

radiation, 88–89

railways. *See* roads and railways

rain, 54, 55, 69, 84

rain forest, 37, 38

rain shadow, 65

Rainbow Warrior, 268

Ramsey, Norman, 132

ranges, 37

recycled rocks, 34–35

Red Sea, 49

refugees, 122

Reich, 192

relative location, 29

relief map, 27

religions, 120, 120 (ill.)

renewable resource, 36

Reno, Nevada, 156

reservations, 165

Revolutionary calendar, 135

Reykjavik, Iceland, 136

Rhode Island, 152

Rhodesia, 257

rice, 218, 218 (ill.)

Richter, Charles F., 80

Richter scale, 80, 81

Ring of Fire, 76

Rio de Janeiro, Brazil, 11, 184, 185

rising sea levels, 43

rivers and lakes. *See also* oceans and seas

Amazon as river with most water, 51

Caspian Sea as world's largest lake, 53

delta, 51–52

drainage basin, 52

flowing direction of rivers, 52

highest rivers in world, 51

Lake Titicaca as world's highest navigable lake, 54

meander, 53, 53 (ill.)

Missouri-Mississippi River as longest river, 51

Nile as longest river, 51

Ojos del Salado as highest lake, 54

oxbow lake, 53

tributary, 52

wadi, 52

watershed, 52

roads and railways

"all roads lead to Rome," 97

Arroyo Seco Freeway as first U.S. freeway, 98

Autobahn, 99

Benz builds first automobile, 100

Century Freeway as last interstate built, 99

Chunnel, 100

Cumberland Gap, 98

Cumberland Road, 97–98

first roads, 97

first train, 101

freeway, 98

highway, 98

interstate highways, 98–99

Lake Ponchartrain Causeway as longest bridge, 99

London Underground as longest subway station, 101

longest Main Street, 100

maps, 26

Mexico City as city with most taxis, 100

most common street names, 100

National Road, 97–98

paved in United States, 99

self-service gas station, 101

taxis in United States, 101

traffic signal, 101

turnpike, 97

Robinson Crusoe, 8

Rock of Gibraltar, 189

rocks

igneous, 34

recycled, 34–35

sedimentary, 34

Rocky Mountains, 171

Rogun dam (Tajikistan), 58

Roman Empire, 209

Romania

orphans, 207–8

overview, 338–39

Transylvania, 208

Rome, 97, 191

Roosevelt, Franklin D., 131

Roosevelt, Theodore, 161 (ill.), 162

Roots (Haley), 259

Ross, James, 23

Royal Flying Doctor Service, 266–67

Royal Observatory, 26

Rub al Khali, Saudi Arabia, 236

Russia. *See also* Soviet Union

Aeroflot, 204

Chechnya, 205

Commonwealth of Independent States, 201–2

driving across, 203

401